Networking and Internetworking with Microcontrollers

Networking and Internetworking with Microcontrollers

By Fred Eady

AMSTERDAM • BOSTON • HEIDELBERG • LONDON
NEW YORK • OXFORD • PARIS • SAN DIEGO
SAN FRANCISCO • SINGAPORE • SYDNEY • TOKYO
Newnes is an imprint of Elsevier

Newnes is an imprint of Elsevier
200 Wheeler Road, Burlington, MA 01803, USA
Linacre House, Jordan Hill, Oxford OX2 8DP, UK

Recognizing the importance of preserving what has been written, Elsevier prints its books on acid-free paper whenever possible.

Library of Congress Cataloging-in-Publication Data

(Application submitted.)

British Library Cataloguing-in-Publication Data
A catalogue record for this book is available from the British Library.

ISBN: 0-7506-7698-1

For information on all Newnes publications
visit our website at www.newnespress.com

03 04 05 06 07 08 10 9 8 7 6 5 4 3 2 1

Printed in the United States of America

Contents

Preface

There are lots of philosophical things I could say here. However, I don't claim to be a philosopher or a poet. My days are spent designing microcontroller hardware, writing code to drive that hardware and then writing about my adventures.

This book is a mean-business document designed to give you the knowledge needed to network microcontroller-based devices successfully. Before you turn the last page of this book, you'll know how to integrate RS-232, I²C and Ethernet into a network device that can be used to communicate via LAN, WAN or Internet. In addition to the knowledge you will gain building the network devices, you'll also walk away with in-depth knowledge of how the code within those network devices works.

Our microcontroller-based network devices will be fabricated using microcontrollers from Atmel and Microchip. To maintain consistency at the coding level, I'll use ImageCraft's ICCAVR Pro C Compiler for the Atmel parts and Custom Computer Service's CCS PIC C Compiler for the PIC parts. Both of these C compilers are moderately priced and easily obtainable via the Internet. You should be able to easily port the C source code from any project in this book to other variants of C.

Our networking adventure will begin with RS-232. We'll build on what we learn in the RS-232 sections and ultimately implement both an I²C-bus and an Ethernet interface. No bit will be left unturned. What you don't see in the pages of this book can be found on the companion CD-ROM. There's also a support web site (http://www.edtp.com) where you can get technical support and purchase parts, kits and assembled units that are discussed in this book.

Both Atmel and Microchip provide a free IDE that you can get for a download from their respective web sites. I'll use Atmel's AVR Studio and Microchip's MPLAB exclusively when working with these microcontrollers. To provide an extra layer of visibility into the microcontrollers, I'll employ the services of a Microchip MPLAB ICE 2000, a Microchip MPLAB ICD 2 and an Atmel AVR JTAG ICE. On the networking side, I'll use a Network Associates Sniffer to show you what's inside the Ethernet packets.

OK...now that you know what this book is about, let's go build some microcontroller-based network devices.

A Quick Look at the Microcontrollers

Atmel's AVR

The Atmel AVR is a very capable and highly networkable microcontroller. There are a number of AVR families, which include the standard AVR line, a low-power AVR microcontroller set, the tinyAVR family and the ATmega AVR microcontrollers. I've chosen to concentrate on networking the ATmega AVR microcontrollers for a number of reasons. Many of the legacy AVRs are being replaced by faster and more powerful ATmega AVRs. For instance, the ATmega16 has replaced the ATmega163 and the ATmega32 has shoved the ATmega323 out. In addition to added functionality, the AVR upgrades fix bugs found in the older silicon they are replacing.

I'm not going to get into internal differences found in the AVR versus other microcontrollers that could be called AVR peers. That's what datasheets are for. However, I will give you my reasons for employing the ATmega AVRs as microcontrollers in network devices. The ATmega AVRs that I will network all have a maximum clock speed of 16 MHz. That may not sound "fast," but the ATmega AVRs execute most instructions in a single cycle and thus are capable of producing 16 MIPS with a 16MHz clock. Basically, the number in the name of a ATmega AVR represents the amount of program Flash in kilobytes. The ATmega16 contains 16K of program Flash while an ATmega32 has 32K words of program flash. The largest ATmega AVR, the ATmega128, contains 128K of program Flash memory. The large portions of program memory are supplemented by big slices of SRAM. The ATmega16 is loaded with 1K of SRAM and the ATmega32 SRAM area doubles the ATmega16 SRAM capacity. Even the ATmega8, the smallest of the ATmega AVRs, has 1K of SRAM. Applying the logic to the rest of the ATmega AVR family reveals the ATmega64, ATmega128 and ATmega8 with 64K, 128K and 8K of program Flash memory, respectively. I think you can see why I've decided to go with Atmel's ATmega AVR line as far as networking is concerned. The high speed and large program Flash memory areas coupled with ample SRAM and EEPROM memory make the ATmega AVR microcontroller a good choice for networking projects.

Atmel's AVR can be obtained from many of the mail order electronic part distributors. AVRs are reasonably priced and come with a tub full of goodies just right for networking. Two 8-bit timers and a 16-bit timer allow the creation of precision delays while an on-chip USART (Universal Synchronous Asynchronous Receiver Transmitter) takes care of the housekeeping chores needed to effect the RS-232 serial protocol. Twenty-one interrupt vectors cover all of the AVR's networking components including the two-wire interface (Atmel's name for I^2C), the SPI subsystem and the USART.

The AVR USART

The ATmega AVR USART is capable of full duplex operation. Like most every other USART in existence, the ATmega AVRs USART supports 5, 6, 7, 8 or 9 data bits plus the standard 1 or 2 stop bits. The USART baud rate generator for the Atmel ATmega AVR is an integral part of the USART hardware. A typical Atmel USART is depicted in the block diagram you see inside Figure 1. All ATmega AVRs contain a USART and the ATmega128 is equipped with a pair of USARTs.

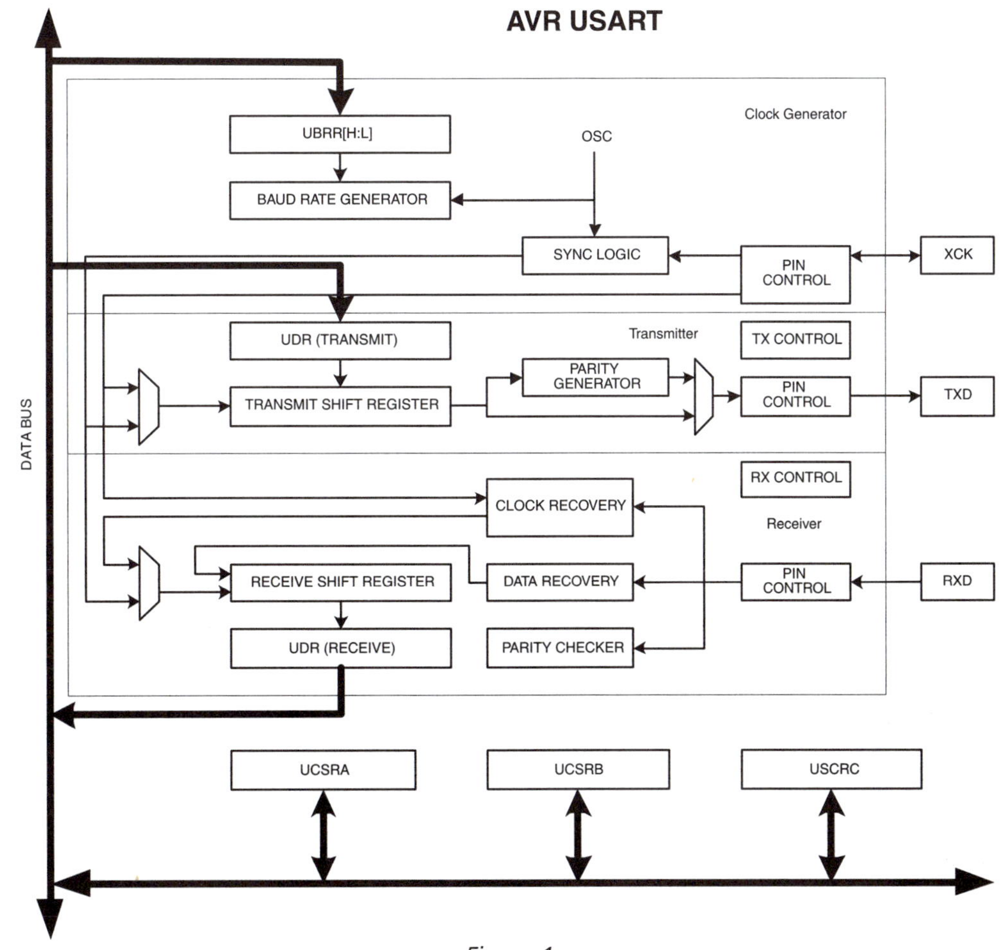

Figure 1

Atmel USARTs allow the ATmega AVRs to enter MPCM (Multi-processor Communication Mode). This mode of operation uses addressing to allow multiple processors to communicate over the same serial bus. MPCM uses the 9-bit character frame format in conjunction with the master/slave paradigm.

The Two-wire Serial Interface

In the Atmel world, I²C is known as TWI, or Two-wire Interface. Other than a name change, TWI looks like and smells like I²C. 128 devices can hang on the two-wire bus and are addressed using the standard I²C 7-bit addressing scheme. Master and slave operation is supported at speeds of up to 400 kHz. To help fight false triggering due to noise, the Atmel TWI module includes noise suppression circuitry. You can even wake up the AVR from sleep with a TWI.

Programming the ATmega AVR

Loading code into an ATmega AVR device is a breeze. There are many ways to accomplish this. There's the AVR ISP (In-System Programmer) programming module that costs less than $40 and hooks up to a personal computer's serial port. Or, AVR programming can be done with the STK500 development board. ATmega AVRs with 16K or more of program memory also support a JTAG interface, which can be used for programming the ATmega AVR program Flash. No matter how you decide to load the code into your ATmega AVR, AVR Studio supports all of the programming devices I've mentioned. AVR Studio is Atmel's front-end IDE software that runs on a personal computer.

I'll complement AVR Studio with ImageCraft's ICCAVR Pro C Compiler. ICCAVR Pro is a true ANSI-based C compiler for the ATmega AVR. I particularly like the code generator and the AVR calculator features of ICCAVR Pro.

Emulating the ATmega AVR

At the helm of emulation for AVRs is AVR Studio. AVR Studio interfaces to the many AVR emulation devices. In this text, the emulation device of choice is the AVR JTAG ICE. The AVR JTAG ICE communicates with an on-chip debug module embedded within the target AVR. The OCD (On-Chip Debugger) module in the ATmega AVRs eliminates the need for a special bondout emulation device.

Microchip's PIC

Most PIC microcontrollers have everything one would need to effect a network application. Larger PICs have on-chip UARTS and USARTS for synchronous and asynchronous communications using the RS-232 protocol. A software UART function can also be implemented for the smaller PICs that don't have the sophistication of a built-in UART or USART module.

In networking, timing is everything. Up to three internal timers can be had on larger PIC devices. Even the tiny 8-pin PIC12F675 has an 8-bit and a 16-bit timer. The timers can be used for generating precision millisecond and microsecond delays or the time of day.

Fortunately, the CCS C Compiler for PIC and its native PIC peripheral routines makes it very easy to assemble a working RS-232 PIC application. In fact, CCS C also has hooks for I²C. The Microchip PIC family complements the CCS C peripheral routines by providing ample Flash memory for program code, scratch pad SRAM and user data storage. The more SRAM the better when it comes to creating buffer areas for interrupt-driven communications applications.

All of our network coding and hardware design and fabrication time will be spent dealing with the Flash-based series of the Microchip PIC family. I've chosen to work with the Flash-based parts because they're inexpensive and easily obtained and don't require the support hardware a standard windowed PIC needs. For instance, using Flash devices eliminates the need for an ultraviolet EPROM eraser. And, since Flash parts can be programmed and reprogrammed in-circuit using ICSP (In-Circuit Serial Programming), fewer microcontroller parts are needed in the development cycle since there is no need to rotate a number of parts through the ultraviolet eraser while you're debugging your code.

For the purposes of networking, I've selected the largest part in the PIC16F87X crew, the PIC16F877. The PIC16F877 can operate with a 20 MHz clock, which gives an instruction cycle of 200 nsec. There are 8K words of program Flash and 368 bytes of SRAM or data memory inside a PIC16F877. Should we decide it's necessary, there is also a block of 256 bytes of EEPROM available for storing constants or whatever else we decide is important to keep even after the power is removed from the part. As we move into putting an RS-232 serial port together on a PIC, you'll see how important interrupts are when it comes to microcontrollers like the PIC. The PIC16F877 can be interrupted in 15 different ways.

I/O pins are also very important in a networking application. Not only do we need enough I/O to perform tasks like monitoring a voltage or turning an external device on or off, there have to be some I/O pins dedicated to the networking task. For instance, a simple micro- controller Ethernet driver application requires at least 16 I/O pins alone. The PIC16F877 has 33 I/O lines we can put to work, which leaves some I/O for things that microcontroller do best—control.

The PIC16F877 offers quite a bit of functionality for things other than effecting networking. However, I'm primarily concerned with giving you the ability to network the PIC16F877. With that, let's start with a look at one of my favorite networking modules, the PIC16F877 USART.

The PIC16F877 USART

USART is short for Universal Synchronous/Asynchronous Receiver/Transmitter. On the PIC, the USART is also called the SCI or Serial Communications Interface. You probably have heard the word UART (Universal Asynchronous Receiver/Transmitter) as for many years that was the only IC used by serial ports in personal computers. Some of today's microcontrollers sport UARTs instead of USARTs.

The PIC16F877 USART takes much of the pain away when it's required to communicate with other serial-based devices. Instead of writing timing routines to produce a specific baud rate, the PIC16F877 USART baud rate is generated by an internal baud rate generator. With a USART or UART, it's not necessary to code routines to look for incoming start bits or time the inter-bit distances to pick up the incoming data. All of that work is done within the USART itself. A USART makes it possible to communicate with other serial devices in full-duplex or half-duplex mode. Full-duplex mode allows communications to flow in both receive and transmit directions simultaneously between two serial devices. Half-duplex mode only allows one device to transmit at a time while the other device listens.

The PIC16F877 MSSP Module

MSSP, or Master Synchronous Serial Port, is yet another PIC16F877 communications subsystem. The MSSP is a serial interface used to bring I^2C applications to life. Like the USART, the MSSP is a register and status bit-oriented module.

I^2C uses six MSSP registers for control, status and buffering. Two PIC16F877 I/O pins are dedicated to I^2C, RC3 for SCL (clock) and RC4 for SDA (data). Like the USART's synchronous function, I^2C is a master/slave communications configuration. Figure 4 is a graphic example of how the MSSP allocates the registers, I/O pins and buffers for I^2C operation.

I^2C is a Philips invention that was designed as a clever way to allow integrated circuits in television sets and stereo rigs to talk to each other. We'll cover I^2C as it pertains to PIC microcontrollers thoroughly in this book. Thanks to Philips, there are hundreds of I^2C-capable devices for us to play with from various manufacturers.

The PIC16F877 lends itself to oddball networking solutions. Using the PIC16F877 precision timers, we can put together a homebrew protocol and bit bang between devices. For instance, in the past I once coded a PIC application that required the PIC to clock data to and from a personal computer's parallel port pin. In addition, in the early days of PIC there were no UARTs or USARTs on the 18-pin PIC16C5X microcontrollers. Therefore, I had to code a "software" UART to emulate the task that today's hardware USARTs perform. You'll find that the software UART is still a good thing to have in your coding toolbox when designing networking and communications applications with the tiny USART-less 8-pin PICs.

The PIC18F452

Another PIC device I'll base networking code on in this book is the PIC18F452. The PIC18F452 is pin-compatible with the PIC16F877. The PIC18F452 is loaded with 16K of on-chip program memory backed up by 1.5K of SRAM. This makes the PIC18F452 a candidate for Ethernet LAN applications. In addition to the increased internal memory area, the PIC18F452 can run twice as fast as the PIC16F877 (40 MHz). All of the PIC16F877 communications peripherals we talked about earlier operate in the same manner on the PIC18F452 and the CCS PIC C Compiler has the capability to generate code for them as well.

The PIC12F675

Sometimes it's more fun to push an economy car to its limits and not drive that performance hot rod with all of the bells and whistles. That's how I feel about the little 8-pin PIC12F675. In comparison, it's as tiny physically as it is logically. The PIC12F675 only has 1K words of program Flash and 64 bytes of SRAM. There are only six I/O pins but inside the PIC12F675 you'll find a couple of timers, an A/D (Analog to Digital) converter and a comparator. Like the big guys, there is on-chip EEPROM but only 128 bytes of it. With some tricky coding, we'll make the tiny PIC do RS-232 with the best of them.

Programming the PIC

The Flash-based PICs that will be featured in this book are all programmed using the ICSP (In-Circuit Serial Programming) method. As this book is focused on microcontroller communications and networking, I won't offer up any made-in-the-garage PIC programming hardware or software. I'm going to stick to the Microchip factory programmers and software. You can use the Microchip MPLAB ICD 2 (In-Circuit Debugger) or the Microchip PRO MATE II for programming the PIC Flash parts.

Emulating the PIC

The Microchip MPLAB ICD 2 and the MPLAB ICE 2000 will be used to debug and display the inner-workings of the PIC code that will be presented in this book. I'll be able to show you all of the code, internal registers and memory areas using the Microchip MPLAB ICE 2000 PIC emulator system. Like the CCS C Compiler and the Microchip PRO MATE II device programmer, the MPLAB ICE 2000 and Microchip MPLAB ICD 2 are natively supported by Microchip's MPLAB. The merger of the Microchip PRO MATE II, the Microchip MPLAB ICE 2000, the Microchip MPLAB ICD 2 and the CCS PIC C Compiler will allow me to show you how things are done inside and outside the PIC using only a single MPLAB IDE screen.

What's on the CD-ROM?

All of the source code and the executable code discussed in this book are on the companion CD-ROM. In addition, all of the Easy Ethernet device schematics are provided in PDF format. Printed circuit board layouts are also part of the CD-ROM package and are included for those readers who wish to build the Easy Ethernet devices from scratch.

CHAPTER 1

The Essence of Microcontroller Networking—RS-232

Let's begin by exploring the RS-232 protocol. Knowing how to manipulate data with RS-232 will help you master more complex communications protocols. You'll also find RS-232 techniques to be invaluable in the development phase of your projects.

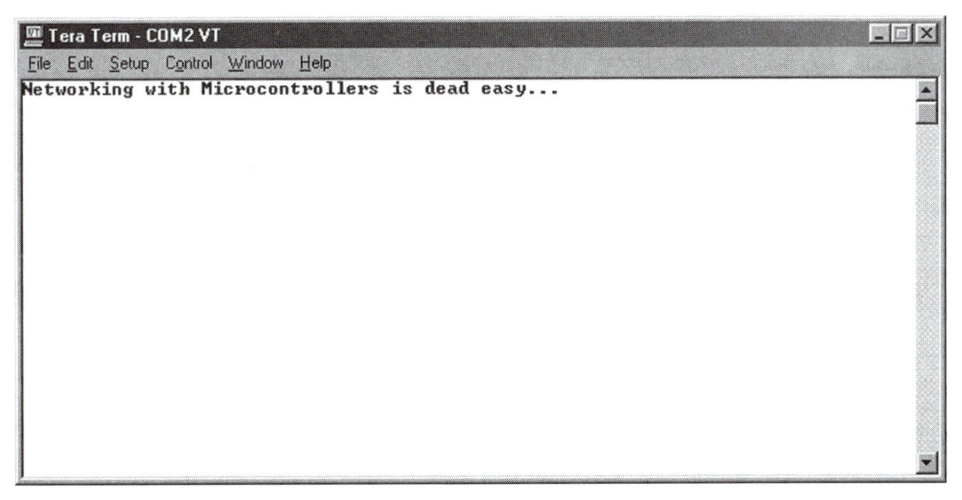

Figure 1.1: Effecting RS-232 communications with a microcontroller is a snap. As you continue reading this book, you will find that knowing how to implement simple RS-232 with a microcontroller can assist you in building and debugging more complex microcontroller projects.

The information you see in the terminal emulator window in Figure 1.1 was generated by some very simple firmware and a not-so-complicated off-the-shelf, two-buck microcontroller. I used a tiny 8-bit microcontroller that does not contain a built-in hardware USART (Universal Synchronous/Asynchronous Receiver/Transmitter), to transfer the ASCII characters you see in Figure 1.1 from one of its I/O pins to an RS-232 converter IC. A serial cable connected between the microcontroller/RS-232 converter IC circuitry and my personal computer's serial port allowed the ASCII characters to flow from the little microcontroller's firmware out of the microcontroller's I/O pin, through the RS-232 converter IC, across the serial cable to the personal computer's USART/RS-232 circuitry and finally end up in the terminal emulator window you see in Figure 1.1.

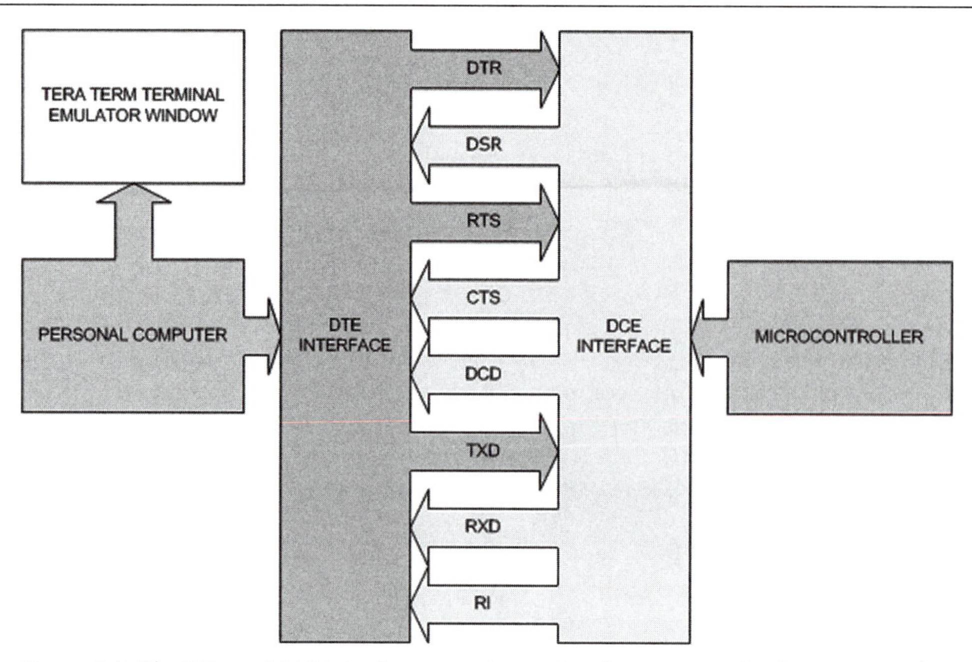

Figure 1.2: The DTE and DCE interfaces usually consist of some sort of voltage-conversion circuitry to translate RS-232 voltage levels to voltage levels that are compatible with the computing equipment on each end of the communications link. The simplest form of an RS-232 link uses only the TXD and RXD signals with a common ground.

What I've just described is one of the simplest forms of microcontroller networking. It is commonly known as serial or RS-232 communications. As you can see in Figure 1.2, RS-232 was designed to tie DTE (Data Terminal Equipment) and DCE (Data Communications Equipment) devices together electronically to effect bidirectional data communications between the devices.

An example of a DTE device is the serial port on your personal computer. Under normal conditions, the DTE interface on your personal computer asserts DTR (Data Terminal Ready) and RTS (Request To Send). DTR and RTS are called modem control signals. A typical DCE device interface responds to the assertion of DTR by activating a signal called DSR (Data Set Ready). The DTE RTS signal is answered by CTS (Clear To Send) from the DCE device. A standard external modem that you would connect to your personal computer serial port is a perfect example of a DCE device.

Some History

In May of 1960, it was evident that a standard was needed to identify the electrical interface between computers and modems. It was decided to establish a standard voltage with standard signal parameters and a standard nomenclature to identify the conductors in the cable that connected computers and data sets. Even today, you will sometimes hear the term data set applied to modems and DCE equipment.

To compete as well as exist in the current communications environment, telecommunications vendors needed common ground to assure that each vendor's equipment set could talk to any other vendor's telecommunications equipment set. In other words, the industry needed a working standard. Without a standard, the whole teleprocessing industry could come to a grinding, nonstandardized halt.

To help establish some harmony, a committee named the Electronic Industries Association was formed. The EIA drafted a standard known as EIA RS-232(X). Though it was a great idea, the original specification was broad in meaning and didn't guarantee compatibility. The new RS-232 specification also had a competitor outside the United States, known as the CCITT, or Consultative Committee on International Telegraphy and Telephony, recommendation V.24.

The RS-232 proposal defined a logical and physical interface between DTE equipment and DCE equipment. The computer's DTE serial port presents both a physical and a logical interface to a modem or data set's DCE port and consists of several conductors for controlling, transmitting and receiving data. Timing and clocking signals are also intermixed within the RS-232 interface. The logical and physical attributes of the RS-232 proposal eventually became a set of standards known today as the EIA RS-232 interface.

Once the signals reach the DCE device, a second interface provides a physical path to the communication channel (RF link, telephone line, fiber-optic link, satellite link, and so forth). For most of you, that second interface is a standard two-conductor analog telephone line, which is terminated inside your modem.

The EIA standard originally identified seven interface conductors and no specific connector. Signal voltages were defined as at least 3 volts but not greater than 20 volts with respect to ground.

In October 1963, RS-232 became RS-232-A and was modified to include a 25-pin connector with a maximum cable length of 50 feet. This revision established fixed relationships between a circuit and specific pin numbers on the 25-pin connector. Also, an alphabetic coding system for each type of interface circuit was presented. The first character of the coding system designated A for ground, B for data, C for control and D for clocking. Table 1.1 lays out the pinout and various names for each RS-232 signal.

Pin	Line Label	Line Name	Signal Direction	Level
1	AA	Positive Ground	N.A.	A,B C
2	BA	Transmitted Data	To DCE	A B,C
3	BB	Received Data	To DTE	A,B,C
4	CA	Request To Send	To DCE	A B,C
5	CB	ClearTo Send	To DTE	A B,C
6	CC	Data Set Ready	To DTE	A B,C
7	AB	Signal Ground	N.A.	A B C
8	CF	Received Line Signal Detector (RS-232); Data Carrier Detect (RS-232A/B)	To DTE	A,B,C
11	N.A.	Select Standby	To DCE	C
12	SCF	Secondary Receive Line Signal Detector	To DTE	C
13	SCB	Secondary Clear To Send	To DTE	C
14	SBA	Secondary Transmitted Data	To DCE	C
14	N.A.	New Sync	To DCE	A,B,C
15	DB	Transmitter Signal Element Timing	To DTE	A B C
16	SBB	Secondary Received Data	To DTE	C
17	DD	Receiver Signal Element Timing	To DTE	A,B,C
18	N.A.	Test	To DCE	C
19	SCA	Secondary Request To Send	To DCE	C
20	CD	Data Terminal Read	To DCE	A,B,C
21	CG	Signal Quality Detector	To DTE	C
22	CE	Indicate Ring/Calling	To DTE	A,B,C

Table 1.1: Specifications list for RS-232 interface.

There are a couple of confusion points. Note the total lack of logic when associating DB-25 pins with DB-9 pins. And, this table is based on the DTE side of the circuit. To get things to work, you must switch the TD and RD pins on the DCE side of the circuit. When you do the switch that puts the DTE TD pin's data into the DCE RD pin and the DCE's TD pin's data into the DTE RD pin. If you're using the modem signals, you have to tie them together properly between the DTE and DCE as well.

The original seven basic circuits and the signal-level definition of –3 volts for mark and +3 volts for space were retained intact, adding ten additional optional circuit definitions. The maximum permissible open-circuit voltage was changed to 25 volts, and a current maximum between any two conductors, including ground, was set at 0.5 ampere. Conductors that permit auto-answer capability were first introduced in this revision.

October 1965 brought about RS232-B, which defined terminating impedances that permitted circuit designers to build hardware with greater reliability. Open-circuit signal levels remained unchanged at –3 to –25 volts as mark and +3 to +25 volts as space, but revision B added an important voltage specification. By specifying that signal ground on pin 7 be tied to frame ground on pin 1 in the DCE equipment, a definite signal reference is established between DTE and DCE devices.

The *Interface Between Data Terminal Equipment and Data Communication Equipment Employing Serial Binary Data Interchange* specification was released in August 1969. It further clarified conductor definitions and stated that properly terminated RS-232 circuits shall not exceed ±15 volts.

RS-232-C came along later and defined the interface between Data Terminal Equipment (DTE) and Data Circuit terminating Equipment (DCE). In the early days, a piece of DTE hardware was usually a dumb terminal. DEC's (Digital Equipment Corporation in those days; Hewlett-Packard/COMPAQ these days) VT100 was and is the most well-known dumb terminal and is still emulated today.

As you would imagine, a standard DTE device should be capable of emitting and receiving a serial data stream. As you have already seen, that includes microcontrollers and personal computers in the "could be a DTE" category. Although DCE equipment can also transmit and receive a serial data stream, the primary purpose of DCE equipment is to receive the DTE-generated bit stream over an RS-232 interface and convert it to a form that's suitable for transmission over a telecommunication medium. In the case of a personal computer modem, that telecommunications medium is most likely a voice-grade telephone line.

Ever noticed that every serial port interface on your personal computer is male and every modem serial port interface you've ever seen is female? There's a reason for that. The RS-232-C standard states that physical DTE port connectors will be male and physical DCE port connectors will be female.

Older personal computers and modems used a 25-pin connector. Today's 9-pin serial connectors aren't really standards although they have become so by proxy. The 9-pin interface first appeared commercially on AT-class PCs in the early 1980s.

RS-232 Standard Operating Procedure

Today, the majority of commercially available equipment is based on the RS-232-C or RS-232-D standard. (The CCITT V.24 and V.28 standards are also common and widely-used.) There are 25 circuits defined in the RS-232 standard. The good news is that most of the 25 RS-232 circuits don't have to be used to effect an asynchronous communications session between a DTE and DCE device. Things could be different for synchronous communications sessions that employ complex communications protocols and that's why the timing and clocking signals are defined in the RS-232 standard. There's a good reason that a 9-pin connector is on your personal computer instead of the standard appointed 25-pin connector. You only need nine RS-232 signal lines to communicate asynchronously using a standard asynchronous modem. Let's look at them from a "commented" standards point of view.

- *Pin 1 (Protective Ground Circuit, AA).* This conductor is bonded to the equipment frame and can be connected to external grounds if other regulations or applications require it.

  Comment: Normally, this is either left open or connected to the signal ground. This signal is not found in the DTE 9-pin serial connector.

- *Pin 2 (Transmitted Data Circuit BA, TD).* This is the data signal generated by the DTE. The serial bit stream from this pin is the data that's ultimately processed by a DCE device.

  Comment: This is pin 3 on the DTE 9-pin serial connector. This is one of the three minimum signals required to effect an RS-232 asynchronous communications session.

- *Pin 3 (Received Data Circuit BB, RD).* Signals on this circuit are generated by the DCE. The serial bit stream originates at a remote DTE device and is a product of the receive circuitry of the local DCE device. This is usually digital data that's produced by an intelligent DCE or modem demodulator circuitry.

  Comment: This is pin 2 on the DTE 9-pin serial connector. This is another of the three minimum signals required to effect an RS-232 asynchronous communications session.

- *Pin 4 (Request To Send Circuit CA, RTS).* This signal prepares the DCE device for a transmit operation. The RTS ON condition puts the DCE in transmit mode, while the OFF condition places the DCE in receive mode. The DCE should respond to an RTS ON by turning ON Clear to Send (CTS). Once RTS is turned OFF, it shouldn't be turned ON again until CTS has been turned OFF. This signal is used in conjunction with DTR, DSR and DCD. RTS is used extensively in flow control.

  Comment: This is pin 7 on the DTE 9-pin serial connector. In simple 3-wire implementations this signal is left disconnected. Sometimes you will see this signal tied to the CTS signal to satisfy a need for RTS and CTS to be active signals in the communications session. You will also see RTS feed CTS in a null modem arrangement.

- *Pin 5 (Clear To Send Circuit CB, CTS).* This signal acknowledges the DTE when RTS has been sensed by the DCE device and usually signals the DTE that the DCE is ready to accept data to be transmitted. Data is transmitted across the communications medium only when this signal is active. This signal is used in conjunction with DTR, DSR and DCD. CTS is used in conjunction with RTS for flow control.

  Comment: This is pin 8 on the DTE 9-pin serial connector. In simple 3-wire implementations this signal is left disconnected. Otherwise, you'll see it tied to RTS in null modem arrangements or where CTS has to be an active participant in the communications session.

- *Pin 6 (Data Set Ready Circuit CC, DSR).* DSR indicates to the DTE device that the DCE equipment is connected to a valid communication medium and, in some cases, indicates that the line is in the OFF HOOK condition. OFF HOOK is an indication that the DCE is either in dialing mode or in session with another remote DCE. When this signal is OFF, the DTE should be instructed to ignore all other DCE signals. If this signal is turned off before DTR, the DTE is to assume an aborted communication session.

Comment: This is pin 6 on the DTE 9-pin serial connector. DSR is sometimes used in a flow control arrangement with DTR. Some modems assert DSR when power to the modem is applied regardless of the condition of the communications medium.

- *Pin 7 (Signal Common Circuit, AB).* This conductor establishes the common-ground reference for all interchange circuits, except Circuit AA, protective ground. The RS-232-B specification permits this circuit to be optionally connected to protective ground within the DCE device as necessary.

  Comment: This is pin 5 on the DTE 9-pin serial connector and is the only ground connection. This is the third wire of the minimal 3-wire configuration. Thus, an RS-232 asynchronous communications session can be effected with only three signals: TX (Transmit Data), RX (Receive Data) and signal ground.

- *Pin 8 (Data Carrier Detect Circuit CF, DCD).* This pin is also known as Received Line Signal Detect (RSLD) or Carrier Detect (CD). This signal is active when a suitable carrier is established between the local and remote DCE devices. When this signal is OFF, RD should be clamped to the mark state (binary 1).

  Comment: This is pin 1 on the DTE 9-pin serial connector. Normally in use only if a modem is in the communications signal path. You will also see this signal tied active in a null modem arrangement.

- *Pin 20 (Data Terminal Ready Circuit CD, DTR).* DTR signals are used to control switching of the DCE to the communication medium. DTR ON indicates to the DCE that connections in progress shall remain in progress, and if no sessions are in progress, new connections can be made. DTR is normally turned off to initiate ON HOOK (hang-up) conditions. The normal DCE response to activating DTR is to activate DSR.

  Comment: This is pin 4 on the DTE 9-pin serial connector. Unless you specify differently or run a program that controls DTR, usually it is present on the personal computer serial port as long as the personal computer is powered on. Occasionally you will see this signal used in flow control.

- *Pin 22 (Ring Indicator Circuit CE, RI).* The ON condition of this signal indicates that a ring signal is being received from the communication medium (telephone line). It's normally up to the control program to act on the presence of this signal.

  Comment: This is pin 9 on the DTE 9-pin serial connector. This signal follows the incoming ring to an extent. Normally, this signal is used by DCE auto-answer algorithms.

That is all that's needed RS-232 signal-wise to establish a session between a DTE and a DCE device. Now that you have a feeling for what each RS-232 signal does, let's review how they react to each other with respect to the transfer of data between a DTE and DCE device.

- Local DTE (personal computer, microcontroller, etc.) is powered up and DTR is asserted.
- Local DCE (modem, data set, microcontroller, etc.) is powered up and senses the DTR from the local DTE.
- Local DCE asserts DSR. If the DCE device is a modem, it goes off-hook (picks up the line). If a dial-up session is to be established, the DTE sends a dial instruction and phone number to the modem.
- If the line is good and the other end (remote DCE) is ready or answers the dial-up from the local DCE, a carrier is generated/detected and the local and remote DCE devices assert DCD. The session is established.
- The transmitting DTE raises RTS.
- The transmitting DCE responds with CTS.
- The control program transmits or receives data.

In our historical review, the DTE or personal computer and DCE or modem took care of converting the RS-232 signal levels to appropriate personal computer circuitry levels. To perform RS-232 asynchronous communications with microcontrollers, we must employ a voltage translation scheme of our own. Fortunately, there are many ways to do this and all of them are easy to implement.

RS-232 Voltage Conversion Considerations

RS-232 converter ICs like those made by Maxim and Sipex convert the negative RS-232 voltages to positive logic voltage levels that microcontroller circuits can understand. The positive RS-232 voltages are converted to a microcontroller's logical 0 (zero) voltage level. If the microcontroller circuitry is powered by +5 VDC, then an RS-232 '1' or mark is converted to a TTL (Transistor Transistor Logic) *high* or '1' and an RS-232 '0' or space is translated into a TTL *low* or '0'. With the advent of 3-volt logic, special RS-232 converter ICs that can operate at the 3-volt power supply levels have been introduced. The bottom line is that the RS-232 marks and spaces must be converted to voltage levels the microcontroller can understand before any communications and data transfer can be realized between devices.

In reality, the full-positive and negative voltage swing called out by the RS-232 standard doesn't have to be employed to effect RS-232 communications links. With the right cable an RS-232 voltage of –3 volts is sufficient to generate a '1' or mark while +3 volts will produce a '0' or space. The area between –3 volts and +3 volts (shown in Figure 1.3) is a transition zone and is where most of the nasty line noise can and should be found. By defining this ±3-volt threshold, the signal-to-noise ratio of the RS-232 physical link is improved. If a high-quality serial cable is used and the distance between stations is relatively short, RS-232 voltages that resemble microcontroller logic voltages can be used to transfer information

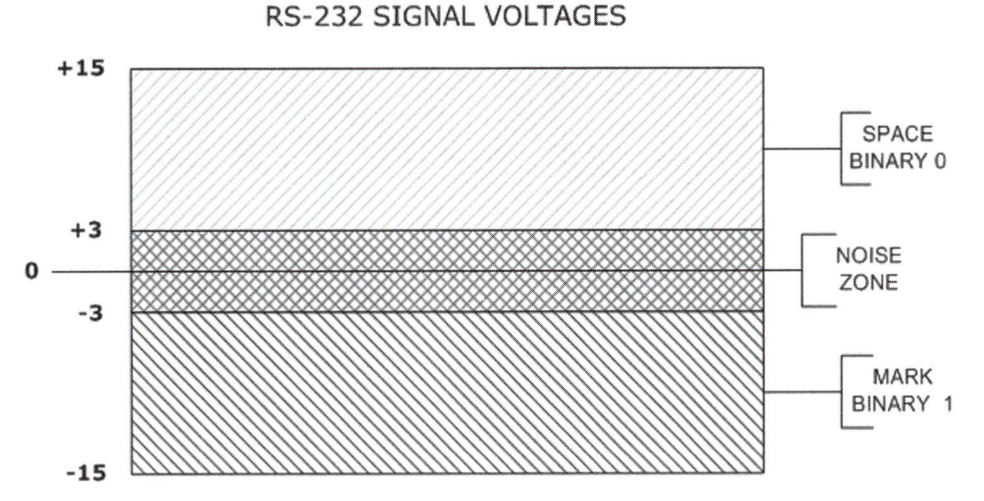

Figure 1.3: Cheap RS-232 implementations dare to use the 0 VDC to +5 VDC region for marks and spaces with 0 VDC being a mark and anything over +3 VDC representing a space. The "NOISE ZONE" I've marked is actually called the transition zone.

between a DTE and DCE device. In addition, using a high-quality cable could extend the 50-foot maximum cable length specified by the RS-232 specification. Reducing the speed of the data transmission can also extend the maximum cable length between a wired set of DTE and DCE devices as well.

The good news is that you don't have to know the nitty-gritty details of the RS-232 specification to use RS-232 as a means of communicating with a microcontroller. In fact, I've already given you more RS-232 history and theory than you really need to know to make a microcontroller talk asynchronously. In this book, we're all about the practical application of RS-232 as it pertains to microcontrollers. So, let's look at some RS-232 hardware and the firmware behind it.

CHAPTER 2

Implementing RS-232 with a Microcontroller

Now that you've completed RS-232 history 101, this chapter will deal with implementing RS-232 on a microcontroller. We'll use the Microchip® PIC12F675 as our RS-232 engine and we'll power our RS-232 engine with code written with the Custom Computer Services C Compiler.

You can build the circuits in this chapter from scratch. I've chosen to use the Microchip PICkit™ 1 as my "breadboard" as it contains circuitry to program the PIC12F675 and an experimenter area that is perfectly suited for additional RS-232 circuitry.

Basic RS-232 Hardware

Let's begin by looking at a simple microcontroller implementation. In its most basic form, an operational microcontroller-based circuit consists of the microcontroller, a simple power supply and a clock source. For this project, I'm going to use the most basic of microcontrollers, an 8-pin Microchip PIC12F675.

The PIC12F675 has an internal clock source but does not contain a USART. That means we will have to implement the functionality of a hardware USART in the PIC12F675's firmware. To do that, we need to know just a bit more about RS-232 signaling. Let's begin by designating the desired RS-232 signaling speed, or baud rate. A common baud rate is 9600 bps (bits per second) and most everything RS-232 can operate at this speed. So, 9600 bps it is.

At 9600 bps, our data packet bit width is the reciprocal of the baud rate, which is 104 μS (104 microseconds). The idea is to try to see if the incoming RS-232 bit is a '1' or '0' by having the PIC12F675 microcontroller USART program check the incoming bit in the dead center of the 104 μS bit width. Since our baud rate is 9600 bps and our bit width for 9600bps is 104 μS, that means we must have the microcontroller check the incoming bit stream every 104 μS.

There are still other things to consider. For instance, how does the microcontroller know when to start and stop the 104μS bit check intervals? For the answer, let's draw again from the RS-232 specification. We assigned a speed of 9600 bps for our data stream. However, we must also specify how many data bits will be transmitted and received in a data packet and how many stop bits will indicate the end of the data packet. We do have a choice as to the number of data bits we can stash into a data packet. The data packet bit length choices are 5 bits, 7 bits, 8 bits and 9 bits. Since the PIC12F675 is an 8-bit device, let's designate a data packet as 8 bits in length. Designating an 8-bit data packet allows the transfer of all readable ASCII characters plus control codes and hexadecimal or BCD (Binary Coded Decimal) data. We could have chosen 7 bits for ASCII transmission as well, but 8 bit data packets are more common and choosing a 7-bit packet inhibits sending a byte of miscellaneous information in a single-data packet.

The PIC12F675's built-in oscillator operates at 4 MHz, which equates to an instruction execution time of 1 µS. That means the PIC12F675 can theoretically execute 104 instructions during a stop-bit width, which is the same as the data-bit width of 104 µS. That time could be used to do some other processing if necessary—104 µS is a long time in microcontroller-land, so for us a single-stop bit will be sufficient.

There's another RS-232 component that can also be defined called *parity*. To keep things easy, we will not assign a parity bit. Parity bits are used to check the integrity of the data packet by inserting an extra bit to make the number of data packet marks even or odd depending on how the user has set up the communications equipment.

Now we have an asynchronous data stream consisting of a start bit, 8 data bits, no parity bit and 1 stop bit. The word asynchronous here means that the data packet can begin at any time without regard to any predetermined timings. If receiving, the presence of a start bit signals the PIC12F675 that a data packet is starting. So far, so good as we haven't done or defined anything out of the RS-232 ordinary.

Let's walk through the voltages that are generated when an RS-232 data packet is sent containing the ASCII representation of the number '2'. A '2' is represented in ASCII by hexadecimal 0x32 or binary 00110010. An idle RS-232 signal is defined as having the voltage on the transmit pin maintain a marking condition for a time that exceeds one data packet bit width. For 9600 bps, the steady marking condition must be greater than 104 µS in length. As you already know, a mark is a negative voltage between –3 and –25 volts and represents a '1' in RS-232 lingo.

To signal the start of a data packet, the transmitting device will drive the RS-232 transmit pin positive into the space voltage region of +3 to +25 volts. This transition from a steady marking state that is greater than or equal to one data packet bit width to a spacing state is called a start bit. Since we are running at 9600 bps, our start bit width is 104 µS, which is equal to our data packet bit width for a baud rate of 9600 bps.

Now here's where things get a bit tricky. Remember that the idea is to sample the incoming bits as closely to their center as possible to determine if the bit is a '1' or a '0'. Under ideal conditions, the start bit is recognized immediately by the receiving microcontroller. If the 104 µS interval begins at the same instant that the start bit is sensed, the microcontroller will sample at the end of the start bit time, which is 104 µS. The first data bit in the incoming data packet will be lost and so will the rest of the data bits, as the microcontroller will be sampling the bits on their leading edges instead of in their centers.

A valid marking condition must exist before a start bit is initiated. So, with that we have a very good idea as to when a start bit should occur. We also know from the RS-232 specification that every valid RS-232 data packet starts with a start bit and ends with at least one stop bit. So, to sync-up with the incoming data bits within the incoming RS-232 data packet, the receiving microcontroller is instructed to wait 1.5 RS-232 data packet bit width times after sensing a valid start bit. This allows the receiving microcontroller to begin the bit sampling in the center of the first incoming data bit. From there all the microcontroller has to do is sample every 104 µS seven more times to get the full 8-bits contained within the incoming RS-232 data packet.

A stop condition is indicated by the transmitting device when the RS-232 voltage being transmitted is returned to the marking state for at least one data bit width time, which is 104 µS for 9600 bps, after the correct number of RS-232 data packet bits are generated. This stop condition, or marking state is actually the stop bit. Everything I just described down to the microsecond is summed up in Figure 2.1

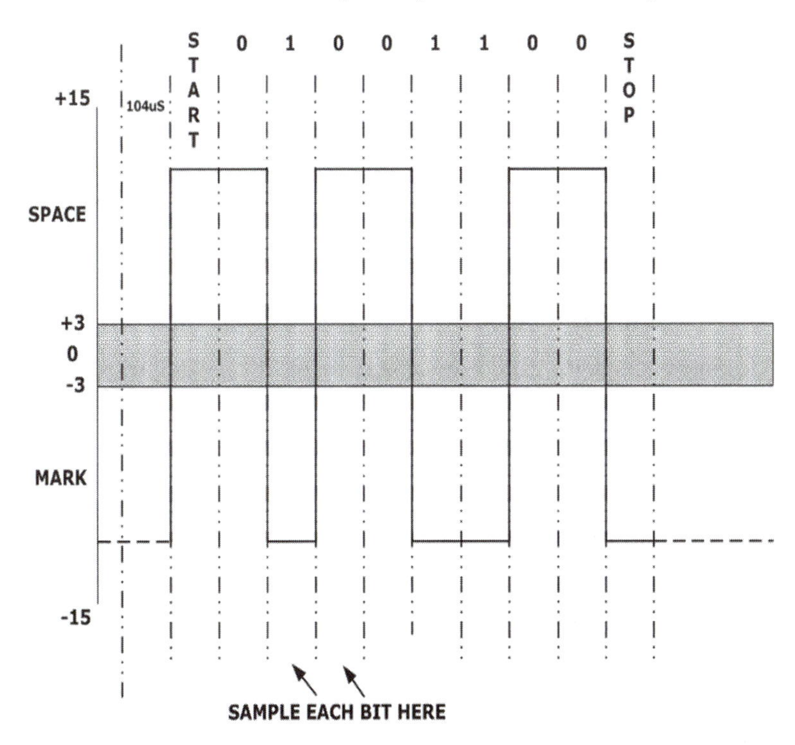

Figure 2.1: This is a graph of a 9600 bps asynchronous RS-232 transmission versus time. The time between each vertical double-dotted line represents 104 µS. Since we are only sampling for each bit one time, the idea is to try to sample as close to the center time of each bit as possible.

Let's run through it again. The transmitting microcontroller is holding it's RS-232 transmit pin in a marking condition. We know that this marking condition must be at least 104 μS in length to satisfy our bit timing for a 9600 bps baud rate. In fact, the marking condition can exist for hours, days or forever as the receiving microcontroller is continually looking for a valid start condition.

The transmitting device drives its transmit pin to a space condition for one data bit time (104 μS for 9600 bps) to indicate the start of an RS-232 data packet. The receiving microcontroller senses the start bit on its receive pin and waits for 156 μS (1.5 × 104 μS). At the 156 μS interval, the receiving microcontroller samples what should be the center of the least significant bit of the incoming RS-232 data packet, bit 0. The microcontroller samples the second bit of the incoming RS-232 data packet 104 microseconds later. The receiving microcontroller samples every 104 μS until the most significant bit of the RS-232 data packet is sampled (bit 7 since we are sending 8-bit data packets).

The receiving microcontroller has 8 bits of data and expects to see its receive line go to a marking condition indicating a stop condition or stop bit. Note that the receiving microcontroller and the sending microcontroller sync-up on every RS-232 data packet using the start bit. From there, every bit inside the RS-232 data packet is expected to be sent and arrive on time according to the baud rate. Later, you'll see that microcontrollers with internal USARTs will perform all of the start bit and receive/transmit timing tasks automatically for you. For now, let's do it caveman style.

Building a Simple Microcontroller RS-232 Transceiver

To convert the RS-232 theory I've presented into real-world events, let's assemble some hardware and implement a simple 3-wire RS-232 session between our PIC12F675 microcontroller and a personal computer.

A personal computer is most always configured as a DTE device. Recalling what we already know about the RS-232 specification, that implies that the personal computer's serial port uses a male 9-pin or male 25-pin connector. From here on out, unless I say otherwise, we'll use the 9-pin connector and pinout for both DTE and DCE devices. So, with that, pin 3 is the DTE transmit pin and pin 2 is the DTE receive pin. For the record, on a 25-pin male serial connector, pin 2 is the DTE transmit pin and pin 3 is the DTE receive pin. The third wire in our 3-wire RS-232 connection is the common ground connection. For a 9-pin male serial connector, the ground pin is pin 5 for both DTE and DCE devices and is designated signal ground in the RS-232 specification. From your history lesson, you know that the 25-pin DTE serial connector's signal ground is found on pin 7.

Applying logic (and your knowledge of the RS-232 specification) to the gender of the personal computer's serial connector would lead one to believe that since a DTE device is represented by a male connector then a DCE device would most likely support a matching female connector. Once again, logic prevails, as that is the real world case. Again, using common sense logic, one would be led to conclude that since the personal computer is a DTE device, our PIC12F675 would be the center of attention in a DCE device. If that is also true, which it is, then that means I can literally plug the personal computer's male DTE serial interface directly into the PIC12F675's female DCE interface and pass data between the personal computer and the PIC12F675. What makes this possible is the DCE serial connector pinout versus the DTE connector pinout. Basically, the DCE device's transmit pin is connected directly to the DTE device's receive pin and the DTE device's transmit pin is wired directly to the DCE device's receive pin with signal ground being common between the DTE and DCE interfaces. Don't confuse this with a "null modem" arrangement as a null modem circuit is intended to attach a DTE device directly to another DTE device by tying complementary modem signals to each other. Therefore, that makes pin 3 on the DCE side the receive pin and pin 2 the DCE transmit pin. Using the standard DTE and DCE pinouts on my connectors means that I can now communicate PIC to personal computer without the need for any special "crossed over" cables. In fact, all I need is three wires.

RS-232 Interface Hardware

As true RS-232 signals are not TTL compatible, the incoming RS-232 voltage levels must be converted to voltage levels compatible with the circuitry behind the serial connector. On the other side of that, the outgoing TTL voltage levels must be shifted to RS-232 signal levels for transmission between the DTE and DCE devices. The easiest way to effect the RS-232 voltage translation process and stay within the RS-232 specification's guidelines is to use a special RS-232 converter IC. One such IC is the industry standard Maxim MAX232CPE.

In the past, if you really wanted to adhere to the RS-232 specification you designed in a ±12-volt or ±15-volt power supply to drive the MC1488 (now called the DS1488) quad line driver. The negative supply voltage coupled with the MC1488 made the marks possible, while the positive 12 volts provided the voltage level necessary to produce a space. On the receiving side, an MC1489 (these days it's called a DS1489) picked up the marks and spaces, converted them to TTL levels and fed them to the device's UART (Universal Asynchronous Receiver/Transmitter).

The DS1488 and DS1489 are still in production and are great choices for low-cost RS-232 interfaces if the power supply voltages are already in the design anyway. However, to really keep it simple and within specification, using a MAX232CPE or similar IC at each end of the RS-232 link is the way to go. The MAX232CPE requires a single +5 VDC and with the help of four common 1 μF capacitors, the MAX232CPE internally generates the voltages necessary to effect marks and spaces on the transmit pin using an internal charge pump. Not only does the MAX232CPE perform the TTL-to-RS-232 conversion duties, it is the "other side" also converting the incoming RS-232 signals into TTL voltages. The MAX232CPE charge pump is capable of producing ±10 VDC when no significant load is present.

A Microcontroller DCE Device

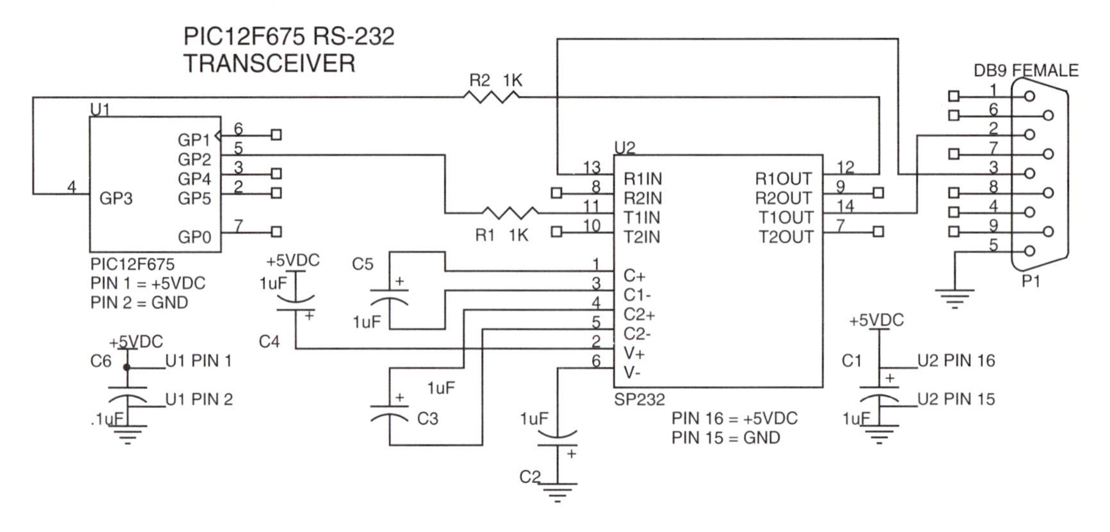

Schematic 2.1: This is the "formal" way to do it. Capacitors C2-C5 help the Sipex SP232ACP's internal charge pump provide the RS-232 voltages that adhere to the RS-232 specification. The PICkit 1 uses this formal approach.

You can build the PIC12F675-based RS-232 transceiver from scratch or you can take a value-added and easier way out by using the Microchip PICkit™ 1 FLASH Starter Kit. Before we move on, let's stop and talk a little about the PICkit 1.

Microchip's PICkit 1 FLASH Starter Kit

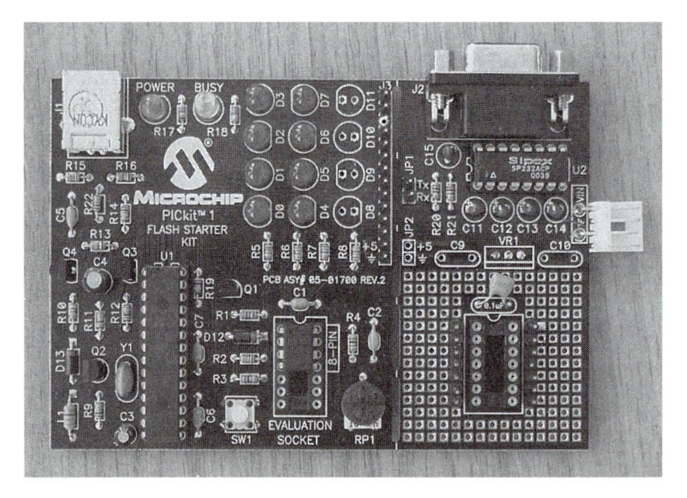

Figure 2.2: Intended for beginners, the PICkit 1 is simple to understand and operate. An 8-pin PIC12F675 is mounted in the evaluation socket. All of the USB circuitry is to the far left of U1, a PIC16C745.

The PICkit 1 FLASH Starter Kit is designed to allow easy and inexpensive evaluation of Microchip's new 14-pin flash-based PICs and some of the legacy 8-pin flash parts like our PIC12F675. The PICkit 1 FLASH Starter Kit programming hardware is centered on the PIC16C745, which contains a USB engine in addition to the normal stuff you would find in a PIC microcontroller.

Along with the hardware and firmware contained in a USB microcontroller, the magic of USB is performed within the Windows operating system. Special programs and drivers running under Microsoft Windows form an alliance between the microcontroller's I/O ports, the microcontroller's USB interface and the application that is running under the Microsoft Windows operating system. In effect, all of the work is done up front and all of the pent up USB programming in the microcontroller and on the personal computer is unleashed when the user plugs a USB device into a personal computer's USB port.

A really neat feature of the PICkit 1 FLASH Starter Kit is that after you have initially downloaded a hex file you can compile the file again and as long as you tell the compiler to always replace the old hex file after a compile, the PICkit 1 will automatically bring in the newly compiled hex file for programming when you click on the Write Device command button. The PICkit 1 FLASH Starter Kit programming interface does this by checking the timestamp of the loaded hex file and loading in the latest time-stamped hex file of the same name.

The target PIC's power is controlled (on or off) by clicking on the Device Power button in the Board Controls box. I used this feature extensively to turn off the PIC12F675 after programming it so I could move it over to the snap-off board socket to run the spin of code I had just compiled and programmed.

The PICkit 1 FLASH Starter Kit hardware communicates with the PICkit 1 FLASH Starter Kit programming interface (Figure 2.3) that runs under Microsoft® Windows®. The PICkit 1 programming interface allows the user/programmer (that's us) to view PIC Program Memory and EEDATA Memory in hexadecimal format. The Program Memory and EEDATA Memory windows contain the contents of a standard Intel hex file the user/programmer loads into the programming interface that has been generated by either a compiler like PicBasic™ Pro Compiler or Custom Computer Services C Compiler or an assembler like PicBasic Pro's PM or Microchip's MPASM™.

The idea is to generate an Intel hex file, load it into the PICkit 1 FLASH Starter Kit programming interface and "burn" or program the binary code into the physical PIC device in the PICkit 1's evaluation socket. A compiled program file (Intel hex file generated by the compiler or assembler) is downloaded into the PICkit 1 FLASH Starter Kit programming interface by using the Import HEX menu item. When the file download is complete, the data contained within the downloaded hex file will appear in the Program Memory and EEDATA windows. At this point, the user/programmer can click on the Write Device button and burn the downloaded code into the target PIC. If all goes well, a green banner will be displayed at the bottom of the PICkit 1 FLASH Starter Kit programming interface window. A red banner signifies that something went wrong in the program cycle.

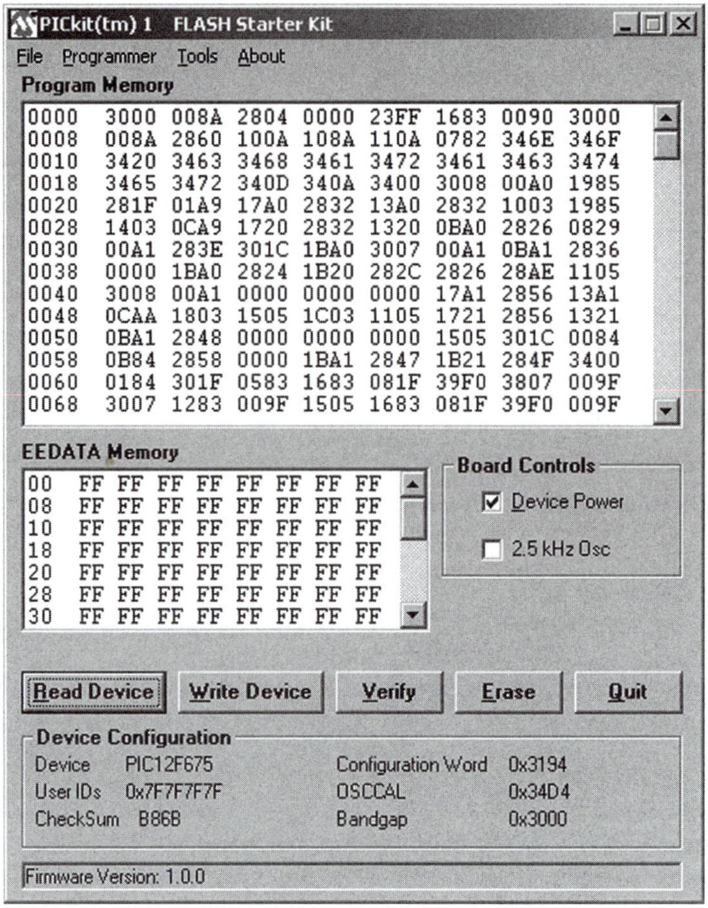

Figure 2.3: Once you load a hex file for programming, each time you issue a Write Device command, the PICkit 1 program finds and reloads the latest version of the hex file you originally specified before programming the PIC.

Providing that the target PIC has not been code protected, the user/programmer can read the contents of the target PIC and save the data as a hex file using the Export Hex menu item. Two other command buttons allow the user/programmer to verify existing code in a PIC mounted in the PICkit 1 program socket with the contents of a hex file and to erase the target PIC part.

The PICkit 1 FLASH Starter Kit shown in Figure 2.2 is a preassembled PIC development board with an unpopulated snap-off experimenter board. The PICkit 1 FLASH Starter Kit is unique in that it doubles as a PIC programmer, but not just any old PIC programmer. A special Visual Basic program that runs on a host personal computer controls the PICkit 1 FLASH Starter Kit. The personal computer is attached to the PICkit 1 FLASH Starter Kit via USB. The bonus is that all of the source code for both the Visual Basic personal computer program and the USB interface is included, in addition to the PIC tutorial and project source code. So, if you're curious about how PIC programmers work and have an interest in how USB works, the PICkit 1 FLASH Starter Kit is a must have device.

I left the snap-off experimenter board attached to the PICkit 1 FLASH Starter Kit and rigged a standard personal computer's diskette drive power connector to get +5 VDC and ground to the snap-off board. These days, personal computer's power supplies are cheap and using a personal computer power supply gave me a power switch and keyed power receptacle for the experimenter board side of the PICkit 1 FLASH Starter Kit while eliminating the need to solder in a 7805 +5 VDC regulator and its supporting circuitry.

I also substituted a pin-for-pin compatible Sipex SP232ACP for the MAX232CPE, as I don't have a through-hole MAX232CPE in my parts inventory. I completed the assembly of my PICkit 1 FLASH Starter Kit experimenter board by installing the TX (transmit) and RX (receive) header pins and the 14 header pins around the PIC socket. Installing the headers will allow easy connections between the Sipex SP232ACP and the PIC12F675.

Even though the pins of the 14-pin socket on the programmer side of the PICkit 1 are connected directly to LEDs, you can still use the pins to run our RS-232 transceiver project. Just solder in the J3 header and use a jumper wire to connect the programmer side TX and RX pins to the snap-off board's TX and RX pins. This allows you to program and execute the programs without having to move the PIC12F675 from the programming socket to the snap-off test socket.

Although the PICkit 1 is nice to have, if you already have a PIC programmer that will burn the PIC12F675 you can build up the "formal" circuit shown in Schematic 2.1 or you can get down and dirty with the "dirty" RS-232 implementation shown in Schematic 2.2.

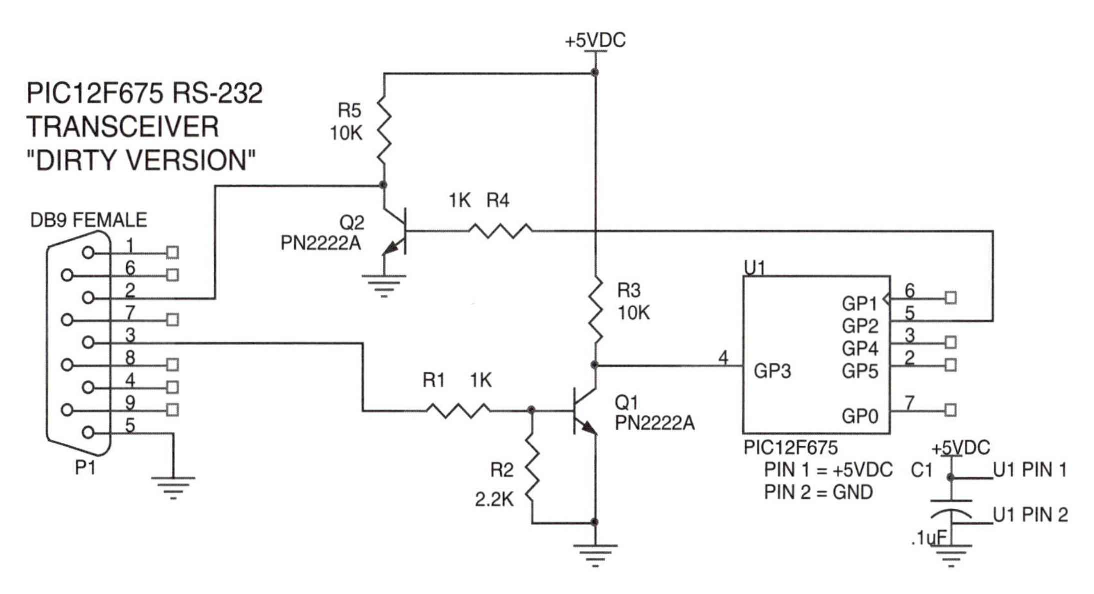

Schematic 2.2: If you don't have a MAX232 or Sipex SP232ACP on hand, or if you want to save some bucks and have some fun at the same time, lash up this "dirty" RS-232 transceiver.

In the “dirty” version, Q1, Q2 and the five resistors perform the RS-232 voltage conversion. Any positive voltage coming in on P1’s pin 3 that is capable of turning on Q1 will be considered “RS-232 OK” and will pass as a binary 0, or space, to the PIC12F675’s GP3 receive pin. If the incoming RS-232 voltages are up to specification and the RS-232 cable is of good quality, this receiver circuit formed by Q1, R1 and R2 will work very well in most instances. The same is true for the transmit circuit, which is driven using Q2, R4 and R5. If the RS-232 cable is not too long and is of a high quality, Q2 will send a “dirty” mark (0 VDC instead of –3 VDC or better) when it is turned on by the PIC12F675’s transmit pin, GP2. A clean space will be transmitted when Q2 is off. If your project can tolerate possible RS-232 bit errors, the “dirty” RS-232 circuitry shown in Schematic 2.2 is a cheap and easy way to implement an RS-232 link.

Writing Some Simple RS-232 Firmware

No matter which direction you took, “dirty” PICkit 1 or homebrew “formal,” I’m sure you’ll agree that the RS-232 hardware was easy to obtain and assemble. The RS-232 code for our minimal RS-232 system is just as easy to write.

There are a variety of C compilers on the market that target the Microchip PIC® family of microcontrollers. I’ve chosen to use the Custom Computer Services C Compiler for Microchip PIC microcontrollers to write the code for the PICkit 1 FLASH Starter Kit RS-232 circuit I’ve assembled. The inexpensive Custom Computer Services C Compiler is easy to use and has features that take the pain out of writing code for PICs. I’ve written a couple of programs that simply send the ASCII character ‘A’ to a HyperTerminal™ or Tera Term Pro™ terminal emulator program.

For those of you that don’t do C, I’ve selected the PicBasic Pro Compiler from microEngineering Labs to represent the RS-232 firmware on the BASIC side of the house. Like Custom Computer Services C Compiler, the PicBasic Pro Compiler from microEngineering Labs is dedicated to producing clean and tight code for Microchip PIC microcontrollers.

Before I describe the code, let’s make sure you have your terminal emulator set up correctly. HyperTerminal is included as an accessory communications program with the Microsoft Windows operating system. It’s fairly easy to prepare HyperTerminal to receive our RS-232 data. Once you open HyperTerminal, the first thing you want to do is name your session. In Figure 2.4, I named my HyperTerminal session “Simple PIC RS-232.”

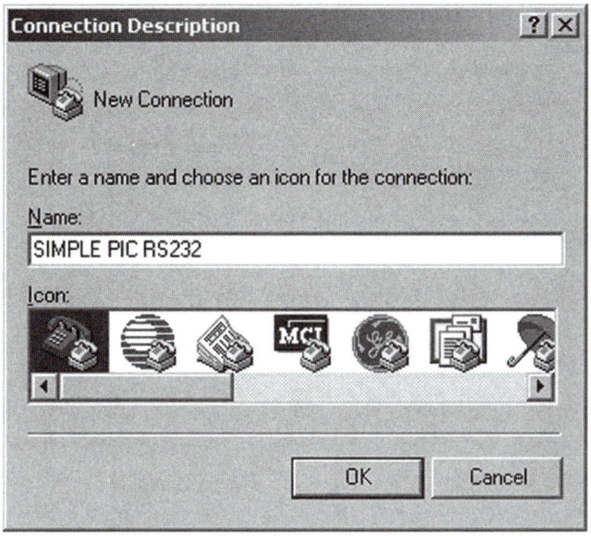

Figure 2.4: Doing this allows you to save the HyperTerminal session with a name for later use.

After you name your session, another window like the one in Figure 2.5 will appear asking which COM port you wish to use. That all depends on what's available on your machine. In my case, I had both COM ports 1 and 2 open and chose COM 1.

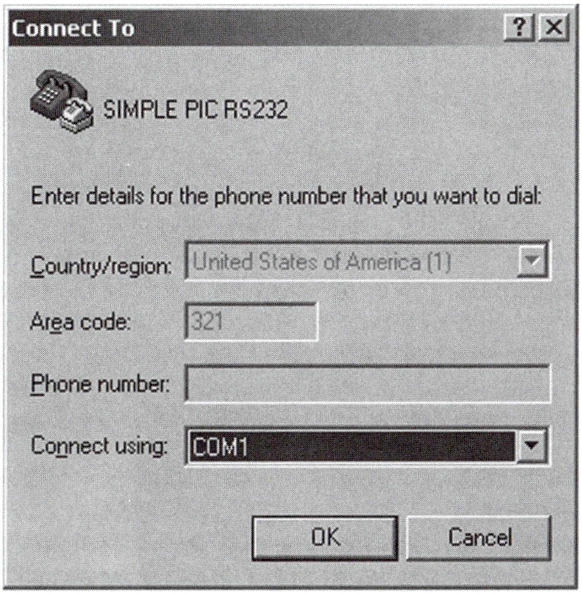

Figure 2.5: Select an open COM port on your personal computer here.

The final step in setting up your HyperTerminal session is the definition of the communications parameters. We defined those earlier as 9600 bps, no parity bit, 8 data bits and 1 stop bit. Set up your serial port as it is shown in Figure 2.6.

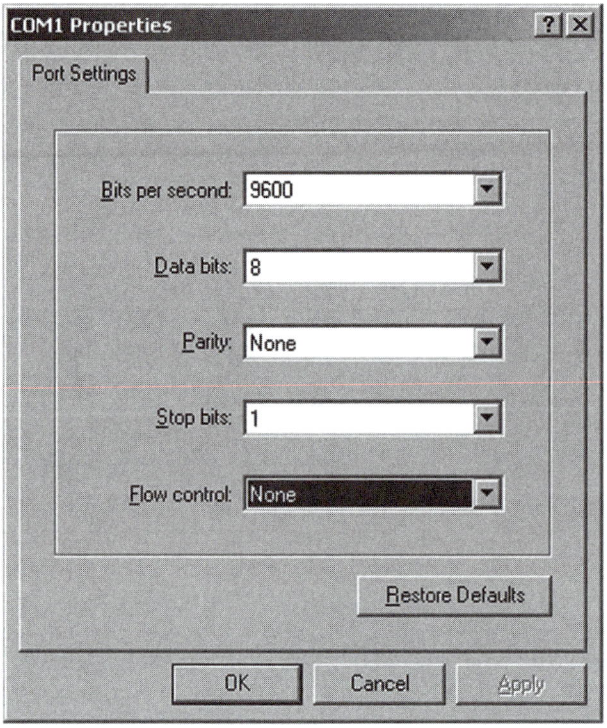

Figure 2.6: No modem control or software signals (flow control) are needed in a simple 3-wire RS-232 connection.

Flow control hasn't been covered yet, and for this project we'll assume there isn't any. Flow control comes in a multitude of flavors. Normally, flow control is implemented by using the modem control signals CTS and RTS. Flow control can also be initiated using software commands like those used to implement XON/XOFF flow control. One could also use a logic signal from a standard I/O pin to effect an unofficial flow control. Flow control excepted, the goal is to end up with a blank terminal emulator window and a blinking cursor in the upper left corner of the terminal emulator window.

Unless you purchase some upgraded HyperTerminal software, you won't be able to do much more than open a HyperTerminal emulator session and send or receive data with the version that is bundled with Windows. Another terminal emulator called Tera Term Pro provides a bit more functionality and flexibility than HyperTerminal and it costs nothing but your time to download it from the Internet. Tera Term Pro setup is similar to that of HyperTerminal, and as you will see in the pull-down menus, there are some things Tera Term Pro can do that the stock HyperTerminal can't. Tera Term Pro's most useful feature is the scripting language that is built into it. Using Tera Term Pro's script commands provides a means of automating the process of transferring and receiving files. We won't need any Tera Term Pro scripting for our simple RS-232 project.

Editing the TERATERM.INI file, which resides inside the Tera Term Pro directory, can be used to set up all of Tera Term Pro's communications parameters. Here, I'll show you how to get a basic Tera Term Pro emulation session to work on your personal computer manually. The first thing you want to do is tell Tera Term Pro that you will be using a serial interface. As you can see in Figure 2.7, Tera Term Pro is capable of doing many other things on differing interfaces.

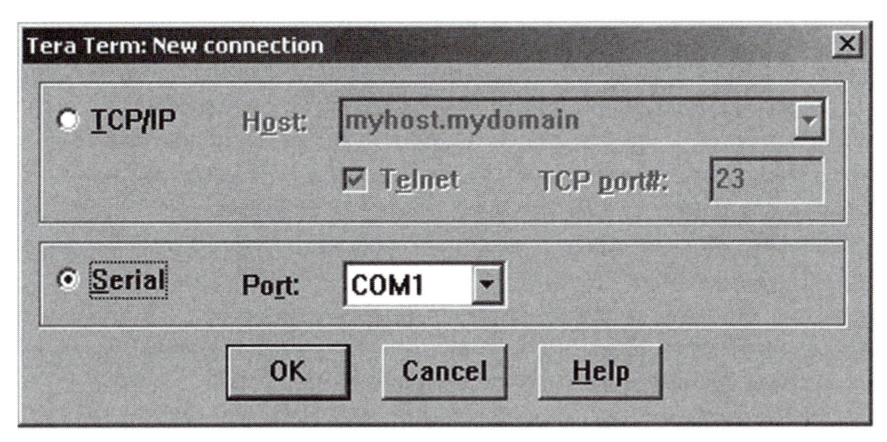

Figure 2.7: I'll show you how to use the TCP/IP part of Tera Term in a later chapter. Right now, use the Serial side and enter a COM port number that's open on your personal computer.

Under the SETUP pull-down menu, you will find an entry for Serial Port. Selecting the Serial Port menu item will bring up a window like the one depicted in Figure 2.8 and allow you to manually set the communications parameters, which are identical to the communications parameters we set in HyperTerminal (9600 bps, 8 data bits, no parity, 1 stop bit).

And again, just like HyperTerminal, you should end up with a blank terminal emulation window with a flashing cursor in the upper left corner. To complete the personal computer and terminal emulator setup, all that's left to do is to attach a pin-for-pin (pin 1 to pin1, pin 2 to pin 2, etc.) 9-pin male-to-female cable between the personal computer's serial (COM port you selected in the setup) port and the PICkit 1 FLASH Starter Kit's 9-pin serial connector on the PICkit 1's snap-off experimenter board. Now, let's pick apart the RS-232 C code.

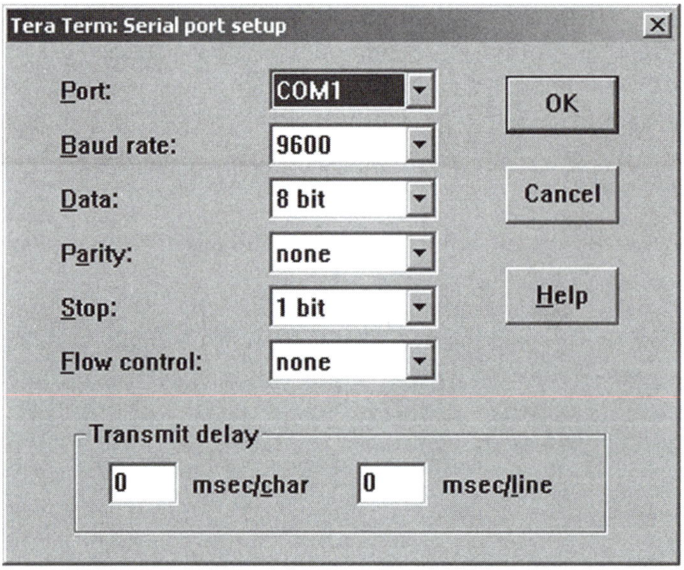

Figure 2.8: The Transmit delay is used to pace the characters. For instance, changing the msec/char field to a 1 would send a character wait 1 ms and then send another character and so on.

I'm not going to assume you know every nuance of C, so this time I'll take us through line by line. The *#include* lines at the top of the listing tell the compiler about the physical attributes of the PIC12F675. The "physical attributes" of a microcontroller device may include the number of I/O pins or the types of special purpose modules that reside inside the microcontroller like analog-to-digital converters or timers. The include files also define associations. For instance, operations that need to express a TRUE or FALSE condition, it's much easier to remember TRUE for 1 and FALSE for 0. Using real words also makes the code easier to read and follow. Another example of what include files do would be equating I/O port names. Instead of having to remember that PORTA is actually address 0x005, the *#include* allow you to simply type in "PORTA" when you are performing tasks against address 0x005. The C include files are readable and you can examine them as you would any other text file. Perusing a microcontroller's datasheet and include files are a good way to learn about what the microcontroller can really do for you. The Custom Computer Services C Compiler comes with an include file for each PIC microcontroller it supports. If there are physical attributes you need to access and they aren't already included in the stock include file there's nothing to stop you from putting together your own include file. I used the PIC12F675 datasheet to build the *f675.h* include file, which includes definitions and associations from the PIC12F675 datasheet that were not included in the canned PIC12F675 include file.

```
#include <12F675.h>
#include <f675.h>
```

The datasheet is the most important tool when working with any microcontroller device. Checking the PIC12F675 datasheet tells us that the PIC12F675 is equipped with an on-chip oscillator that does not require an external crystal or resonator. Another look at the PIC12F675 datasheet tells us the internal clock speed of the internal oscillator is a nominal 4 MHz. Another plus in using the Custom Computer Services C Compiler is that once the clock speed is defined to the compiler, things like delays and baud rates are automatically calculated and applied inside the compiler routines that rely on the microcontroller's clock speed. So, the line *#use delay (clock = 4000000)* sets the PIC12F675 clock rate at 4 MHz and tells the compiler to use 4 MHz for its delay and baud rate calculations.

```
#use delay(clock=4000000)
```

Bits inside fuse words are used to turn on or turn off certain special purpose modules, functions or features that the PIC12F675 offers to the programmer. Again, checking the PIC12F675 datasheet, we know that the PIC12F675 can be instructed to use the internal oscillator or depend on an external crystal arrangement. The *INTRC_I/O* fuse instruction sets a fuse bit that activates the PIC12F675's internal 4 MHz oscillator. In addition to selecting the clock type, the *INTRC_I/O* bit deactivates the clock signal from being accessible via a PIC12F675 I/O line.

```
#fuses INTRC_IO,NOWDT,NOMCLR,NOPROTECT,NOCPD,NOBROWNOUT
```

The next fuse instruction, *NOWDT*, deactivates the PIC12F675 watchdog function. Watchdog timers are commonly used to monitor the microcontroller's execution of instructions. If the microcontroller "hangs" or "loops" and the watchdog timer doesn't get reset, the microcontroller is forced to reset itself and restart the application that is programmed into it. For simple programs like this one, the watchdog timer function is not necessary.

As you've probably already figured out, the "NO" in front of the rest of the fuse instructions turns off a particular PIC12F675 function. *NOMCLR* saves an I/O line on the PIC12F675 by not requiring the MCLR reset pin to be offered to the programmer externally. Instead, the MCLR pin function is performed internal to the PIC12F675.

Activating code protection makes reading the PIC12F675's program memory with a PIC programmer impossible. Since I haven't written any code that would stop an alien attack, *NOPROTECT* and *NOCPD* allow the code loaded into the PIC12F675 program memory to be accessed by the standard methods.

I'm also not anticipating my personal computer power's supply voltages to dip or "brownout" under load, so there is no need for brownout protection, and *NOBROWNOUT* is pretty obvious as to how I feel about that.

While we're on the fuse bit subject, the Custom Computer Services C Compiler has a really nice pull-down *View* menu feature that describes and lists the valid fuses for the microcontroller you're writing code for. In that same pull-down *View* menu, the compiler also gives you access to the microcontroller datasheets, which are stored in a directory as standard PDF files. The scope of this book isn't really about teaching you C or tutoring you on how to use the Custom Computer Services C Compiler. However, as we continue on this networking hop, I'll point out goodies inside the compiler packages that will help you write the best code with the least effort. If you're not a C person, who knows, you may pick up enough C to become proficient with the language.

The Custom Computer Services C Compiler does many things behind the scenes to assist you but sometimes it comes at the expense of extra code that is generated by the compiler. If you're a control freak like I am, I want to be in command as much as possible. So, the *#use fast_io(A)* code line tells the Custom Computer Services C Compiler to allow me and not the compiler to determine the direction (input or output) of each PIC I/O line.

```
#use fast_io(A)
```

Our simple RS-232 C program actually consists of three subprograms: TX_program_1, TX_program_2 and TX_program_3. Each program does the same thing—transmits the ASCII character 0x41 or 'A'. By simply placing each subprogram between a set of *#ifdef* and *#endif* preprocessor statements, I can compile one of the subprograms at a time by "defining" which program is active during the compilation time. The subprogram to compile is chosen by "commenting out" the other subprograms I don't want to be compiled. For instance, to select TX_program_1 in Code Snippet 2.1, I comment out *#define TX_program_2* and *#define TX_program_3*. When I run the Custom Computer Services C Compiler, all that will be included in the final output file will be the common code plus all of the code between *#ifdef TX_program_1* and its corresponding *#endif* preprocessor statement. I've used the Custom Computer Services C Compiler to write more complex programs and you'll get a taste of that as we progress.

```
//*******************************************************
// COMMENT OUT THE PROGRAMS YOU DON'T WANT TO RUN
//*******************************************************
#define   TX_program_1   //this program will be compiled
//#define  TX_program_2 //this program will not be compiled
//#define  TX_program_3 //this program will not be compiled
```

Code Snippet 2.1: When you begin to write larger C microcontroller programs, you'll use the // to comment out parts of code instead of deleting them.

All C programs have a *main* function like the one shown in Code Snippet 2.2. The main microcontroller application program actually flows inside the *main* function braces. In our RS-232 code, any code that is not fenced in by *#ifdef TX_program_x* and a related *#endif* is always compiled and can react with the selected TX_program_x code segment.

```
//*******************************************************
// MAIN PROGRAM STARTS HERE
//*******************************************************
// This code fragment will always be compiled
void main() {

   setup_adc_ports(0);
   setup_adc(ADC_OFF);
   setup_timer_1(T1_DISABLED);
   setup_comparator(NC_NC_NC_NC);
   setup_vref(FALSE);

                         //PORTA pin 2 = TX line
   SET_TRIS_A(0b00001000); //PORTA pin 3 = RX line
```

Code Snippet 2.2: The Custom Computer Services C Compiler program wizard generated all of the setup statements.

In addition to the on-chip analog-to-digital converter, the PIC12F675 also contains an analog comparator, a voltage reference and some timers. Since we won't be using any services provided by these modules, the *setup_xxxxx* lines of code are there to turn off TIMER_1, the analog-to-digital converter, the comparator and the voltage reference. Executing the "setup" lines will also free up any I/O pins that the service modules may have wanted to use.

All of the subprograms have a few things in common; each subprogram transmits the letter 'A' and each subprogram uses the same PIC12F675 I/O pins for transmitting and receiving. That means that I can set the I/O direction of the PIC12F675's receive and transmit pins in the common code. The *SET_TRIS_A(0b00001000)* code line completes the manual I/O direction task and feeds my control freak animal as I, not the compiler, set the PIC12F675's I/O pin direction.

A Bit of RS-232 Transmit Code

Earlier I talked about how each of the data bits inside a data packet must be 104 μS in duration to be recognized as a 9600 bps bit stream. The first program, TX_program_1, is a crude 9600 bps algorithm that uses delays and bit voltage levels to transmit the ASCII character 'A'. To make things a bit easier to read in the TX_program_1 main code, I've defined the TX (transmit) pin, PIN_A2, and the RX (receive) pin, PIN_A3, in the PIN DEFINITIONS area before the main program code as seen in Code Snippet 2.3.

```
//*******************************************************
// TX_PROGRAM_1 PIN DEFINITIONS
//*******************************************************
#ifdef TX_program_1
#define TX  PIN_A2
#define RX  PIN_A3
#endif
```

Code Snippet 2.3: It's best to keep the C code human readable.

TX_program_1 begins by placing the TX line in a marking state for 1 ms. The *output_high(TX)* instructs the PIC12F675 to present a TTL high (binary 1) to the Sipex SP232ACP's TTL input. The Sipex SP232ACP inverts that to present a RS-232 mark on pin 2 (DCE transmit pin) of the communications cable. The *while(1)* statement says that while the tested condition is 1 or while the tested condition is TRUE, the code between the braces ({}) will run. Since 1 never changes value and 1 represents TRUE, the code will run in this loop inside the braces forever. This is one way of creating a continuous loop. I could have also used *for(;;)* to accomplish the same thing. I've included both statements in the source code and Code Snippet 2.4 for you to try.

```
//*******************************************************
//  TRANSMIT PROGRAM 1
//*******************************************************
#ifdef TX_program_1

   output_high(TX);        //mark for more than 104uS
   delay_ms(1);

   while(1)
   //for(;;)
   {
      output_low(TX);      //send 0  START BIT
      delay_us(104);
      output_high(TX);     //send 1  LSB of 'A'
      delay_us(104);
      output_low(TX);      //send 0
      delay_us(104);
      output_low(TX);      //send 0
      delay_us(104);
      output_low(TX);      //send 0
      delay_us(104);
      output_low(TX);      //send 0
      delay_us(104);
      output_low(TX);      //send 0
      delay_us(104);
```

```
        output_high(TX);        //send 1
        delay_us(104);
        output_low(TX);         //send 0  MSB of 'A'
        delay_us(104);
        output_high(TX);        //send 1  STOP BIT
        delay_us(104);

        delay_ms(1000);         //pace the transmission
    }
#endif
```

Code Snippet 2.4: As you'll see in the code in the upcoming chapters, I like to use while(1).

The first *output_low(TX)* is a start bit. The TTL low (binary 0) from the PIC12F675 I/O pin is inverted by the Sipex SP232ACP and comes out as a space on the RS-232 side. Note that the ASCII 'A' is transmitted to the personal computer's least significant bit first. Eight bits and eight *output_XXX/delay_us (104)* sequences later, Tera Term Pro displays the 'A' it received in the terminal emulator window I've captured in Figure 2.9.

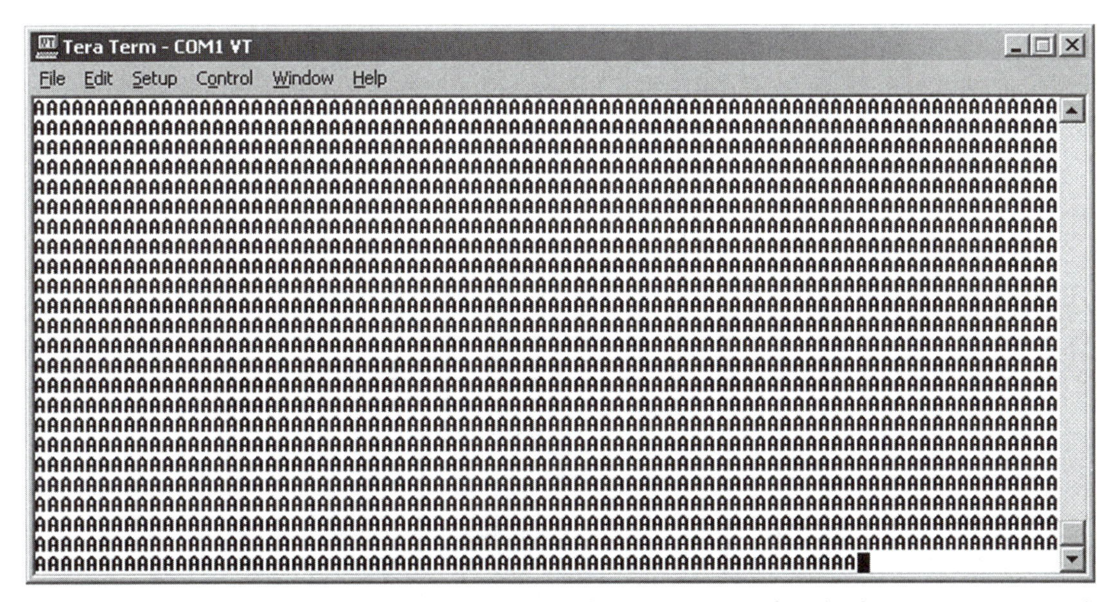

*Figure 2.9: Notice I "paced" the transmission in Code Snippet 2.4. After the first 'A' was sent, each 'A' thereafter was sent one per second (*delay_ms(1000)*).*

I've put a pacing statement at the end of the loop. This will allow you to see the characters as they appear in 1-second (1000 milliseconds = 1 second) intervals in the Tera Term Pro emulator window. You can comment this statement out to see the 'A's zip by.

Let's comment out *#define TX_program_1* and *#define TX_program_3* to select TX_program_2. Note the *#use rs232* statement. This is Custom Computer Services' way of having the compiler set the baud rate and assign the RS-232 I/O pins for you without having to consult the datasheet to make the adjustments manually on a bit-by-bit basis. Remember, the baud rate here is calculated based on the microcontroller's clock speed which is defined at the beginning of the program using the *#use delay(clock=4000000)* statement. Since the PIC12F675 has no internal USART, we can choose almost any pair of I/O pins to be TX and RX. Note that I said that "almost" any pair of PIC12F675 I/O pins could be chosen. The PIC12F675 has an input only pin (GP3) and since this is an input only pin, it can't be used as an output and thus can't be used as a transmit pin. You're probably also wondering where I'm getting these PORTA definitions when the PIC12F675 datasheet states that GPIO is used to define the PIC12F675 I/O port names. That's a Custom Computer Services C compiler thing. It uses PORTA designations instead of GPIO names. GPIO and PORTA are both located in their respective data memory maps at location 0x05. So, it's only a name difference. The whole of TX_program_2 is shown in Code Snippet 2.5.

```
//*******************************************************
//  TRANSMIT PROGRAM 2
//*******************************************************
#ifdef TX_program_2

#use rs232(BAUD=9600, XMIT=PIN_A2, RCV=PIN_A3)

   while(1)
   {
      printf("A");
      //printf("Your first name here");
      delay_ms(1000);
   }
#endif
```

Code Snippet 2.5: Wow! Consider doing this in PIC assembler. Are you beginning to like C?

TX_program_1 consists of 23 lines of C statements (25 if you include the defines for the TX and RX lines). TX_program_2 is comprised of only three C statements and does the exact same thing as TX_program_1. What gives? The trick is the plenty powerful *printf* statement. I'm not going to explain the coding in detail, but you can see for yourself that using *printf* has more advantages than drawbacks. Replace the 'A' with your first name and compile and run the program again. Cool, huh? That's what the C compiler *printf* services buys you. Of course, in the embedded world nothing is free. So, to gain the ease of use of the *printf* function, you pay in the increased amount of code the function generates and the additional amount of program memory that is consumed. To get an idea of how much extra code is generated, the Custom Computer Services C Compiler allows you to view the assembler code it generates. Compile TX_program_2 yourself and take a look at the list file to get an idea of what I'm talking about. Even though more code is generated, it's only generated once and placed in memory for use by other calls to the *printf* code. So in the long run, for the price of a little additional code, you get increased functionality with a minimum of coding effort.

The TX_program_3 in Code Snippet 2.6 is a simplified version of TX_program_1. However, it is very similar to TX_program_2 as it is short and sweet and it sends a single 'A' to the Tera Term Pro emulator window. Compile and run TX_program_3 to see the 'A's sequence through. Then comment out the putc (put character) line and try to compile and run with the *You can't put but 1 character here* line. The compiler will choke and tell you that you can't do this. Why? Because putc is an abbreviation for *put character*. That means a single character and not a string of characters.

```
//*******************************************************
//  TRANSMIT PROGRAM 3
//*******************************************************
#ifdef TX_program_3

#use rs232(BAUD=9600, XMIT=PIN_A2, RCV=PIN_A3)
   while(1)
   {
      putc('A');
      //putc('You can't put but 1 character here');
      delay_ms(1000);
   }
#endif
```

Code Snippet 2.6: Use putc when you want to conserve program memory and have small canned messages or single characters to send.

As you can see from the example code, using C for RS-232 work removes most of the housekeeping hassles associated with setting up RS-232 hardware and let's you concentrate on getting your data transferred from point A to point B. If I haven't convinced you that C is the easier road to RS-232 happiness, and if you just really have to write some assembler to transmit a byte, Code Snippet 2.7 is a working example of C-less assembler RS-232 transmit routine:

```
;*********** RS-232 TRANSMIT SUBROUTINE
;
SENDIT
   MOVWF     XMTREG       ;LOAD BYTE TO TRANSMIT
XMTR
   MOVLW     8            ;LOAD NUMBER OF BITS TO SEND
   MOVWF     COUNT
   BCF       RS232,TX     ;WRITE 0 TO SERIAL PORT
```

```
   CALL      DELAY1       ;WAIT 1 BIT PERIOD
XNEXT
   BCF       STATUS,C     ;CLEAR CARRY
   RRF       XMTREG,F     ;ROTATE TRANSMIT REGISTER RIGHT THRU CARRY
   BTFSC     STATUS,C     ;CHECK CARRY STATUS AFTER THE ROTATE
   BSF       RS232,TX     ;IF CARRY IS SET, WRITE A 1 TO SERIAL PORT
   BTFSS     STATUS,C     ;CHECK CARRY STATUS AFTER THE ROTATE
   BCF       RS232,TX     ;IF CARRY IS CLEAR, WRITE A 0 TO SERIAL PORT
   CALL      DELAY1
   DECFSZ    COUNT,F      ;DECREMENT THE COUNT REGISTER
   GOTO      XNEXT        ;NOT DONE, GO GET NEXT BIT AND SEND IT
   BSF       RS232,TX     ;Send Stop Bit
   CALL      DELAY1       ;WAIT ONE BIT PERIOD
   RETLW     0            ;DONE, RETURN TO CALLER
DELAY1
   MOVLW     BAUD         ;104uS for 9600 BAUD
STARTUP
   MOVWF     DLYCNT
REDO1
   NOP
   NOP
   NOP
   DECFSZ    DLYCNT,F
   GOTO      REDO1
   RETLW     0
```

Code Snippet 2.7: This homegrown code was all I had when I started writing microcontroller RS-232 communications functions.

To make the assembler transmit routine work, all you have to do is calculate the bit delay time (number of cycles to expend) versus the clock frequency your project is using and plug your results into the BAUD variable. Remember, if you choose to do this as a C program, the C compiler and its related RS-232 libraries perform the automagic RS-232 setup work.

Now that you have an idea of the hows and whys of sending data with a minimal microcontroller like the PIC12F675, let's figure out how to make that PIC12F675 receive RS-232 data.

Some RS-232 Receive Code

One would believe that we could take what we know about data packet timing and write a few lines of C code akin to TX_program_1 to receive some characters from our Tera Term Pro session. That cannot easily be done, however, even though your RS-232 receive C code will consist of mostly C statements, you'll probably still end up writing the time critical routines in assembler. We actually got lucky in TX_program_1, as our delay loop overhead was small enough to not disrupt our data packet bit timing. Why reinvent the wheel by writing RS-232 receive code from scratch? Let the C compiler and RS-232 libraries do the work. Code Snippet 2.8 is an example of writing a receive routine in Microchip assembler for our PIC12F675.

```
;*********** RS-232 RECEIVE SUBROUTINE
;
GETBYTE
   CLRF      RCVREG
   BTFSC     RS232,RD      ;LOOK FOR A START BIT
   GOTO      GETBYTE

   CALL      STARTBIT      ;go do start bit delay
RCVR
   MOVLW     8             ;load W with 8
   MOVWF     COUNT         ;load w to count
R_NEXT
   BCF       STATUS,C      ;clear the carry bit
   BTFSC     RS232,RD      ;look for data bit
   BSF       STATUS,C      ;if 0..skip this instruction
   RRF       RCVREG,F      ;ROTATE BIT FROM CARRY INTO RECREG
   CALL      DELAY1        ;go wait 104 uS
   DECFSZ    COUNT,F       ;decrement COUNT..skip if 0
   GOTO      R_NEXT        ;skip this instruction if COUNT=0

   RETLW     0

STARTBIT

   MOVLW     STARTDLY      ;DELAY FOR 156uS
   GOTO      STARTUP

DELAY1
   MOVLW     BAUD          ;104uS for 9600 BAUD
STARTUP
   MOVWF     DLYCNT
REDO1
   NOP
   NOP
   NOP
   DECFSZ    DLYCNT,F
   GOTO      REDO1
   RETLW     0
```

Code Snippet 2.8: Timing is very critical in this code and the faster the baud rate, the more critical the timing becomes.

Again, to make this code return a character you have to calculate the value of the BAUD variable, which depends on the microcontroller's clock frequency and the amount of loop overhead in the code. In short, you have to count instruction cycles and translate them to elapsed time to set the BAUD value correctly. This is how I used to do it before the introduction of C for PIC microcontrollers. I can tell you that if you don't have a way to view the register values in the debugging process, you will be forced to use time-consuming, trial-and-error coding techniques.

What if you wanted to transmit a random character and not just the character 'A'? I ask this question because if we are to continue with our building of simple RS-232 routines, we must be able to view the results of our receive algorithms. Assuming we would want to test the assembler RS-232 receive code you were just introduced to, how would we transmit the received character to our Tera Term Pro emulator session?

What if we chose to use TX_program_1 to echo the character received by our RS-232 receive assembler program? TX_program_1 would need some heavy-duty modifications to scan the received character's bits and translate them to *output_low* or *output_high* states used in the TX_program_1 algorithm. The overhead of the code needed for the TX_program_1 modification would most likely interfere with the RS-232 data packet bit timing and cause the RS-232 transmit character code to fail or operate erratically. In that case, incorporating the assembler transmit routine would be a better choice than modifying the TX_program_1 code.

Although there is nothing wrong with either the assembler transmit code or the assembler receive code, a couple of simple C statements can eliminate a truckload of RS-232 coding grief. Those little C statements are *putc* and *getc*. The *getc* instruction performs the same task as our RS-232 assembler receive routine. Code Snippet 2.9 an example of how the *getc* function is written in a C program.

```
#use rs232(BAUD=9600, XMIT=PIN_A2, RCV=PIN_A3)
int8 character_in;
//Receive a character
character_in = getc();
```

Code Snippet 2.9: This simple concept will take you far when writing your own microcontroller RS-232 communications programs.

The variable *character_in* is a byte, which is defined by the *int8* (8-bit integer) data type descriptor. The *getc* function returns a character, which in this code snippet's case is placed in the *character_in* memory location.

Let's write a C program called RX_program_1 that receives a keyboarded character from our Tera Term Pro session and echoes it back to the same Tera Term Pro session. Don't blink or you'll miss it. The whole program consumes three lines of actual code in Code Snippet 2.10.

```
//**************************************************
//  RECEIVE PROGRAM 1
//**************************************************
#ifdef RX_program_1

#use rs232(BAUD=9600, XMIT=PIN_A2, RCV=PIN_A3)

   while(1){
   putc(getc());
   }
#endif
```

Code Snippet 2.10: The getc function is called first and as soon as a character is received, the putc part of the statement pushes the character out of the microcontroller's serial port.

Ah—the beauty of C! In RX_program_1, the *getc* function is executed first and returns an 8-bit character. The *putc* function sends the results of the *getc* function, which is the keyboarded and received ASCII character, out to the Tera Term Pro session. The *while(1)* statement assures that this get and put operation will continue until power is removed from the PIC12F675.

You are trained and can now write and execute a basic RS-232 routine in either C or assembler using the smallest of microcontrollers. It's also evident (I hope) that C is the easier choice. Notice I used the word "easier" and not the word "better" because there may be situations where C is "too big" for your application. In those cases, assembler can be more efficient and more compact. What if neither programming C nor assembler is comfortable for you? Keep reading. Most of you will be in for a pleasant surprise.

CHAPTER 3

Writing RS-232 Microcontroller Routines in BASIC

In the preceding chapter, I've attempted to convert those of you that are still writing your microcontroller code in assembler to writing your microcontroller code using the C programming language. However, I learned personal computer assembler first, and then as the personal computer BASIC language evolved I moved to that as my primary personal computer programming language. After getting a grip on just what programming was, I finally ended up using C for most of my personal computer programming needs. Note that I said "most of," not "all of." If the personal computer application fits, I will revert to using some form of the BASIC programming language as BASIC is still a viable and powerful programming tool. This chapter will prove to you that the BASIC language is just as meaningful and just as powerful in the microcontroller programming world as it still is today in the personal computer programming environment.

At first, this wasn't as "BASIC" as I would have liked. It took some effort to understand the PicBasic Pro system, but once I had a grasp of what was going on, things got better in a hurry.

BASIC RS-232

Like the Custom Computer Services C Compiler, microEngineering Labs' PicBasic Pro has very good intentions about making things easy for the PIC programmer. For instance, in PicBasic Pro the watchdog timer is enabled by default and the PicBasic Pro compiler automatically inserts clear watchdog timer commands into the code at the appropriate locations. Akin to Custom Computer Services C Compiler, some PicBasic Pro instructions actually change the PIC port pin to an input pin or output pin automatically to effect their function. The PicBasic Pro built-in functions *SerIn* and *SerOut* are examples of PicBasic Pro instructions that automatically set the I/O pin direction (input for *SerIn* and output for *SerOut*) when called.

PicBasic Pro comes with its own special IDE, CodeDesigner Lite, and also melds seamlessly with the latest version of MPLAB IDE. Let's put together a simple PicBasic Pro program using CodeDesigner Lite™ that receives a character from Tera Term Pro and then echoes that character back to Tera Term Pro.

Just like before, the first thing we must do is provide a means of defining the RS-232 baud rate, parity setting, stop bit setting and bit inversion setting. In PicBasic Pro, the bit inversion setting is used to emulate an RS-232 converter IC when connecting the serial I/O pins directly to another serial device like your personal computer's serial port. Remember that the TTL bits in the data packet are inverted and voltage-shifted after passing through the Sipex SP232ACP RS-232 converter on our PICkit 1. A binary 1 becomes a mark or negative voltage and a binary 0 becomes a positive voltage space. If you look at the voltage levels of an RS-232 signal, you'll see that it is possible to fool a serial interface using the PIC's TTL I/O levels. Most serial ports will sense a TTL low (1.8 volts or below) as a mark and a TTL high (3 volts and above) as a space. If we use a PIC microcontroller to send an 'A' without using an RS-232 converter IC, the TTL binary sequence LSB (least significant bit) to MSB (most significant bit) would look like this:

```
START BIT (0) 10000010 (1) STOP BIT
```

If an RS-232 converter IC is used, we know that a binary 1 becomes a mark and a binary 0 translates to a space on the RS-232 side of the converter IC. In this case, we don't have the inversion (and voltage conversion) at the sending serial port because the RS-232 converter IC is not in the circuit. If the receiving side of our serial link uses an RS-232 converter IC, the incoming data will be presented to the receiving device's application inverted. So, after its RS-232 converter at the receiver does its thing, the receiving device would actually see:

```
START BIT (1) 01111101 (0) STOP BIT
```

What a mess. First of all, the start bit will not be recognized and will be seen as the first data bit. Second, the binary pattern is an inverse of the character 'A' and because the start bit was detected one bit too late, the data bit are shifted as a result. Finally, the stop bit is wrong and if this bit pattern gets through to the application at all, the stop bit will be incorrectly recognized as a data bit. The receiving application will most likely throw this RS-232 data packet (and any similar to this that follow) in the trash. PicBasic Pro's bit inversion option is used to avoid situations like the one we just discussed that stem from not having an RS-232 converter IC at one end of the link. As you can see, the bit inversion takes the place of an RS-232 converter IC and allows direct connection to a "true" serial port. Although you can hook raw microcontroller port I/O pins directly to an RS-232 port, be careful as the RS-232 voltages will damage your microcontroller.

PicBasic Pro's *SerIn* and *SerOut* functions use predefined modes to set up the baud rate. The modes are defined in an include file that comes with the PicBasic Pro Compiler package called *modedefs.bas*. For the *SerIn* and *SerOut* functions, 8N1 (8 data bits, no parity and 1 stop bit) is the default setting for an RS-232 data packet and can't be changed. We'll choose 9600 bps as our baud rate and since we do have a Sipex SP232ACP RS-232 converter IC in our PICkit 1 circuit there is no need to specify bit inversion. Thus, our mode will be specified as T9600, where the "T" stands for True. If inversion were required, our mode specification would change to N9600, with "N" signifying that the TTL data will be inverted.

Another automatic feature of PicBasic Pro is its assumption that the target PIC is running a 4 MHz clock. Although other clock speeds can be defined, the 4 MHz clock default is a good thing for us since our PIC12F675 is running on its internal 4 MHz clock. To maintain as accurate an internal clock as possible, the PIC12F675 uses an oscillator calibration value called OSCCAL. The OSCCAL value is kept in program memory space. The PicBasic Pro "DEFINE OSCCAL_1K" definition automatically moves the OSCCAL value that resides in program memory to the OSCCAL register every time the program is run. If you're not careful, you can erase the OSCCAL value. No worries. The PICkit 1 host program has an option that will rebuild the OSCCAL value for you.

There's another way to generate serial I/O using PicBasic Pro. The *DEBUG* and "DEBUGIN" functions allow most any I/O pin to become a serial transmitter or serial receiver. As you may ascertain from their names, the *DEBUG* functions are primarily intended to help you debug your code by allowing you to insert the *DEBUG* statements at various points inside your code to send variable values to a Tera Term Pro or HyperTerminal session. To use the *DEBUG* functions the I/O port, the I/O pin, the baud rate and the bit inversion mode must be specified. I'll stick with the PIC12F675 RS-232 port, pin, baud and bit inversion values we've used throughout our discussion, which makes GPIO our *DEBUG* and *DEBUGIN* port with GP2 acting as the transmit pin and GP3 doing the receiver duty. Our selected baud rate is 9600 bps and there is no bit inversion. Like the "*SerIn*" and "*SerOut*" functions, the number of bits in the RS-232 data packet, the parity and the number of stop bits is set at 8N1 and cannot be altered by the programmer. Code Snippet 3.1 is our PicBasic Pro BASIC code up to this point:

```
INCLUDE "modedefs.bas"
DEFINE OSCCAL_1K  1
DEFINE DEBUG_REG GPIO
DEFINE DEBUGIN_REG GPIO
DEFINE DEBUG_BIT 2
DEFINE DEBUGIN_BIT 3
DEFINE DEBUG_BAUD 9600
DEFINE DEBUG_MODE 0
DEFINE DEBUGIN_MODE 0
DEFINE NO_CLRWDT 1
```

Code Snippet 3.1: Hmmm...Looks kind of like C.

I went ahead and threw in the last DEFINE line as it's important if you want to change one of PicBasic Pro's automatic features such as whether the watchdog timer runs or not.

How a PicBasic Pro source file is compiled depends on the assembler and some configuration fuse settings found in the PicBasic Pro PIC12F675 include file. The code snippet below is the PicBasic Pro PIC12F675 include file that determines what microcontroller fuses are active, versus which assembler is invoked. CodeDesigner Lite uses the native PicBasic Pro assembler, PM. Notice that I have modified the original code turning the watchdog timer off (*wdt_off*) and internalizing the PIC12F675 MCLR function freeing the GP2 pin for I/O use with *mclr_off*. The last DEFINE statement, DEFINE NO_CLRWDT 1, instructs the PicBasic Pro Compiler not to insert the clear watchdog statements into the final code.

MPLAB IDE allows the use of either the PicBasic Pro's PM or Microchip's MPASM assembler. The code we generate with CodeDesigner Lite will compile without modification under the MPLAB IDE. Notice that I have made the same configuration fuse adjustments in the MPASM header code in Code Snippet 3.2.

```
;*****************************************************************
;* 12F675.INC
*
;*
*
;* By        : Leonard Zerman, Jeff Schmoyer
*
;* Notice    : Copyright (c) 2002 microEngineering Labs, Inc.
*
;*            All Rights Reserved
*
;* Date      : 09/27/02
*
;* Version   : 2.43
*
;* Notes     :
*
;*****************************************************************
      NOLIST
   ifdef PM_USED
      LIST
      INCLUDE 'M12F675.INC'    ; PM header
      device  pic12F675, intrc_osc_noclkout,bod_off, wdt_off,
pwrt_on, mclr_off, protect_off
      XALL
      NOLIST
   Else
      LIST
      LIST p = 12F675, r = DEC, w = -302
      INCLUDE "P12F675.INC"    ; MPASM  Header
      __config _INTRC_OSC_NOCLKOUT & BODEN_OFF & _WDT_OFF & _PWRTE_ON
& _MCLRE_OFF & _CP_OFF
      NOLIST
   EndIF
      LIST
```

Code Snippet 3.2: The trick is to recognize that both the PM and MPASM configuration code is included in this file.

Unlike the Custom Computer Services C Compiler, the PicBasic Pro Compiler does not have preprocessor directives like *#ifdef*. So, I'll sacrifice a byte to simulate the C preprocessor directives and add some PIC12F675 setup information to our PicBasic Pro code. Two bits are assigned to *debug_prog* and *serio_prog* and I've allocated a byte, *chr*, to hold the ASCII character our program will send and receive. Equating a '1' to a bit allows the program represented by that bit to be compiled. Conversely, a '0' assigned to a bit effectively turns that program off to the compiler. Normally, this kind of code would not be something you'd want to include in a professional project. I'm incorporating it here to make the PicBasic Pro Compiler serial communications example easier for you to understand and compile by stuffing two programs into one as we did with the Custom Computer Services C Compiler C source.

Recall that we had to generate some C code to turn off the analog section of the PIC12F675 so we could use those dual-purpose pins for digital I/O. Well, we're using the same microcontroller hardware, a PIC12F675, and we must perform the same analog deactivation process using PicBasic Pro code (adcon0 = 0). Our C code example also included a "TRIS" statement to setup the input or output status of the PIC transmit and receive I/O lines. We don't need to code any BASIC "TRIS" statements here as the *DEBUG/DEBUGIN* and "*SerIn/SerOut*" functions automatically set the selected PIC's I/O pin for input or output depending on the function call. The conveniences provided by the PicBasic Pro Compiler make for a very tidy set of RS-232 echo routines as shown in Code Snippet 3.3 below.

```
INCLUDE "modedefs.bas"
DEFINE OSCCAL_1K  1
DEFINE DEBUG_REG GPIO
DEFINE DEBUGIN_REG GPIO
DEFINE DEBUG_BIT 2
DEFINE DEBUGIN_BIT 3
DEFINE DEBUG_BAUD 9600
DEFINE DEBUG_MODE 0
DEFINE DEBUGIN_MODE 0
DEFINE NO_CLRWDT 1

debug_prog VAR BIT
serio_prog VAR BIT
chr VAR BYTE

adcon0 = 0

;PROGRAM TO RUN = 1
debug_prog = 1
serio_prog = 0
```

```
loop:
   ;CHARACTER ECHO USING DEBUG FUNCTIONS
   IF debug_prog Then
   DebugIn [chr]
   Debug chr
   EndIF

   ;CHARACTER  ECHO USING SER_IO FUNCTIONS
   IF serio_prog Then
   SerIn GPIO.3,T9600,chr
   SerOut   GPIO.2,T9600,[chr]
   EndIF

   GoTo loop

End
```

Code Snippet 3.3: A bit wordier than its C counterpart, but that's BASIC no matter where you encounter it. The bottom line is that it works just like the C code.

To me, writing code in BASIC is fun. The neat thing about BASIC is that it's easy to learn no matter what microcontroller or personal computer you're writing an application for. The microEngineering Labs PicBasic Pro package is easy to use and powerful in function. No matter which programming language you choose to write your RS-232 code with, you'll need some hardware that is capable of turning your typing into reality. Let's round up some RS-232 components, a PIC microcontroller and turn on the soldering iron.

CHAPTER 4

Building Some RS-232 Communications Hardware

Enough theory and coding already—let's solder some stuff. Before we can start connecting components together to form our physical hardware, there are a few more things you need to know about microcontrollers and RS-232.

Up to now we've been sending and receiving under the guidance of the RS-232 standard with a microcontroller that doesn't contain any internal serial communications circuitry. The PicBasic Pro Compiler and the Custom Computer Services C Compiler compensate for this lack of circuitry and allow us to emulate the missing serial hardware using the compiler's firmware. The firmware implementation of a serial port works fine until you have to do other things and look for incoming serial data simultaneously.

A Few More BASIC RS-232 Instructions

When other tasks are being serviced by the microcontroller, it may be possible to miss an incoming RS-232 message. PicBasic Pro handles this situation by allowing the user/programmer to "time out" after checking for incoming serial data. Let's add another module to our PicBasic Pro source code to demonstrate how that would work.

```
INCLUDE "modedefs.bas"        ;get mode defs
DEFINE OSCCAL_1K  1
DEFINE DEBUG_REG GPIO
DEFINE DEBUGIN_REG GPIO
DEFINE DEBUG_BIT 2            ;serial out GP2
DEFINE DEBUGIN_BIT 3          ;serial in GP3
DEFINE DEBUG_BAUD 9600        ;9600,N,8,1
DEFINE DEBUG_MODE 0           ;no inversion
DEFINE DEBUGIN_MODE 0         ;no inversion
DEFINE NO_CLRWDT 1            ;don't add clrwdt

debug_prog VAR BIT
serio_prog VAR BIT
serio_wait_prog VAR BIT

chr VAR BYTE                  ;for ASCII character

adcon0 = 0                    ;turn off analog I/O
```

```
;PROGRAM TO RUN = 1
debug_prog = 0
serio_prog = 0
serio_wait_prog = 1

loop:
      ;CHARACTER ECHO USING DEBUG FUNCTIONS
      IF debug_prog Then
      DebugIn [chr]
      Debug chr
      EndIF

      ;CHARACTER ECHO USING SER_IO FUNCTIONS
      IF serio_prog Then
      SerIn GPIO.3,T9600,chr
      SerOut   GPIO.2,T9600,[chr]
      EndIF

      ;CHECK FOR A CHARACTER EVERY SECOND
      IF serio_wait_prog Then
      SerIn GPIO.3,T9600,1000,no_data,chr
      SerOut GPIO.2,T9600,[chr,13,10]
      GoTo loop
no_data:
      SerOut   GPIO.2,T9600,["no character",13,10]
      EndIF

      GoTo loop

End
```

Code Snippet 4.1: The microEngineering Labs PicBasic Pro Compiler has as many built-in tricks up its sleeve as the Custom Computer Services C Compiler does.

As you can see in Code Snippet 4.1, I've added a third module and a corresponding third bit, *serio_wait_prog*. The *serio_wait_prog* code looks for an incoming character for 1000 ms (1 second). If no character is received after waiting for 1 second, the program jumps to the "no_data" label and prints "no character" followed by a carriage return (decimal 13) and a linefeed (decimal 10). If a character is detected within the 1-second window, the received character is sent followed by a CRLF (carriage return/linefeed) sequence.

Things are a bit different on the C side in Code Snippet 4.2, but the results are the same.

```
#include <12F675.h>
#include <f675.h>
#use delay(clock=4000000)
#fuses INTRC_IO,NOWDT,NOMCLR,NOPROTECT,NOCPD,NOBROWNOUT

#use fast_io(A)

int32 timeout;

//*******************************************************
// COMMENT OUT THE PROGRAMS YOU DON'T WANT TO RUN
//*******************************************************
//#define  TX_program_1
//#define  TX_program_2
//#define  TX_program_3
//#define  RX_program_1
#define  SERIO_program

////////////////////////////////////////////
//*******************************************************
//  SERIO WAIT PROGRAM
//*******************************************************
#ifdef SERIO_program

#use rs232(BAUD=9600, XMIT=PIN_A2, RCV=PIN_A3)
   while(1)
   {
   timeout=0;
   while(!kbhit()&&(++timeout<50000))
      delay_us(10);
   if(kbhit())
         printf("%c\r\n",(getc()));
   else
         printf("no character\r\n");
   }
#endif
```

Code Snippet 4.2: This C code is a bit more complicated than the BASIC version.

I added *int32 timeout* to allocate a 32-bit area of PIC memory to hold a timeout value and another #define statement to point at our new C source module, *SERIO_program.* Instead of having the PicBasic Pro luxury of inserting a timeout value in the function call, our C source code uses the *kbhit* function to signal the presence of a character. Since we have no "real" serial hardware functionality inside our PIC12F675, the *kbhit* function returns a TRUE after detecting a valid start bit on the PIC12F675's GPIO receive line. So, every 10 μs our C program looks for a start bit and increments the timeout value. After about a second or so, if no character has been detected, our C program sends "no character" followed by a carriage return/line feed sequence (\r\n). If a valid start bit is detected and followed by a valid character, the character is retrieved using the *getc* function and sent to the Tera Term Pro terminal emulator using the *printf* function, which also appends the sent character with a CRLF sequence.

If you're working with a microcontroller like the PIC12F675, the serial I/O routines I've just described will work well for you. The only drawback is that you have to continually run the routines in conjunction with your main application to "poll" for an incoming character. There will be times when you won't have enough processing time or processor resources to do the polling. That's when you call in some bigger guns.

CHAPTER 5

Using Microcontroller USARTs

Before we move on and delve into microcontroller USARTs, the very first thing we must do is set our PICkit 1 FLASH Starter Kit aside and move to another development platform that will support our USART-laden microcontroller. We'll develop our PIC USART code using a device you will come to know in later chapters as the Easy Ethernet CS8900A. The partially assembled Easy Ethernet CS8900A you see in Photo 5.1 is based on 40-pin PIC microcontrollers that adhere to the pinouts used by the PIC6F87X and PIC18FXX2 families.

Photo 5.1: Don't worry; you'll get a full play-by-play on the components you see in the photo that aren't yet mounted. Right now, we want to concentrate on the RS-232 portion of the Easy Ethernet CS8900A.

The microcontroller we will be working with in this part of the RS-232 project is the PIC18F452. The PIC18F452 is full of goodies but we're only interested in its USART hardware at this point. As you already know, we will use the RS-232 circuitry of the Easy Ethernet CS8900A for the RS-232 engine that will be driven by the code we write in this section. Schematic 5.1 is a representation of the Easy Ethernet CS8900A's RS-232 circuitry.

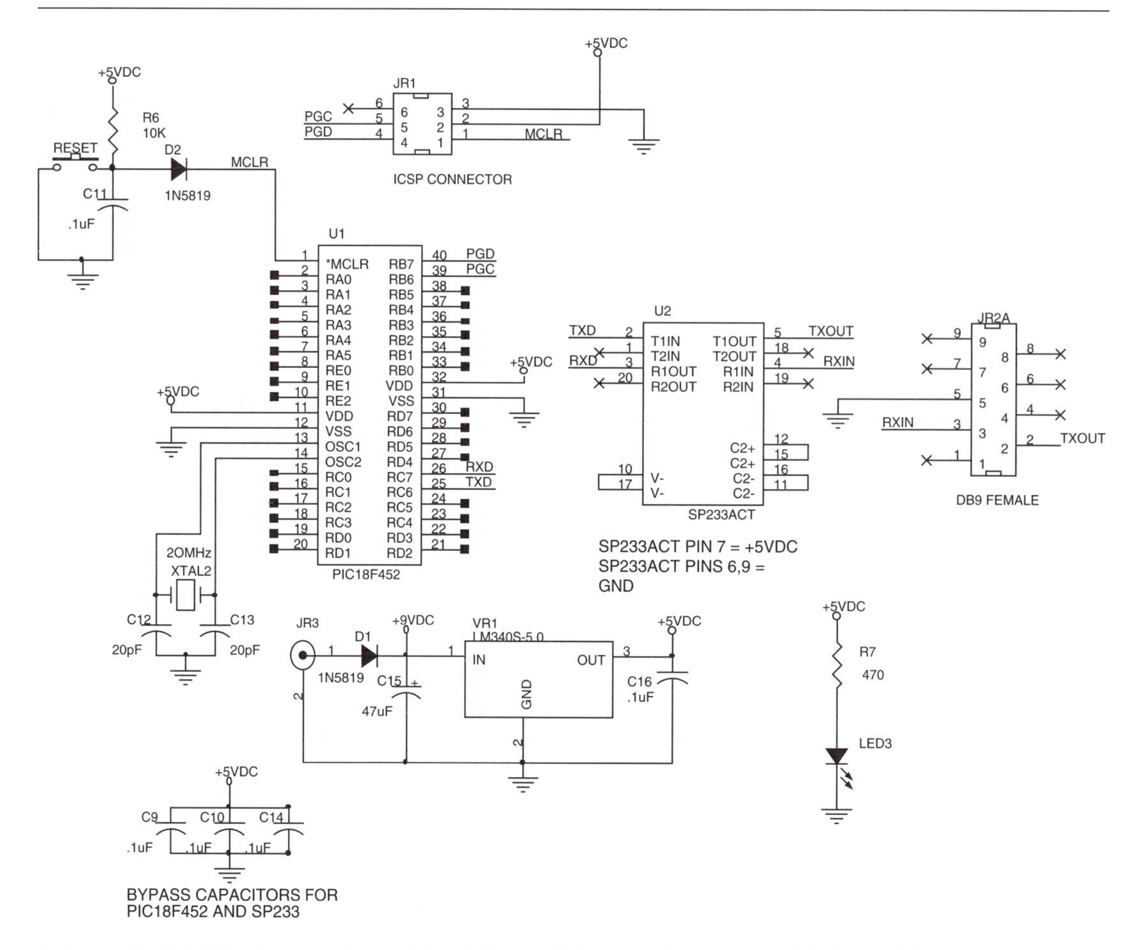

Schematic 5.1: This is a portion of the full Easy Ethernet CS8900A, which you'll learn more about later. We have just enough hardware here to program and power a microcontroller-based RS-232 engine.

From segments of our previous discussion, you already have a basic idea of what a USART can do. So, let's take a close-up look of what's inside a typical microcontroller USART.

If you consult the Microchip PIC datasheets, you'll find that some of the PICs contain internal USART circuitry, which in the datasheet is also referred to as the SCI, or Serial Communications Interface. Although there are numerous ways to use the PIC USART, as you already know, the most common application of the PIC USART is to communicate to a personal computer's serial port using the RS-232 protocol.

A microcontroller USART is a collection of special purpose registers. For instance, the USART complex inside our PIC18F452 is made up of status registers, data registers, and interrupt registers. Rather than try to educate you on every bit in every register, I'll cover only the registers and bits that we will need to be concerned with in the writing of our code.

TXSTA REGISTER

7	6	5	4	3	2	1	0
CSRC	TX9	TXEN	SYNC	–	BRGH	TRMT	TX9D
–	0	1	0	0	0	1	0

Figure 5.1: For standard 8-bit operations, TXSTA would contain X0100010. CSRC is a "don't care" bit for asynchronous mode. 8-bit transmission is selected by clearing bit TX9. Setting TXEN enables the asynchronous transmitter and a 0 in the SYNC bit position selects asynchronous mode. BRGH is set for low speed and TRMT indicates that at this moment the Transmit Shift Register is empty. Since we're not using 9-bit transmission, TX9D is a "don't care" and is arbitrarily clear in our example.

Many of the C and BASIC serial communications functions use the USART registers behind the scenes to do their magic. The Transmit Status and Control Register depicted in Figure 5.1, TXSTA, is used to select asynchronous or synchronous mode, enable the transmit function, provide transmit shift register status and perform some 9-bit tasks that have to do with address bits, data bits or parity bits. Right now, we're not concerned with 9-bit transmissions or synchronous data sessions, which leaves the transmit enable and transmit shift register functions as those of possible interest to us.

RCSTA REGISTER

7	6	5	4	3	2	1	0
SPEN	RX9	SREN	CREN	ADDEN	FERR	OERR	RX9D

Figure 5.2: Setting SPEN and CREN while clearing RX9 enables an 8-bit serial port. SREN is a "don't care" bit in asynchronous mode. The level of bit ADDEN decides whether RX9D is used as an address bit or parity bit in 9-bit receive mode. FERR and OERR are the framing and overrun error bits.

To complement the TXSTA, the PIC USART complex houses yet another status register component, the Receive Status and Control Register, or RCSTA, which is shown in Figure 5.2. In addition to the 9-bit receive duties, the RCSTA contains bits that provide status for framing and overrun errors. We've already seen an example of an "inverted" data packet that could generate a framing error. A framing error occurs when the STOP bit is detected as a zero. The USART is counting the incoming bits and knows that a STOP bit should always be a one (a mark).

Overrun errors have to do with the FIFO (First In, First Out) buffer within the USART. The typical PIC USART can buffer two characters in its FIFO. When the third byte is received and there are still two bytes in the FIFO (the FIFO is full), the FIFO is "overrun" and the overrun error is signaled. To avoid overrun errors, we must read the USART FIFO as quickly as possible once a character has entered it. In addition to signaling framing and overrun errors, the RCSTA is in control of the microcontroller serial port and has bits to enable or disable the receiver and microcontroller serial port.

The TXREG and RCREG are the transmit and receive data registers. TXREG and RCREG hold the actual data being transported on the link. Most of the time, C and BASIC compilers shield the user/programmer from having to deal directly with these registers.

However, despite the cloaking done by the compilers, these registers are easily accessible to the programmer.

Baud rate is set in a PIC USART by loading a value into the SPBRG register. There is a formula for calculating the SPBRG value and the PIC datasheets contain a table in the USART area listing many of the common baud rate settings for a particular microcontroller clock speed. The Custom Computer Services C Compiler uses the "#use RS232" directive to set the baud rate while PicBasic Pro provides preset modes for setting the baud rate.

Until now, we have only studied "polled" serial I/O routines. Polling in this sense means that we must have our program periodically check for incoming serial data. If the time between polling cycles is too long, there is a possibility that incoming characters could be lost if there isn't any code that is performing receive buffering. To improve our chances of receiving every character that comes in, the USART employs the services of interrupt registers. The interrupt mechanism is also used by the USART to transmit data as well. Using serial I/O interrupts, the main program can be running continuously without having to poll for characters on a regular basis. Using interrupt routines also greatly reduces the chance of getting overrun errors as the character is serviced as soon as it hits the USART's FIFO. Another advantage of using serial I/O interrupts is that we can now transmit and receive at the same time. This is called "full duplex" operation. Half-duplex sessions are only allowed to receive or transmit in one direction at any time. All of my examples thus far run as half-duplex.

By using the TX9 bit in the TXSTA register, our PIC USART can be forced to transmit nine data bits. This is accomplished by placing the ninth bit in the TX9D bit location of the TXSTA register before writing an 8-bit data packet to the TXREG register. The ninth bit could be a parity bit or an address bit. Our application doesn't require parity or addressing. So, we will use the PIC USART to send 8-bit data packets in normal asynchronous mode. Once data has been written to TXREG, the eight (or nine) bits are moved into the Transmit Shift Register. From there, they are clocked out onto the TX pin preceded by a START bit and followed by a STOP bit.

Some Interrupt-Driven USART Code

Our goal here is to generate some C code that will interrupt the microcontroller when a character is received so it can be processed as quickly as possible. Also, we want our interrupt code to handle the housekeeping associated with transmitting a character as well.

The beginning of our USART C code looks much like the starting portion of our polled serial I/O code. Don't get all wound up about the fuses. I've simply opened up the PIC18F452 memory areas for unrestricted reading and writing. I'm sure you've noticed that there is one unfamiliar option in our header code. The "#device ICD=TRUE" is a Custom Computer Services C Compiler directive that instructs the C compiler to generate code that is compatible with Microchip's MPLAB ICD 2 debugging hardware. An NOP (no operation) assembler instruction must reside at program memory location 0 to allow proper operation of the MPLAB ICD 2 hardware, and the "#device ICD=TRUE" makes that possible too. In our polling code using the PIC12F675, we arbitrarily chose our transmit and receive I/O pins as there was no hardware USART involved. This time around, we will not be emulating a USART in software. So, we must use the hardware USART pinout for our transmit and receive pins that is dictated by the microcontroller. Pins C6 and C7 are the dedicated USART transmit and receive pins on the PIC18F452. The USART pins are specifically called out in Code Snippet 5.1.

```
#include <18F452.h>
#device ICD=TRUE
#use delay(clock=20000000)
#use rs232(baud=9600,parity=N,xmit=PIN_C6,rcv=PIN_C7,bits=8)
#fuses HS,PUT,NOWRTB,NOEBTR,NOWRT,NOWDT,NOLVP,NOPROTECT
```

Code Snippet 5.1: The MPLAB ICD 2 will allow us to see into the PIC18F452. The ICSP connector on the Easy Ethernet CS8900A is the link between the MPLAB ICD 2 and the PIC18F452 microcontroller.

Our interrupt routines (Code Snippet 5.2) will be responsible for actually handling the raw incoming and outgoing data. Once the data is processed, we will need some routines to manipulate it. The *recvchar* function removes a byte of data from the receive buffer and adjusts the receive buffer pointer to point to the next character in the queue. The *sendchar* function inserts a character to be transmitted into the transmit buffer, adjusts the transmit buffer pointer and enables the transmit interrupt. Even though we will build a very efficient interrupt-driven and buffered serial I/O engine, we will still need to know if there are characters in the receive buffer that need our attention. The *CharInQueue* function returns a TRUE if there is a character in the buffer and a FALSE if the receive buffer is empty.

```
int8 recvchar(void);
int8 sendchar(int8 data);
int8 CharInQueue(void);
```

Code Snippet 5.2: These routines are used in conjunction with the RS-232 interrupt structure of the PIC18F452.

Before we take a look at each of the serial I/O functions, I want to show you how the transmit and receive buffers work. I'll use the PIC18F452 microcontroller and serial circuitry found on the Easy Ethernet CS8900A and the MPLAB ICD 2 in debugger mode to illustrate the receive and transmit buffer structure and functionality.

Our Easy Ethernet CS8900A is equipped with a Sipex SP233ACT RS-232 converter, and I have mounted a PIC18F452 in the Easy Ethernet CS8900A's 40-pin target socket. The PIC18F452 is being clocked at 20 MHz.

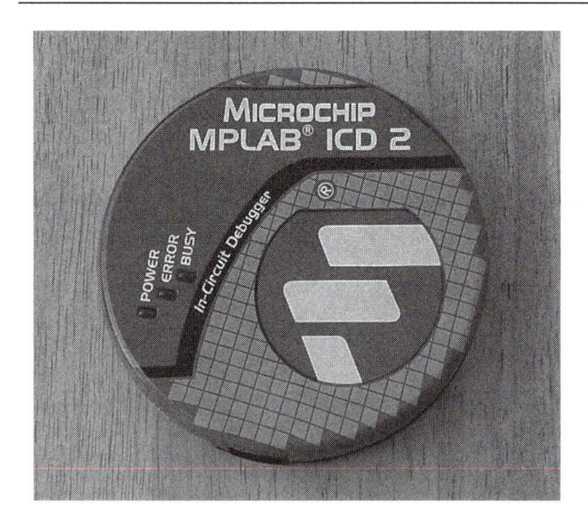

Photo 5.2: This is the MPLAB ICD 2; otherwise known as the hockey puck. The MPLAB ICD 2 uses debug facilities built into the PIC18F452 to give us a marvelous view of the PIC18F452's internals.

I've setup the MPLAB ICD 2 within the MPLAB IDE as a programmer and debugger for the PIC18F452. The MPLAB ICD 2 is affectionately known as the "hockey puck" and I've attached the puck to my personal computer and the MPLAB IDE using the puck's USB interface. The Easy Ethernet CS8900A has an ICSP™ (In-Circuit Serial Programming™) socket that allows the puck to be connected to it as a programmer/debugger device via a 6-pin flat cable. The puck is equipped with a wall wart power jack and can also be configured to provide power to the circuit it is attached to. Since the Easy Ethernet CS8900A has the wall wart power interface and circuitry required to power itself and the MPLAB ICD 2, I've configured the puck to obtain its power from the Easy Ethernet CS8900A.

```
#define USART_RX_BUFFER_SIZE 16   // 1,2,4,8,16,32,64,128 or 256 bytes
#define USART_RX_BUFFER_MASK ( USART_RX_BUFFER_SIZE - 1 )
//#if ( USART_RX_BUFFER_SIZE & USART_RX_BUFFER_MASK )
//#error RX buffer size is not a power of 2
//#endif
```

Code Snippet 5.3: Things will really be busy if the buffer has to actually buffer 16 incoming bytes.

Let's start by defining the receive buffer area. As you can see in Code Snippet 5.3, the receive buffer is 16 bytes in length (#define USART_RX_BUFFER_SIZE 16), and can be as long as 256 bytes. The USART_RX_BUFFER_MASK is used to calculate an index into the buffer and is set to a value of one less than the receive buffer size. Let's stop and think logically about the relationship between the receive buffer size and the receive buffer mask. If the receive buffer is 16 bytes long and begins with an index of zero (0), then the number of bits it would take to go from 0 to 15 (actually 16 byte locations) is 4. Here is the 16-byte progression of the 4 bits in a binary representation starting at index 0 and ending at index 15:

0000	0
0001	1
0010	2
0011	3
0100	4
0101	5
0110	6
0111	7
1000	8
1001	9
1010	10 or hexadecimal A
1011	11 or hexadecimal B
1100	12 or hexadecimal C
1101	13 or hexadecimal D
1110	14 or hexadecimal E
1111	15 or hexadecimal F

Thus, our 4-bit receive mask can address 16 bytes of data. Let's check our theory using the mask table and a lower receive buffer count. Let's say our receive buffer size was 8 bytes in length instead of 16 bytes in length. That means our receive buffer mask would only have to use 3 bits to index 8 bytes of data (index 0 through index 7, which is actually 8 bytes including the index 0 value). If you were to check all of the valid receive buffer sizes, you would notice that every time the receive buffer size increases it doubles, and every time the receive buffer size doubles, you add an extra bit to the left of your receive buffer mask. For instance, for a buffer size of 256 bytes, the receiver mask value would be one less than the receive buffer size, 255, which equates to 8 receive mask bits. Note in the commented out area that the receive buffer size must be a power of 2. With that, 2 to the 8th power (2^8) is equal to 256 or *100000000* in binary. Subtracting 1 from 2^8 to account for index 0 is equal to *011111111* or *11111111* in binary notation, which is our 8 bits needed to cover 256 byte indexes.

Just for grins, I selected a transmit buffer size of 128 bytes in Code Snippet 5.4. The idea of the mask helping to determine an index is the same for both the receive and transmit buffers. Doing the math tells us that we will need enough bits to represent 128 minus 1, or 127 in binary (01111111 or 1111111). I've commented out the power of 2 check in both the receive and transmit definitions, but left them as a reminder as to how the receive and transmit buffer sizes should be selected.

```
#define USART_TX_BUFFER_SIZE 128 // 1,2,4,8,16,32,64,128 or 256 bytes
#define USART_TX_BUFFER_MASK ( USART_TX_BUFFER_SIZE - 1 )
//#if ( USART_TX_BUFFER_SIZE & USART_TX_BUFFER_MASK )
//#error TX buffer size is not a power of 2
//#endif
```

Code Snippet 5.4: Again, things inside the PIC18F452 would be steaming if it had to buffer 128 outgoing RS-232 bytes. The PIC18F452 is fast enough to keep both the transmit and receive buffers near empty at all times under normal loads with high baud rates.

Now that we've defined our transmit and receive buffer sizes and mask values, let's allocate some memory for the actual transmit and receive buffers in Code Snippet 5.5.

```
int8 USART_RxBuf[USART_RX_BUFFER_SIZE];
int8 USART_TxBuf[USART_TX_BUFFER_SIZE];
int8 USART_TxHead,USART_TxTail,USART_RxHead,USART_RxTail;
```

Code Snippet 5.5: These are simple memory arrays with their associated array pointers.

Both the receive and transmit buffers use a head and tail scheme to indicate whether or not there is any data inside the buffer structure that needs attention. The easiest way for me to explain this is to reference the *CharInQueue* function. Let's assume both the USART_RxHead and USART_RxTail variables are both initialized to zero and no data has entered the receive buffer. When *CharInQueue* is called, the USART_RxHead and USART_RxTail values will be equal at zero, and a FALSE will be returned by the *CharInQueue* function. A FALSE means the receive buffer is empty. Thus, if the buffer head and tail are equal, there is no data between them, which equates to an empty buffer. Let's walk through the USART receive interrupt processing an incoming character.

We begin by making sure our receive and transmit buffer heads and tails are all at the index 0 point. This is done in the C *main* function with the code shown in Code Snippet 5.6.

```
   USART_RxTail = 0x00;
   USART_RxHead = 0x00;
   USART_TxTail = 0x00;
   USART_TxHead = 0x00;
```

Code Snippet 5.6: Never assume a register or memory location is clear.

Once all of the microcontroller hardware is initialized, we can use the code in Code Snippet 5.7 to enable the receive interrupt.

```
enable_interrupts(INT_RDA);
enable_interrupts(global);
```

Code Snippet 5.7: The global *parameter can enable or disable the entire set of PIC18F452 microcontroller interrupts with a single statement.*

Earlier, I mentioned that the C and BASIC compilers usually hide their access of the real receive and transmit data registers. Here, we will actually access them directly in our interrupt routines. So, if we want to use them, we must define them and their locations in the PIC microcontroller memory map as shown in Code Snippet 5.8.

```
#byte  RCREG  =  0x0FAE
#byte  TXREG  =  0x0FAD
```

Code Snippet 5.8: These values could also be stuffed into an include *file or header file and referenced in the RS-232 source code. I've included them in the main source code for clarity.*

OK...Our USART has been enabled and is actively looking for a start bit and a following character. Before we start our simulation, I'm going to take some poetic license here and add a few lines of code to put a spotlight on our transmit and receive buffer areas. The MPLAB ICD 2 allows me a single hardware breakpoint, and I'll place it immediately following the unofficial code snippet, Code Snippet 5.9.

```
    for(x=0;x<USART_RX_BUFFER_SIZE;++x)
       USART_RxBuf[x] = 'R';
    for(x=0;x<USART_TX_BUFFER_SIZE;++x)
       USART_TxBuf[x] = 'T';
//breakpoint is inserted here
```

Code Snippet 5.9: Writing and debugging code without the right tools is like driving a car with no gauges.

In Figure 5.3, the R's denote the bytes in the receive buffer, and the T's do the same for the transmit buffer. If you count carefully, you'll see that there are exactly 16 contiguous R's and 128 contiguous T's. We'll be keeping up with the data using the head and tail index pointer values. So, we can keep the R and T designators in the transmit and receive buffers as we run the program. What you will see is characters coming in to replace the pre-positioned T and R markers.

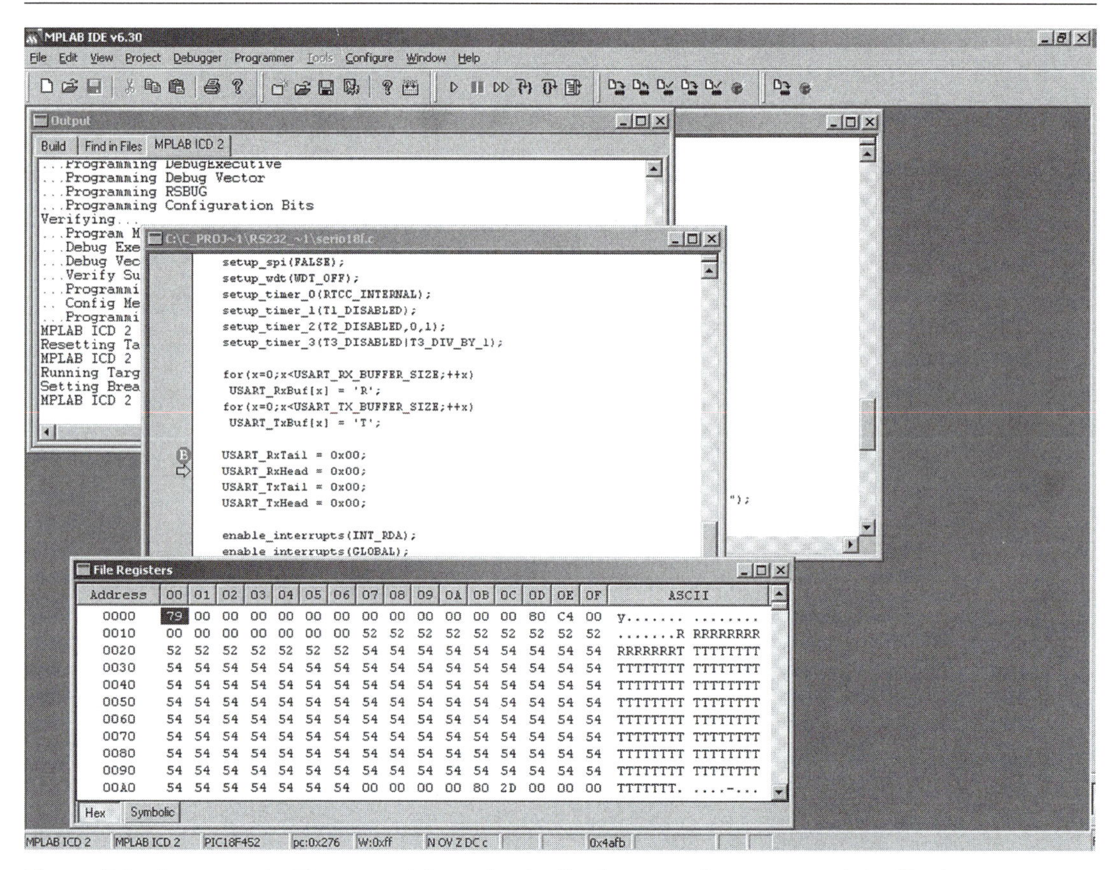

Figure 5.3: Count em'...There are 16 receive buffer bytes and 128 transmit buffer bytes.

To get a complete view of what's going on in the buffer areas, we'll also need to keep up with the transmit and receive buffer head and tail values. Since the head and tail values reside in and change in the same memory location throughout the program, it's easier to add them to a "watch" list for observation. I've brought the watch list to the front in Figure 5.4.

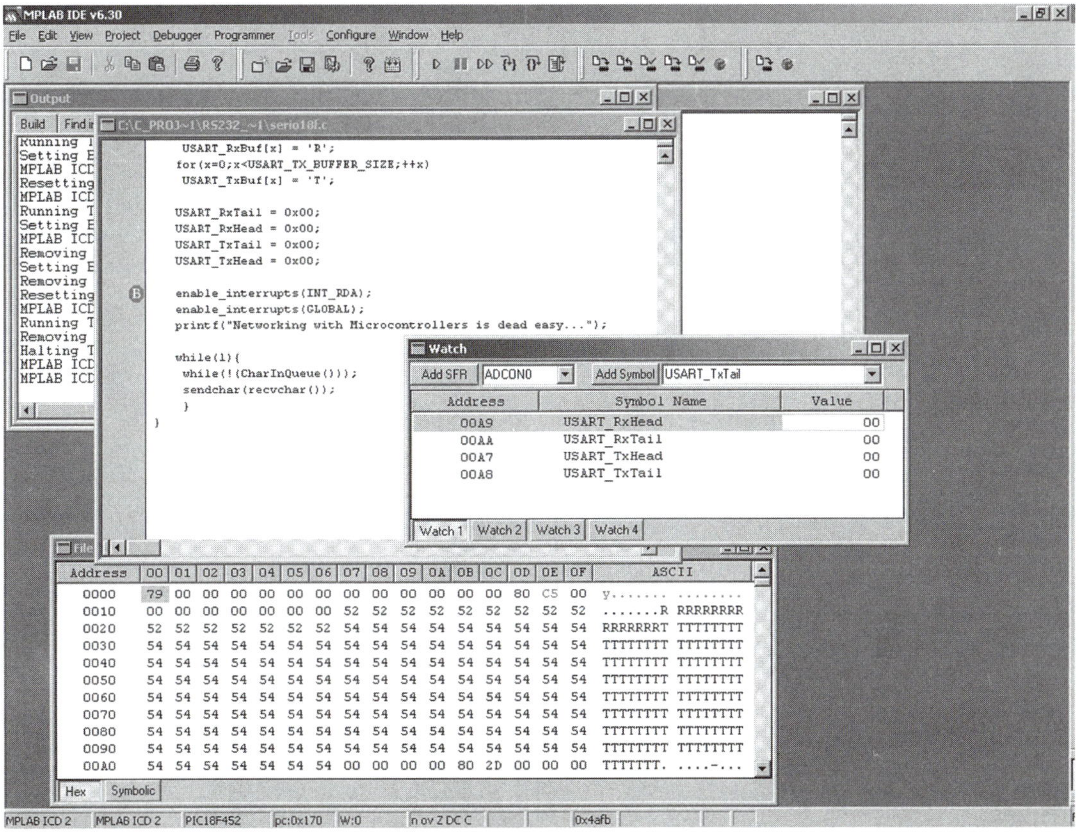

MPLAB IDE v6.30

File Edit View Project Debugger Programmer Tools Configure Window Help

Output

Build | Find in

```
Running T
Setting E
MPLAB ICD
Resetting
MPLAB ICD
Running T
Setting E
MPLAB ICD
Removing
Setting E
Removing
Resetting
MPLAB ICD
Running T
Removing
Halting T
MPLAB ICD
MPLAB ICD
```

C:\C_PROJ~1\RS232_~1\serio18f.c

```
  USART_RxBuf[x] = 'R';
 for(x=0;x<USART_TX_BUFFER_SIZE;++x)
  USART_TxBuf[x] = 'T';

 USART_RxTail = 0x00;
 USART_RxHead = 0x00;
 USART_TxTail = 0x00;
 USART_TxHead = 0x00;

 enable_interrupts(INT_RDA);
 enable_interrupts(GLOBAL);
 printf("Networking with Microcontrollers is dead easy...");

 while(1){
  while(!(CharInQueue()));
  sendchar(recvchar());
  }
}
```

Watch

Add SFR | ADCON0 | Add Symbol | USART_TxTail

Address	Symbol Name	Value
00A9	USART_RxHead	00
00AA	USART_RxTail	00
00A7	USART_TxHead	00
00A8	USART_TxTail	00

Watch 1 | Watch 2 | Watch 3 | Watch 4

File Registers

Address	00	01	02	03	04	05	06	07	08	09	0A	0B	0C	0D	0E	0F	ASCII
0000	79	00	00	00	00	00	00	00	00	00	00	00	00	80	C5	00	y....... ....
0010	00	00	00	00	00	00	00	52	52	52	52	52	52	52	52	52	.......R RRRRRRRR
0020	52	52	52	52	52	52	52	54	54	54	54	54	54	54	54	54	RRRRRRRT TTTTTTTT
0030	54	54	54	54	54	54	54	54	54	54	54	54	54	54	54	54	TTTTTTTT TTTTTTTT
0040	54	54	54	54	54	54	54	54	54	54	54	54	54	54	54	54	TTTTTTTT TTTTTTTT
0050	54	54	54	54	54	54	54	54	54	54	54	54	54	54	54	54	TTTTTTTT TTTTTTTT
0060	54	54	54	54	54	54	54	54	54	54	54	54	54	54	54	54	TTTTTTTT TTTTTTTT
0070	54	54	54	54	54	54	54	54	54	54	54	54	54	54	54	54	TTTTTTTT TTTTTTTT
0080	54	54	54	54	54	54	54	54	54	54	54	54	54	54	54	54	TTTTTTTT TTTTTTTT
0090	54	54	54	54	54	54	54	54	54	54	54	54	54	54	54	54	TTTTTTTT TTTTTTTT
00A0	54	54	54	54	54	54	54	00	00	00	00	80	2D	00	00	00	TTTTTTT. ....-...

Hex | Symbolic

MPLAB ICD 2 | MPLAB ICD 2 | PIC18F452 | pc:0x170 | W:0 | n ov Z DC C | 0x4afb

Figure 5.4: You can also see the USART buffer pointer values in the File Register window just beyond the last T.

The receive and transmit buffer head and tail memory locations will be cleared to 0x00 (the "x" means hexadecimal) just before the USART receive interrupt is enabled. Note that the watch window gives us the address of the head and tail variables so we can find them easily in the File Register window if we want to.

Since the MPLAB ICD 2 only allows me a single breakpoint, I'll move that breakpoint to a point just inside the receive interrupt routine after I manually step through the instructions to clear the receive and transmit buffer heads and tails. After setting the new breakpoint, I'll run the program using the MPLAB ICD 2 debugger inside of MPLAB IDE and send an ASCII '1' from my Tera Term Pro session.

After entering an ASCII '1' in the Tera Term Pro session, the breakpoint stopped the execution of our program after the *data = RCREG* statement in Code Snippet 5.10.

```
#int_RDA
void USART_RX_interrupt(void)
{
   int8 data;
   int8 tmphead;

   data = RCREG;            /* read the received data */
```

Code Snippet 5.10: This is the entry point for the RDA (Received Data Available) interrupt service routine. Note that the very first thing we do is grab that incoming RS-232 data byte.

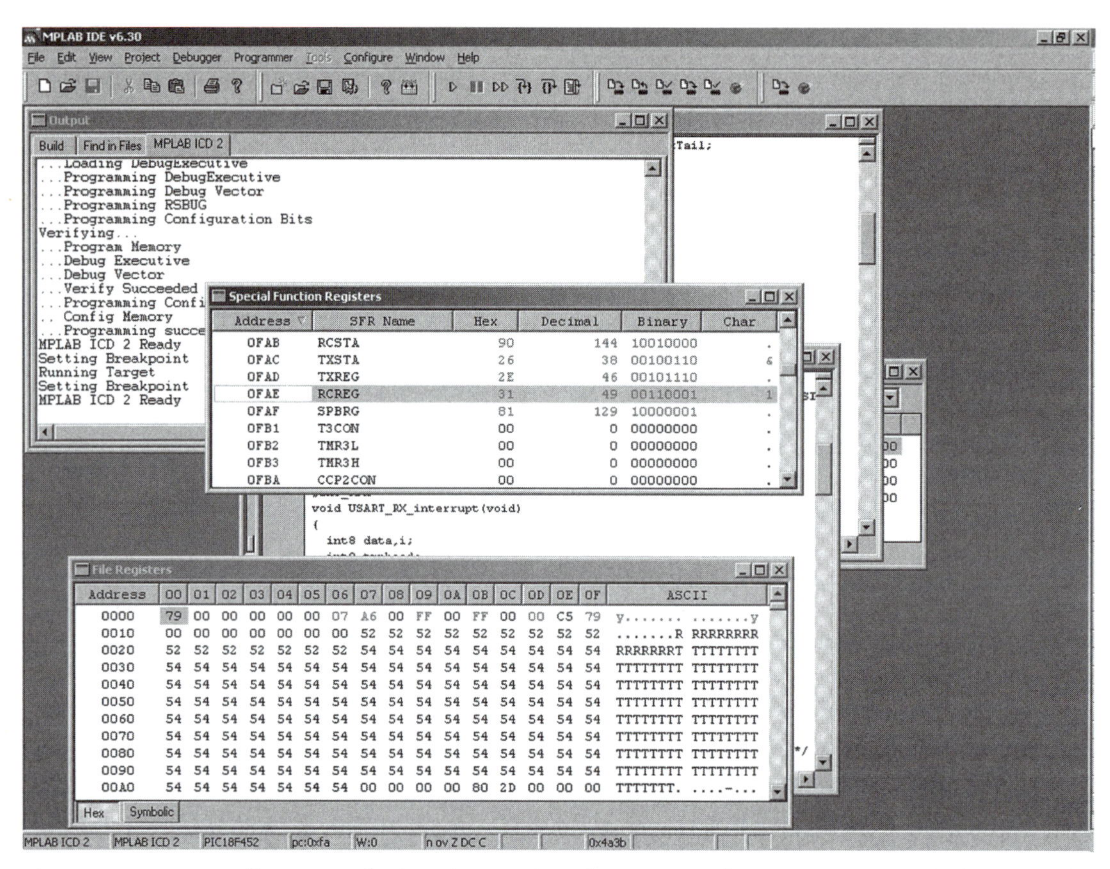

Figure 5.5: You can figure out the baud rate using the SPBRG value. At 20 MHz with the BRGH bit set (BRGH is bit 2 of the TXSTA register), 129 in the SPBRG equates to 9600 bps. You can also easily find this correlation in the PIC datasheet's USART section.

I've added the SFR (Special Function Registers) window to our set of MPLAB ICD 2 debugger windows in Figure 5.5 to show you the transmit and receive data registers, TXREG and RCREG. The last line of RS-232 interrupt C code that was executed picked up the contents of the RCREG and put them in the *data* variable. An ASCII '1' (0x31 or 00110001) is not the same as a binary '1' (00000001), as you can see in the RCREG value inside the SFR window in Figure 5.5. You can also check out the USART receive status by interpreting the bits within the RCSTA (Receive Status and Control Register), which is also located in the SFR window. I'll save you some datasheet time and tell you the seventh bit of the 0x90 value (1XXXXXXX) is the SPEN bit, which enables the serial port and configures the RX and TX pins of the PIC18F452 as serial port pins that are dedicated to the internal USART. The fourth bit that completes the 0x90 value (XXX1XXXX) is the CREN bit. CREN is short for Continuous Receive Enable bit and enables the USART's receiver. If an overrun error occurs, it can be cleared by clearing and setting the CREN bit. This doesn't recover any of the corrupted data but you may be able to read the USART's FIFO to retrieve and salvage what you can. What you do with framing and overrun errors depends on how you code your serial application to handle the possible error situations.

So far, thanks to our PIC18F452 microcontroller's USART, we have an ASCII '1' in the RCREG and the '1' has been transferred to a variable called *data*. In Figure 5.6 I've added *data* to our watch window for a positive confirmation of the contents of the *data* memory location. Notice also that I've changed the size of the watch values from 16 bits to 8 bits to make it easier to read as sometimes the 16-bit value includes another adjacent and unrelated value. When you are using the MPLAB IDE and you don't like the way something "looks," click on it to see if you can alter the item by tweaking its properties. A "right click" on the "31" in Figure 5.6 provided me with the ability to change the format of the Value display (hex, binary, and so forth), and gave me the power to change the number of bits that display in the Value column.

We've got to move the contents of the *data* variable to our receive buffer as fast as we can to avoid a possible USART FIFO overrun condition. However, we can't just stuff the *data* variable's contents into any buffer location. There's got to be some order to this and that's why the head and tail index calculations are incorporated. Let's work through calculating the index for our '1'.

We know that this is our first incoming character and our watch window tells us that *USART_RxHead*, or receive buffer head value, is set to 0x00. Since our *USART_RxTail* is also at 0x00, we would need to put a byte of distance between the buffer head and tail to cause it not to be "empty." That's easy. We simply increment our head value by 1.

Now, we only have 16 bytes of buffer space and if we run over that, we're into our transmit buffer space. That wouldn't be good as you'd have to learn to smoke and drink while trying to figure out who's doing what to whom in the buffer space. To keep you sober and our received characters within their allocated receive buffer memory, we will always bounce our receive buffer head index value off of our receive buffer mask value (USART_RX_BUFFER_MASK).

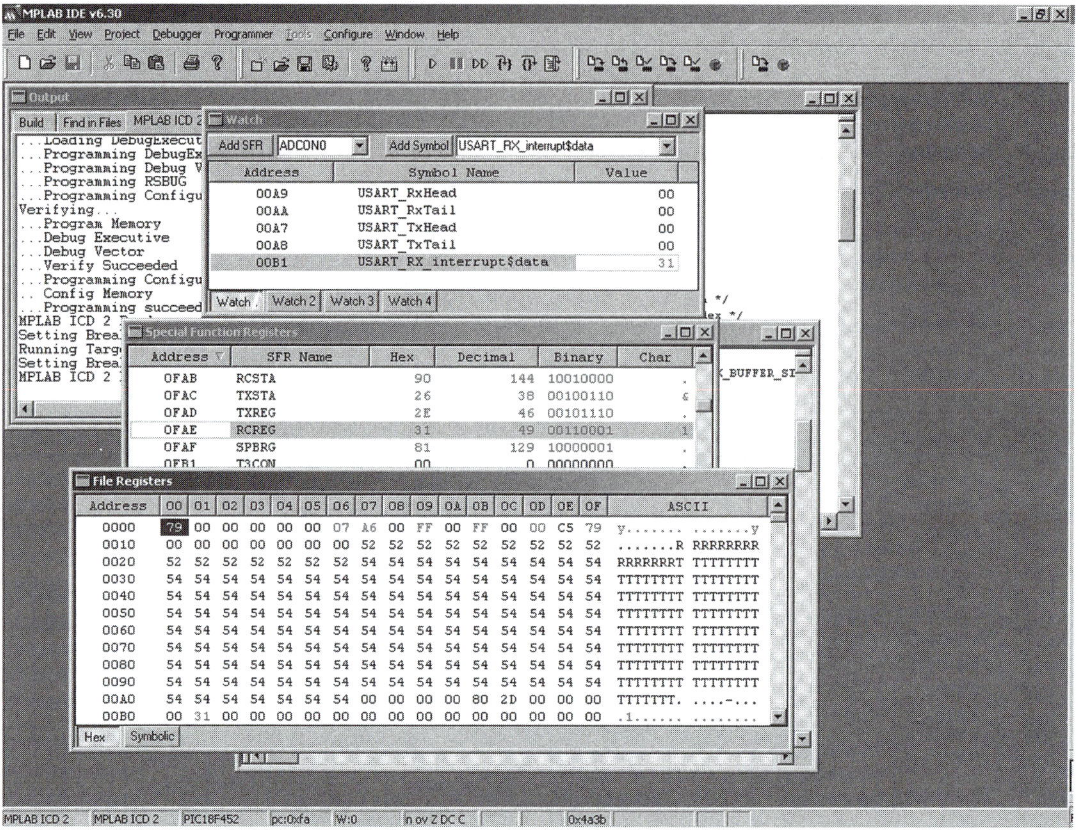

Figure 5.6: I've pulled in just enough File Register data to show you the *data* variable memory location at 0x00B1.

Code Snippet 5.11 shows us how the receive buffer mask and the receive buffer head value work together to store the received byte *data*:

```
   // calculate buffer index
   tmphead = ( USART_RxHead + 1 ) & USART_RX_BUFFER_MASK;

tmphead = (0x00 +1) & 0x0F;
or  tmphead = 0x01 & 0x0F;
performing the bitwise AND (&) operation: 00000001 AND 00001111
results in tmphead resolving to: 00000001 or 0x01
```

Code Snippet 5.11: Math can be fun, even if it's binary math.

The results of our calculations are confirmed in Figure 5.7.

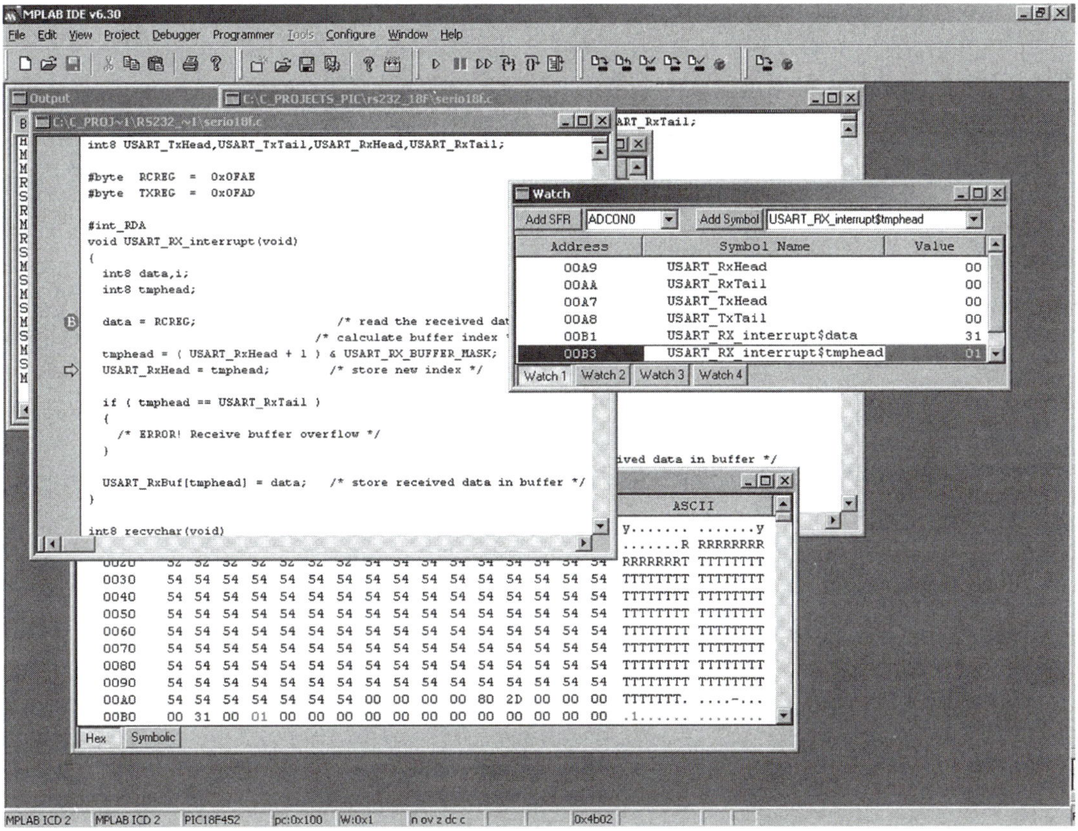

Figure 5.7: Once the incoming byte is seized, it is put into a buffer slot as quickly as possible.

Once the new receive head index value is calculated, we can store it as such by executing the next line of C source in Code Snippet 5.12.

```
// store new index
USART_RxHead = tmphead;
```

Code Snippet 5.12: The first byte of incoming RS-232 data will be stored in the buffer at the index location we just calculated.

Figure 5.8 confirms the store operation.

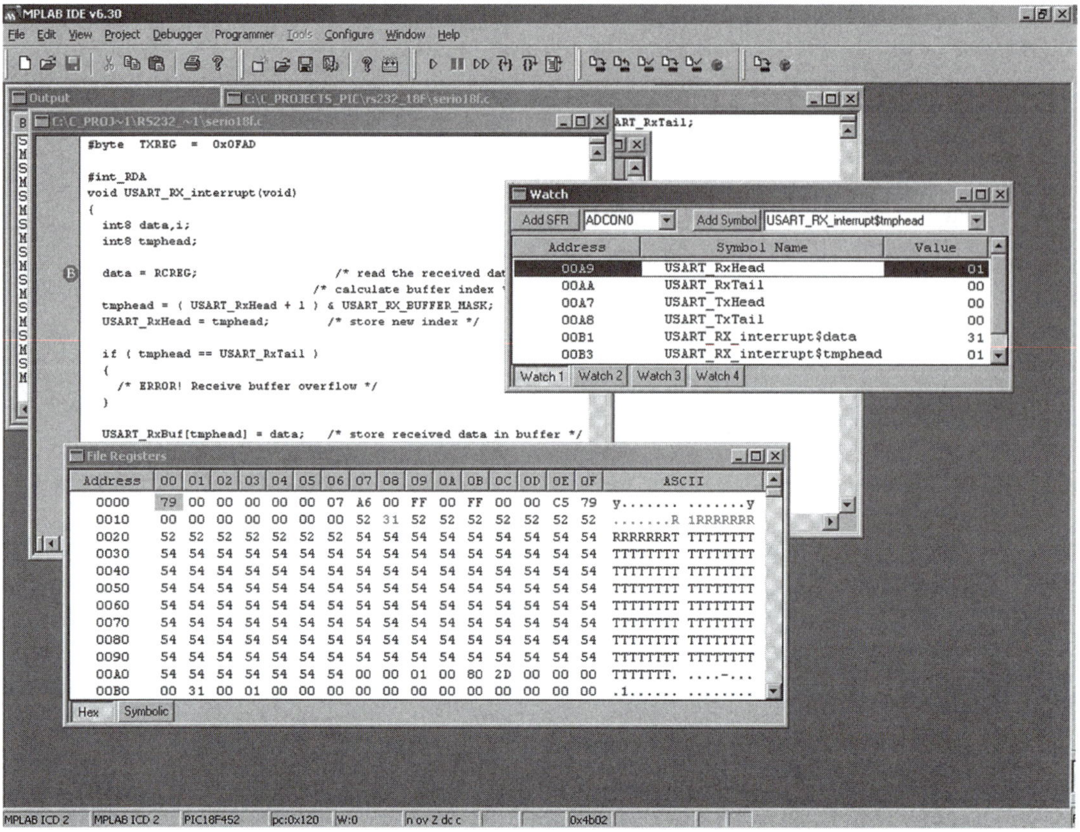

MPLAB IDE v6.30

File Edit View Project Debugger Programmer Tools Configure Window Help

Output

C:\C_PROJECTS_PIC\rs232_18F\serio18f.c

C:\C_PROJ~1\RS232_~1\serio18f.c

```
#byte  TXREG  =  0x0FAD

#int_RDA
void USART_RX_interrupt(void)
{
  int8 data,i;
  int8 tmphead;

  data = RCREG;                     /* read the received dat
                                 /* calculate buffer index *
  tmphead = ( USART_RxHead + 1 ) & USART_RX_BUFFER_MASK;
  USART_RxHead = tmphead;         /* store new index */

  if ( tmphead == USART_RxTail )
  {
    /* ERROR! Receive buffer overflow */
  }

  USART_RxBuf[tmphead] = data;   /* store received data in buffer */
```

ART_RxTail;

Watch

Add SFR | ADCON0 | Add Symbol | USART_RX_interrupt$tmphead

Address	Symbol Name	Value
00A9	USART_RxHead	01
00AA	USART_RxTail	00
00A7	USART_TxHead	00
00A8	USART_TxTail	00
00B1	USART_RX_interrupt$data	31
00B3	USART_RX_interrupt$tmphead	01

Watch 1 | Watch 2 | Watch 3 | Watch 4

File Registers

Address	00	01	02	03	04	05	06	07	08	09	0A	0B	0C	0D	0E	0F	ASCII
0000	79	00	00	00	00	00	07	A6	00	FF	00	FF	00	00	C5	79	y....... ....y
0010	00	00	00	00	00	00	00	52	31	52	52	52	52	52	52	52	.......R 1RRRRRRR
0020	52	52	52	52	52	52	52	54	54	54	54	54	54	54	54	54	RRRRRRRT TTTTTTTT
0030	54	54	54	54	54	54	54	54	54	54	54	54	54	54	54	54	TTTTTTTT TTTTTTTT
0040	54	54	54	54	54	54	54	54	54	54	54	54	54	54	54	54	TTTTTTTT TTTTTTTT
0050	54	54	54	54	54	54	54	54	54	54	54	54	54	54	54	54	TTTTTTTT TTTTTTTT
0060	54	54	54	54	54	54	54	54	54	54	54	54	54	54	54	54	TTTTTTTT TTTTTTTT
0070	54	54	54	54	54	54	54	54	54	54	54	54	54	54	54	54	TTTTTTTT TTTTTTTT
0080	54	54	54	54	54	54	54	54	54	54	54	54	54	54	54	54	TTTTTTTT TTTTTTTT
0090	54	54	54	54	54	54	54	54	54	54	54	54	54	54	54	54	TTTTTTTT TTTTTTTT
00A0	54	54	54	54	54	54	54	00	00	01	00	80	2D	00	00	00	TTTTTTT. ....-...
00B0	00	31	00	01	00	00	00	00	00	00	00	00	00	00	00	00	.1...... ....

Hex | Symbolic

MPLAB ICD 2 | MPLAB ICD 2 | PIC18F452 | pc:0x120 | W:0 | n ov Z dc c | 0x4b02

Figure 5.8: Our ASCII '1' takes up residence in the second slot of receive buffer memory.

Bad things can happen if we inadvertently stash our current head value into our current tail location. You can't type fast enough to do that in our application. And, the watch window tells us that the next C *if* statement will certainly fail and bypass the buffer overflow error code between the braces in Code Snippet 5.13.

```
if ( tmphead == USART_RxTail )
{
    /* ERROR! Receive buffer overflow */
}
```

Code Snippet 5.13: We could put some code here to attempt to reset the buffer pointers or we could simply issue a return *and do nothing. The code between the braces would have to be tailored to your application.*

Instead, as shown in Figure 5.8, our incoming ASCII '1' is stored at index position 1 inside the receive buffer memory area.

```
USART_RxBuf[tmphead] = data;    /* store received data in buffer */
```

Meanwhile, our C *main* function in Code Snippet 5.14 is constantly executing the *CharInQueue* function on what was up until now an empty receive queue.

```
while(1){
   while(!(CharInQueue()));
   sendchar(recvchar());
}
```

Code Snippet 5.14: The CharInQueue *function returns a TRUE when our ASCII '1' is received and kicks off the next C statement,* sendchar(recvchar()).

Now that we have a character in the receive buffer, the *CharInQueue* function will return a TRUE when it is called. The TRUE condition kicks off the *recvchar* function inside the *sendchar* call.

The *recvchar* function in Code Snippet 5.15 is not an interrupt handler and can be called from within the application at any time. So, an internal line of *CharInQueue* emulation code is included within the *recvchar* function.

```
int8 recvchar(void)
{
   int8 tmptail;
                              /* wait for incoming data */
   while ( USART_RxHead == USART_RxTail );
```

Code Snippet 5.15: If the recvchar *function is called, there had better be some data in the receive queue.*

If the *USART_RxHead* value is equal to the *USART_RxTail*, the receive buffer is empty and there's nothing for the *recvchar* function to do. The downside to this is that the program will stay here until a character is received by the USART and put into the receive buffer. Using the interrupt routine assures us that we won't get stuck here.

Data is inserted into the receive buffer at the head index location and removed from the receive buffer at the tail index location. We have to keep the tail inside the receive buffer area too, and the wrangler is once again the receive buffer mask value.

```
    // calculate buffer index
tmptail = ( USART_RxTail + 1 ) & USART_RX_BUFFER_MASK;
USART_RxTail = tmptail;         // store new index

return USART_RxBuf[tmptail];    // return data
```

Code Snippet 5.16: The tail index is calculated exactly like the head index.

This code works in Code Snippet 5.16 just like the interrupt code, except instead of using the current head index value to determine the new head index, the current tail index value is used to find a new tail index value. If all goes as coded, the *tmptail* value will resolve to 0x01 and point to our ASCII '1' residing at index location 1 inside our receive buffer.

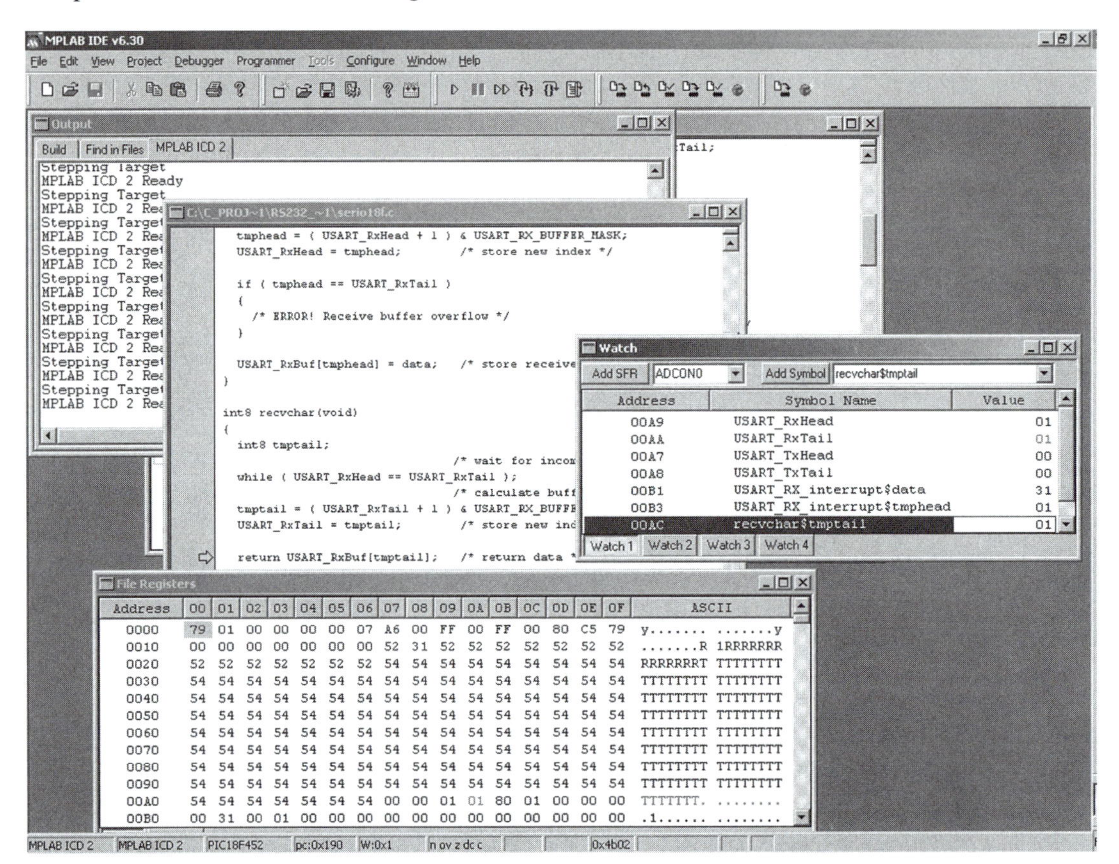

Figure 5.9: The recvchar function removes data from the receive buffer's tail.

I added *tmptail* to our watch window list in Figure 5.9. Note that *USART_RxTail* is now our new receive buffer official tail index value now that the data at receive buffer index location 1 has been processed. At this point, the receive buffer is once again empty as the *USART_RxHead* and *USART_RxTail* occupy the same positions in the receive buffer memory space.

With the *recvchar* function returning the ASCII '1' from the receive buffer, the *sendchar* function now has a value to act upon. The *sendchar* function's job is to insert a character into the transmit buffer, adjust the transmit buffer index value and enable the transmit interrupt mechanism. The code within the *sendchar* function is very similar to the receive code we've already examined.

Like the *recvchar* function, the *sendchar* function in Code Snippet 5.17 is not an interrupt handler and can be called at will. If by chance the transmit buffer is full when the *sendchar* function is called, the code *while (tmphead == USART_TxTail);* will stall the *sendchar* function until a character is removed from the transmit buffer. If the transmit interrupt mechanism is operating correctly, the byte would eventually be removed from the transmit buffer allowing the *sendchar* function to continue.

```
int8 sendchar(int8 data)
{
   int8 tmphead;
                            /* calculate buffer index */
   tmphead = ( USART_TxHead + 1 ) & USART_TX_BUFFER_MASK;
                            /* wait for free space in buffer */
   while ( tmphead == USART_TxTail );
                            /* store data in buffer */
   USART_TxBuf[tmphead] = (int8)data;
   USART_TxHead = tmphead;     /* store new index */

   enable_interrupts(INT_TBE);

   return data;
}
```

Code Snippet 5.17: Sending the received character is like firing an artillery piece. Once an expended cartridge is removed from the gun, a new shell is pushed into its place, the breech is closed and the gunner waits for the fire order. In our RS-232 transmit routine, we wait for a free buffer location, throw the data into that location and wait for the transmit interrupt to come along and fire off the data byte.

Here's how everything looks just before the code enables the transmit interrupt.

Our roving ASCII '1' is now positioned in the index 1 position of the 128-byte transmit buffer. Now that our transmit buffer head value doesn't match the transmit buffer tail value, the transmit buffer is no longer considered empty. I've added the *sendchar* function's *tmphead* and *data* variables to the watch window list in Figure 5.10 so we can keep up with them.

A peek at the TXSTA (Transmit Status and Control Register) bits in Figure 5.10 tells us that the TXEN (Transmit Enable) bit (XX1XXXXX) is active, and thus, so is the USART transmitter. The BRGH (High Baud Rate Selector; more logically Baud Rate Generator High (Speed)) bit (XXXXX1XX) indicates that the USART is set for high-speed asynchronous mode, and the TRMT (Transmit Shift Register Status) bit (XXXXXX1X) is set telling us that the transmit shift register (TSR) is empty.

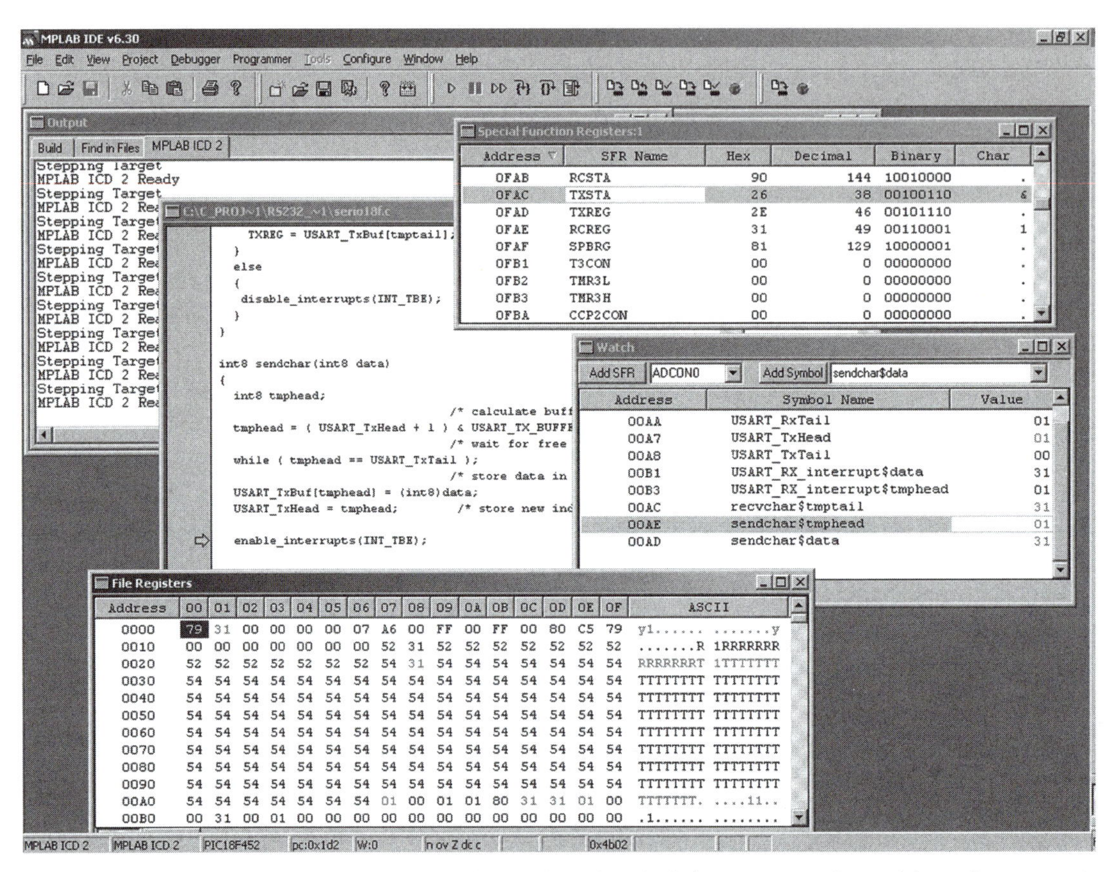

Figure 5.10: We've aimed our gun and slapped in the shell (our ASCII '1'). Enabling the transmit interrupt mechanism will close the breech.

Looks like the USART transmitter is just waitin' on a friend. In reality, it's a shark that's always hungry. I'm willing to bet my book royalties that every tooth is in place on the jaws of the transmit interrupt mechanism and they're ready to bite.

Once the USART transmit interrupt is enabled, a simple feed to the TXREG is all that's necessary to get the byte out of the TX pin. As soon as the last STOP bit is transmitted and if there's any data to send, the TSR is loaded with new data to send from TXREG. When the data transfer from TXREG to TSR is complete, TXREG is considered to be empty and the TXIF ((USART) Transmit Interrupt Flag) bit in the PIR1 (Peripheral Interrupt Request (Flag) Register 1) register is set. If the transmit interrupt is enabled and the TXIF flag is set, the shark is in the water. Let's take a dip and find out if the shark is on patrol.

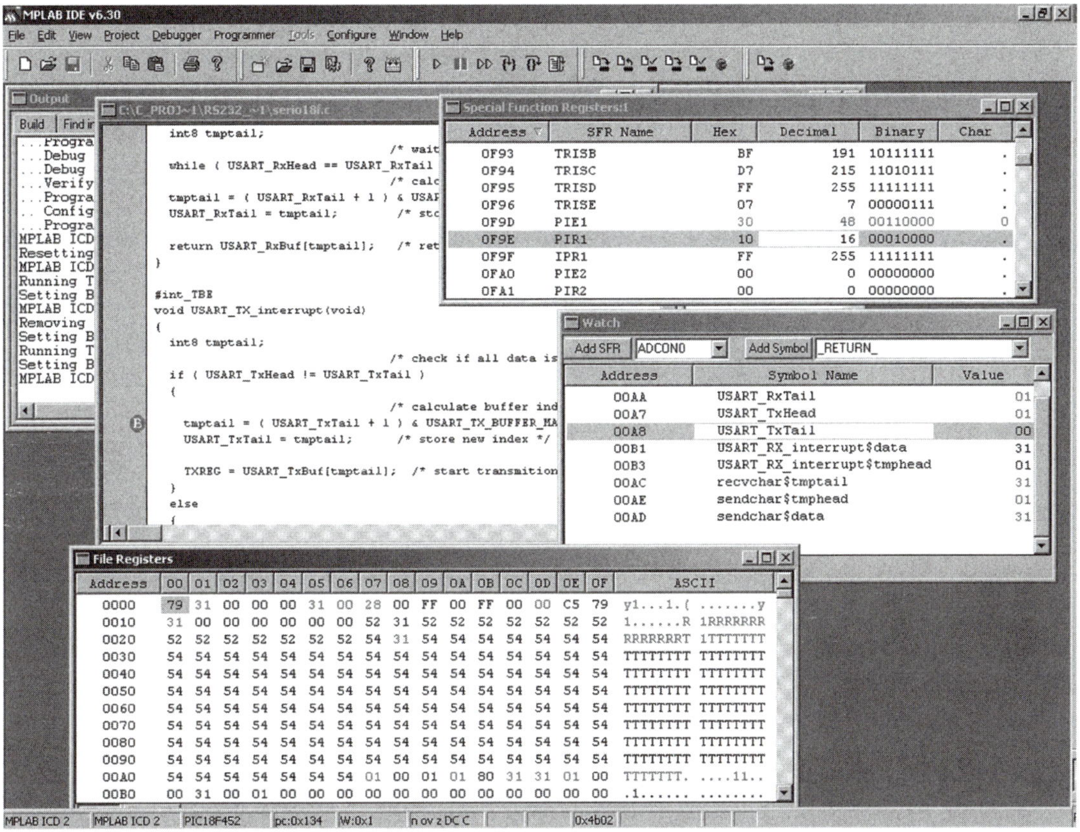

Figure 5.11: Guns and sharks...What a way to describe a software routine. Everything is poised for a transmission in this figure. All we need is some blood in the water or the order to fire.

There she is swimming around in Figure 5.11. Bit 4 of the PIR1 register is the TXIF bit. Bit 4 and 5 of the PIE1 (Peripheral Interrupt Enable Register 1) register tell us that the transmit and receive interrupts are enabled respectively.

We already know that there is data to be transmitted inside the transmit buffer. So, our *if (USART_TxHead != USART_TxTail)* statement will evaluate as TRUE, and the code between the braces will be executed. The Watch window in Figure 5.11 confirms this as well.

Just like the receive routines, the transmit routines in Code Snippet 5.18 feed the transmit buffer area with the head and feed the microcontroller application from the tail of the buffer. The data pointed to by the transmit buffer's tail index value is loaded into TXREG. From the TXREG register, the ASCII '1' makes its way to the TSR and out the USART's TX pin.

```
#int_TBE
void USART_TX_interrupt(void)
{
   int8 tmptail;
                          /* check if all data is transmitted */
   if ( USART_TxHead != USART_TxTail )
   {
                          /* calculate buffer index */
      tmptail = ( USART_TxTail + 1 ) & USART_TX_BUFFER_MASK;
      USART_TxTail = tmptail;        /* store new index */

      TXREG = USART_TxBuf[tmptail];     /* start transmission */
   }
   else
   {
      disable_interrupts(INT_TBE);
   }
}
```

Code Snippet 5.18: The gun is fired and the breech is opened, then the disable_interrupts(INT_TBE) statement is executed. If you're following the marine analogy, the shark bites and swallows your leg when the disable_interrupts(INT_TBE) statement is executed.

Figure 5.12 tells us many things. According to the *USART_TxHead* and *USART_TxTail* values, the transmit buffer is empty as is the receive buffer. Bit 1 of the TXSTA has cleared indicating that the TSR is full. The TSR is full because an ASCII '1' (0x31) was loaded into the TXREG register. After allowing the program to resume normally, that ASCII '1' we just followed through the receive and transmit buffers appeared in my Tera Term Pro terminal emulator window.

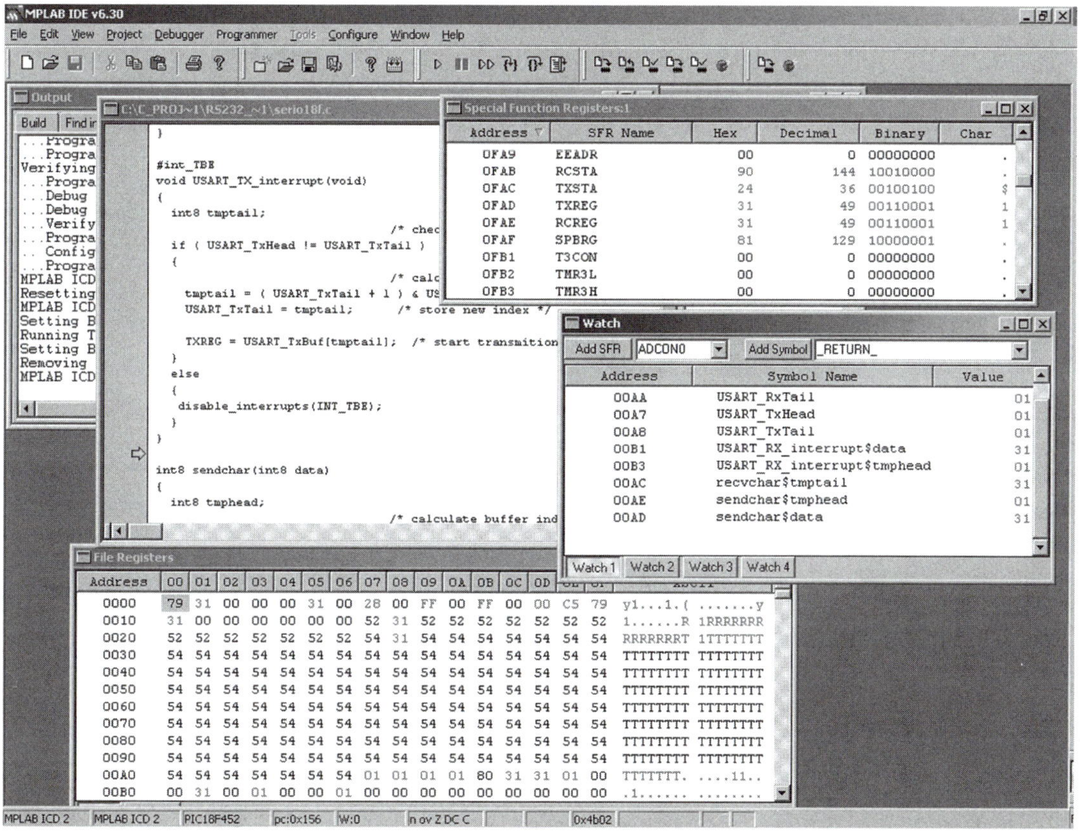

Figure 5.12: You can write and debug all of the code we've discussed without using an MPLAB ICD 2, but it wouldn't be as much fun.

You're probably wondering when and how the very first index position in the buffers, zero (0), is used. Actually, index 0 is the last to be accessed. Let's assume that we have 15 characters in our 16-byte receive buffer, and a new 16th character is ready to be inserted into the receive buffer memory area. Code Snippet 5.19 tells us how it would go:

```
    // calculate buffer index
    tmphead = ( USART_RxHead + 1 ) & USART_RX_BUFFER_MASK;

tmphead = (0x0F +1) & 0x0F;
or  tmphead = 0x10 & 0x0F;
performing the bitwise AND (&) operation: 00010000 AND 00001111
results in tmphead resolving to: 00000000 or 0x00
```

Code Snippet 5.19: You can use this logic on any microcontroller you come into contact with.

As you can see, the receive buffer mask will never let the index go out-of-bounds and wraps the index around to zero to accommodate the 16th received character. I'll leave the PIC portion of our RS-232 discussion with a view of an actual Tera Term Pro session and the contents of the PIC18F452's receive and transmit buffers, courtesy of the MPLAB ICD 2 and the RS-232 portion of the Easy Ethernet CS8900A in Figure 5.13.

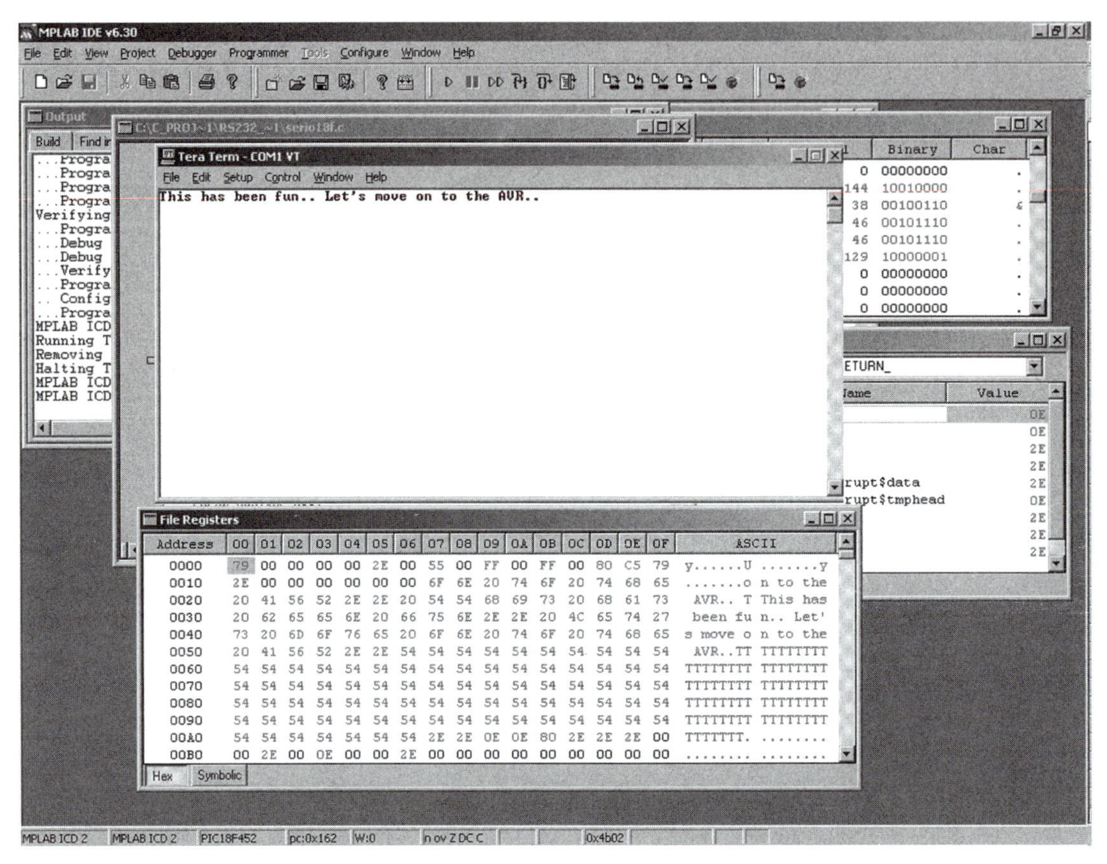

Figure 5.13: In reality, all of this seemingly complex code is really simple once you break it down into digestible pieces. You can't make it simple if you don't have the tools. The MPLAB ICD 2 is a must if you want to get down to the bit and byte level of your PIC18FXXX-based programs.

Applying What We Know about RS-232 to the Atmel AVR

The RS-232 engine you see in Photo 5.3 is based on what you will come to know as the Easy Ethernet AVR and is powered by an Atmel® ATmega16 AVR. In this chapter, I'll show you how to get an AVR RS-232 engine running using virtually the same RS-232 code as that used by the Microchip PIC.

Photo 5.3: The same "don't worry, you'll see this again" goes for the obviously partless area of this device. In later chapters, you'll come to know the device in this photo as the Easy Ethernet AVR.

Everything you know about the PIC18F452 USART applies logically to the ATmega16 USART. Of course, Microchip and Atmel do things differently inside the bowels of their respective USARTs, but in the end, both the Atmel and Microchip USARTs do the same thing and that's send and receive asynchronous data frames. Also, everything you already know about RS-232 as it relates to the PIC18F452 applies to the ATmega16. If it didn't, the Microchip and Atmel parts wouldn't be able to communicate using the RS-232 standard.

The RS-232 converter and power supply circuitry for the Easy Ethernet AVR and Easy Ethernet CS8900A are identical. The difference in hardware between the two platforms lies in the ATmega16 and its supporting circuitry. The ATmega16 uses a different programming interface and is supported by a bevy of Atmel programming tools. The serial circuitry for the partially assembled Easy Ethernet AVR is shown in Schematic 5.2.

We'll be using a different C compiler to write the code for the AVR device. The AVR C compiler is produced by ImageCraft™ and is called ICCAVR. The Easy Ethernet AVR RS-232 firmware was written with the professional version of ICCAVR. The ImageCraft AVR C compiler has its own IDE. We'll use AVR Studio as our debugging platform and write the C code inside the ImageCraft IDE.

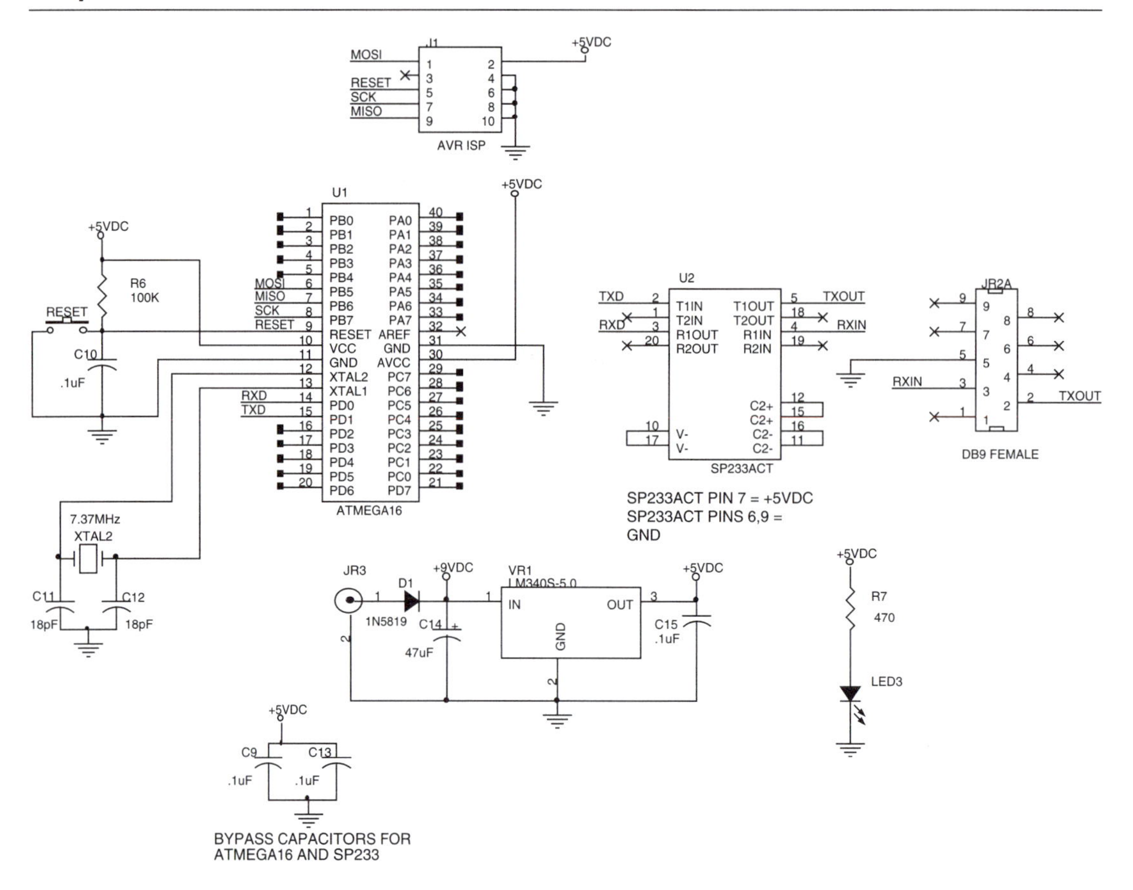

Schematic 5.2: We could run the ATmega16 much faster but there's no need to here. The AVR JTAG ICE will use the AVR's PORTC pins while we're examining the AVR RS-232 code.

You can download AVR Studio free from the Atmel web site. AVR Studio is very similar to Microchip's MPLAB IDE, which is also a free download. AVR Studio is an IDE (Integrated Development Environment) that consists of an editor, a project manager, an assembler and compiler interface, a simulator and a debugger. The AVR Studio editor is capable of being coupled with almost any AVR compiler and C or assembler source code can be edited in the AVR Studio debugger source window.

ImageCraft's ICCAVR comes complete with an integrated project manager, C editor, ANSI Terminal emulator and an Application Builder to generate peripheral initialization code. ICCAVR also supports symbolic debugging in the AVR Studio IDE. That means we can write our code in the ICCAVR project environment and debug it C statements and all in the AVR Studio IDE. It gets better. We can also use ICCAVR to generate a debug file that is used by AVR Studio to drive a piece of debugging/programming hardware called JTAG ICE.

The AVR JTAG ICE allows a programmer's eye view of the ATmega16 internals. The ability to "look inside" of an AVR microcontroller makes the AVR JTAG ICE very similar logically to the MPLAB ICD 2 that I used to show you what was going on inside the Microchip PIC18F452 microcontroller. You can use the AVR JTAG ICE to debug your code using ImageCraft's ICCAVR C source or the AVR assembler generated by the ImageCraft AVR C Compiler.

Atmel's AVR JTAG ICE is based on a concept called On-chip Debugging or OCD. A majority of the microcontroller emulators you can buy today use a specialized "bond out" integrated circuit that contains the CPU and I/O cores of the microcontroller it is emulating. Thus, the code is actually running on the "bond out" device and the supporting emulator hardware and software have the ability to reach into the "bond out" and pull out the microcontroller internals for inspection and debugging. The Microchip MPLAB ICE 2000 uses "bond out" technology. Instead of depending on a "bond out" device, the AVR JTAG ICE interfaces with the target AVR's internal OCD system via a JTAG IEEE 1149.1 compliant interface.

Every Atmel AVR microcontroller that pins out a JTAG pin set houses OCD logic. That includes the Atmel ATmega16, which I've chosen to be the command microcontroller for our AVR RS-232 project. The AVR JTAG ICE takes control of the ATmega16 and controls the execution of the AVR firmware using the ATmega16's OCD logic by way of the ATmega16's JTAG interface pin set. I purposely left the Easy Ethernet AVR's JTAG pins open to allow the use of the AVR JTAG ICE in this project and the AVR Easy Ethernet AVR project.

Coding the AVR RS-232 Routines

I'll begin by telling you the AVR RS-232 code is logically identical to the Microchip PIC18F452 RS-232 code. In fact, the only differences in the physical code lie in the way the interrupt handler definition code is structured and some USART register naming conventions. The best way to convey the similarities and differences is to put the AVR code side by side with the PIC code.

A *#pragma* directive controls the actions of the ImageCraft ICCAVR C Compiler. In Code Snippet 5.20 the *#pragma interrupt_handler* declares functions as interrupt handlers so that the compiler will generate a *reti* (return from interrupt) instead of a standard *ret* (return) instruction. The other neat thing the #pragma directive does is to make sure that all the registers that the interrupt handler functions use are automatically saved and restored. Also, the *#pragma* directive instructs the compiler to generate the interrupt vectors based on the vector numbers. For all of this magic to work, the *#pragma* directive must precede the interrupt handler function definitions.

```
//********************************************************************
//*   PIC
//********************************************************************
#include <18F452.h>
#device ICD=TRUE
#use delay(clock=20000000)
#use rs232(baud=9600,parity=N,xmit=PIN_C6,rcv=PIN_C7,bits=8)
#fuses HS,PUT,NOWRTB,NOEBTR,NOWRT,NOWDT,NOLVP,NOPROTECT

#int_RDA
void USART_RX_interrupt(void)
#int_TBE
void USART_TX_interrupt(void)

//********************************************************************
//*   AVR
//********************************************************************
#include <iom16v.h>
#include <macros.h>

#pragma interrupt_handler USART_RX_interrupt:iv_USART_RX
#pragma interrupt_handler USART_TX_interrupt:iv_USART_UDRE
```

Code Snippet 5.20: You're already familiar with #include *statements. The* #pragma *directives immediately follow the* #include *statements, which are at the top of the AVR RS-232 source code.*

All of the additional *#device, #use and #fuses* information that is in the PIC header code is entered into the AVR mix in different ways and different places. The *#device ICD=TRUE* statement does many things to prepare the PIC for use with the MPLAB ICD 2 including setting up debugger activation bits within the PIC18F452. In a similar fashion, fuses within the ATmega16 determine if an AVR JTAG ICE can be attached to the ATmega16's JTAG I/O pins. There are no AVR counterparts for the PIC's *#use delay(clock=20000000)* and *#use rs232(baud=9600,parity=N,xmit=PIN_C6,rcv=PIN_C7,bits=8)* directives. So, we must calculate and code the AVR baud rate and delay routine values manually.

Minor differences in syntax exist when the AVR and PIC function prototypes are declared in Code Snippet 5.21. The AVR *init_USART(unsigned int baud)* function takes the place of the *#use rs232* directive used by the Custom Computer Services C Compiler.

```
//*****************************************************************
//*    PIC FUNCTION PROTOTYPES
//*****************************************************************
int8 recvchar(void);
int8 sendchar(int8 data);
int8 CharInQueue(void);

//*****************************************************************
//*    AVR FUNCTION PROTOTYPES
//*****************************************************************
int recvchar( void );
int sendchar( int );
unsigned char CharInQueue(void);
void init_USART(unsigned int baud);
```

Code Snippet 5.21: The Custom Computer Services C Compiler uses the int8 naming convention. This allows an integer to be defined with specific bit lengths. For instance, using the built-in functionality that the Custom Computer Services C Compiler provides allows you to define a 16-bit integer using the int16 directive. ICCAVR follows the ANSI standards, and an int is 16-bits in length no matter what.

```
//*****************************************************************
//*    BAUD RATE NUMBERS FOR UBRR
//*****************************************************************
#define  b9600   47     // 7.3728MHz clock
#define  b19200 23
#define  b38400 11
#define  b57600 7
```

Code Snippet 5.22: These baud rates came directly from the baud rate table in the ATmega16 datasheet. Later in this book, I'll introduce you to a program that is bundled with ICCAVR that calculates baud rate and delay register values.

The Custom Computer Services C Compiler *#use rs232* directive also uses a parameter within the body of the directive to set the baud rate. Since 7.3728 MHz is a common microcontroller oscillator frequency, I was able to use a table in the ATmega16 datasheet to extract the baud rates in Code Snippet 5.22.

ICCAVR has some built-in tricks up its sleeve as well. The ICCAVR Project Wizard generated the bulk of the code in Code Snippet 5.23.

```
//********************************************************************
//*   USART Function
//********************************************************************
void init_USART(unsigned int baud)
{
   UCSRB = 0x00; //disable while setting baud rate
   UCSRA = 0x00;
   UCSRC = 0x86;
   UBRRL = baud; //set baud rate lo
   UBRRH = 0x00; //set baud rate hi
   UCSRB = 0x98;
}
```

Code Snippet 5.23: I added the baud variable. I also answered some simple questions and the ICCAVR Project Wizard did the rest.

None of the AVR USART register names look like anything we used in the PIC USART code. So, to gain a perspective as to what's going on here, let's take a closer look at these ATmega16's USART registers. We'll begin with the UCSRB in Figure 5.14.

USART CONTROL AND STATUS REGISTER B

7	6	5	4	3	2	1	0
RXCIE	TXCIE	UDRIE	RXEN	TXEN	UCSZ2	RXB8	TXB8
1	0	0	1	1	0	0	0

Figure 5.14: We use the heck out of this register in the AVR USART code. The UCSZ2 bit participates in the determination of the asynchronous frame length. RXB8 and TXB8 are used in 9-bit RS-232 sessions.

The USCRB is the AVR USART register we will use to enable and disable the USART transmitter and receiver and enable and disable the AVR transmit and receive interrupts. The USCRB is initially cleared and after the baud rate registers are loaded, its value is set to 0x98. Setting the USCRB to 0x98 enables both the AVR USART transmitter and receiver and enables the USART'S receive complete interrupt.

Some of the flag bits associated with the transmit and receive interrupt bits in UCSRB are located in the UCSRA, which is shown graphically in Figure 5.15. You'll also find the framing (FE), parity (PE) and overrun (DOR) error flags in the UCSRA. The AVR USART is able to run at 2X speed by setting the U2X bit of the UCSRA and multiprocessor mode by setting the MPCM bit of the UCSRA.

USART CONTROL AND STATUS REGISTER A

7	6	5	4	3	2	1	0
RXC	TXC	UDRE	FE	DOR	PE	U2X	MPCM
0	0	0	0	0	0	0	0

Figure 5.15: When the AVR USART is running in asynchronous mode, the 2X mode halves the baud rate divisor value and doubles the baud rate.

The UDRE bit of the UCSRA is set after a reset to indicate that the USART transmitter is ready. We're using transmit interrupts and don't want any vectors to the transmit interrupt handler to occur while we're setting up the USART. So, the UCSRA is cleared to indicate that the transmitter is not ready.

The UCSRC in Figure 5.16 shares the same I/O location as the UBRRH. The URSEL bit must be set to access the bits in the UCSRC. Looking at our baud rate value, we'll never have to access the UBRRH. You don't see the *URSEL* bit toggled in the AVR RS-232 code because we take advantage of the fact that the UBRRH is cleared at reset.

USART CONTROL AND STATUS REGISTER C

7	6	5	4	3	2	1	0
URSEL	UMSEL	UPM1	UPM0	USBS	UCSZ1	UCSZ0	UCPOL
1	0	0	0	0	1	1	0

Figure 5.16: This AVR USART register shares I/O space with the UBRRH register. If your application uses a baud rate value that is larger than 8 bits, be sure to toggle URSEL clear before writing to the UBRRH register. An example of a baud rate that would require the UBRRH is 2400 bps with the U2X bit set and a clock of 7.3728 MHz.

The *UMSEL* bit of the UCSRC selects asynchronous mode when it is cleared. The 'S' in USART stands for synchronous, and a set *UMSEL* bit puts the AVR USART in synchronous mode.

I like the AVR USART register naming conventions as you can almost guess what each bit does by its name. You've probably already figured out that the 'U' in each bit name is short for USART. Can you deduce the nature of the *UPM* bit set? Let's give it a whirl; U = USART – P = parity – M = mode. The *UPM* bit defines the parity setting. With both of the *UPM* bits cleared, parity is disabled. A clear *USBS* (USART Stop Bit Select) bit instructs the USART logic to recognize the end of an asynchronous frame using a single stop bit.

The *UCSZ2* bit in the UCSRB is cleared and plays along with the *UCSZ1* and *UCSZ0* bits of the UCSRC to set the asynchronous frame data length to 8 bits. The *UCPOL* bit is used in synchronous mode and is a "don't care" bit to us right now.

The only things left that are different are just a few lines of code. I've highlighted them in Code Snippet 5.24.

```
//*****************************************************************
//*   USART Receive Interrupt Handler
//*****************************************************************
// AVR
   data = UDR;              /* read the received data */
// PIC
   data = RCREG;            /* read the received data */

//*****************************************************************
//*   USART Transmit Interrupt Handler
//*****************************************************************
// AVR
   else
   {
      UCSRB &= ~(1<<UDRIE);   /* disable UDRE interrupt */
   }
// PIC
   else
   {
      disable_interrupts(INT_TBE);
   }

//*****************************************************************
//*   USART Transmit Character Function
//*****************************************************************
// AVR
   UCSRB |= (1<<UDRIE);       /* enable UDRE interrupt */
// PIC
   enable_interrupts(INT_TBE);

//*****************************************************************
//*   MAIN FUNCTION
//*****************************************************************
// AVR
   CLI(); //disable all interrupts
   SEI(); //re-enable interrupts
// PIC
   enable_interrupts(GLOBAL);
   disable_interrupts(GLOBAL); //not used in our code
```

Code Snippet 5.24: This is just a difference in the underlying design of each C compiler. ImageCraft C tends to lean more toward the ANSI standards, while the Custom Computer Services C Compiler favors the PIC.

The Custom Computer Services C Compiler PIC statements in Code Snippet 5.24 are pretty simple to understand. Some of you may be lost in what seems to be extra dribble of the AVR code. Actually, it's easy to figure out if you break it down. Let's decipher the line *UCSRB &= ~(1<<UDRIE).*

Do the stuff inside the parentheses first:

1. UDRIE = 5 (value kept in iom16v.h)
2. Shift binary 00000001 left *UDRIE* (5) times
3. Shift results in binary result in 00100000

Then apply the '~' complement operator. Complementing binary 00100000 results in 11011111.

Finally, AND (&) the shifted and complemented binary number with the contents of UCSRB. UCSRB would have a value of 0xB8 at this point.

UCSRB	10111000
Binary number	11011111
Result	10011000

The idea was to clear the *UDRIE* bit and disable the USART Data Register Empty interrupt and that's just what we did. The same logic applies for setting the *UDRIE* bit except an OR (|) is used instead of the AND (&), and the shifted binary value representing *UDRIE* is not complemented.

Photo 5.4: I added standard .1-inch center header posts to allow the inclusion of the AVR JTAG ICE into the AVR Studio RS-232 debugging mix.

I modified the partially assembled Easy Ethernet AVR by adding a pin here and there to allow easy hookup of my AVR JTAG ICE (Photo 5.4). After lashing in the AVR JTAG ICE, I compiled our AVR RS-232 code using ICCAVR and loaded the debug file into AVR Studio and the AVR JTAG ICE. I then opened up a 9600 bps Tera Term Pro terminal emulator window, started the AVR JTAG ICE debug session and typed some ASCII text into the Tera Term Pro session. The ASCII text was echoed back to the Tera Term Pro session, and I stopped the AVR Studio debug session so I could show you Figure 5.17.

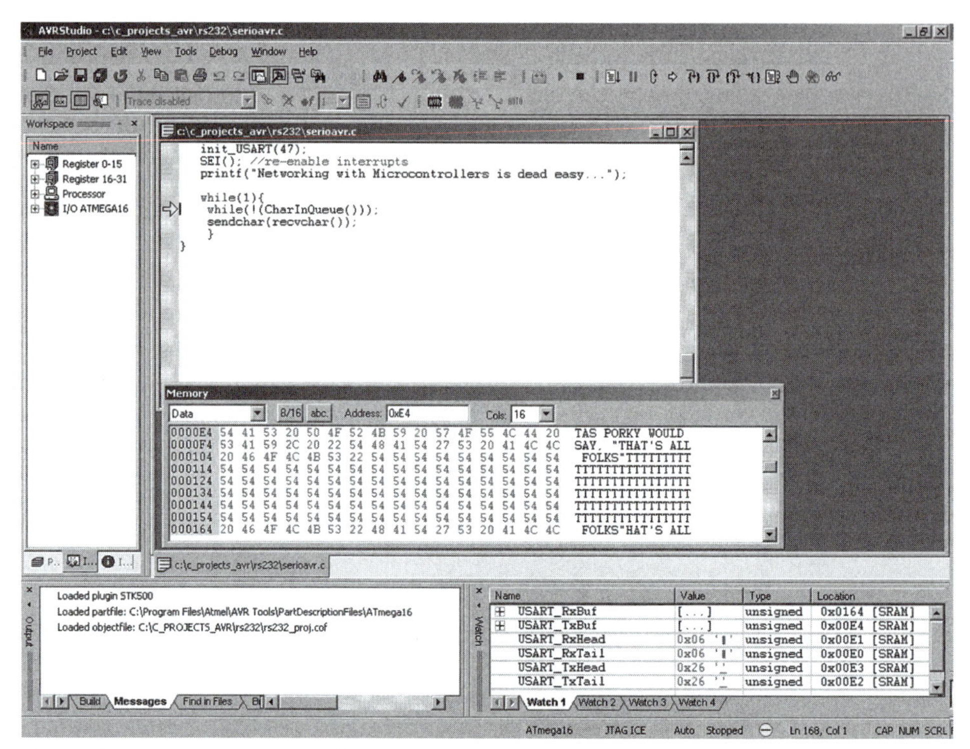

Figure 5.17: Now that's one famous quote!

The ability to build and code an RS-232 interface gives you, the microcontroller programmer, a leg up when it comes to debugging most any microcontroller-based project. Having the ability to send data from the microcontroller's serial port allows you to add checkpoints in your code to assist in debugging. Once you've learned the way of RS-232 for one microcontroller, you can use that knowledge to assist you when you move to a project on a different manufacturer's microcontroller. The rudiments of RS-232 do not change from microcontroller to microcontroller. The basic tenets of RS-232 ring true throughout the vastness of microcontroller-dom.

Porky has said goodbye for us from RS-232 land. So, let's explore yet another popular microcontroller serial protocol in the next chapter.

CHAPTER 6

I^2C...The Other Serial Protocol

RS-232 is a great point-to-point protocol when communicating between two distinct and sometimes distant pieces of equipment. However, there are times that you'll need to be able to talk to multiple electronic modules across a communications link that only spans the distance of a single printed circuit board. It would be possible to "network" the board-sharing modules using the RS-232 9-bit addressing protocol but there are lots of caveats in that approach. Even though you could eliminate the RS-232 voltage conversion circuitry, you would find yourself doing a tremendous amount of USART transmit and receive line housekeeping. For instance, you would have to generate an algorithm to handle collisions between modules attempting to transmit at the same time, or collisions that occur in the middle of a message another module is already transmitting. The USART transmit and receive lines are not automatically passive or tristated when inactive. Thus, you would also have to write some code to make sure the transmit and receive lines are inactive when they're supposed to be. If you want an RS-232 LAN, it can be done but there is a better way.

Initially designed for use in commercial audio and video systems, the inter-IC or I^2C bus is a Philips Semiconductor creation. Just as its name implies, the I^2C bus is a bidirectional 2-wire bus that is used to transport data between ICs (integrated circuits). Unlike RS-232, the I^2C bus doesn't need any voltage converters or special interface parts. If an IC is I^2C-bus compatible, everything needed to operate on the I^2C bus is incorporated on-chip within the IC.

If you take another look at our RS-232 schematic (Schematic 6.1), you'll see that there are two bus lines integral to the PIC18F452: a serial data line (SDA) and a serial clock line (SCL). The SDA and SCL bus lines make up the I^2C interface, and since these lines are designated as an integral part of the PIC18F452 that makes the PIC18F452 an I^2C-compatible device. Being I^2C-compatible, the PIC18F452 has provisions for a unique I^2C bus address. Using the built-in I^2C functionality, the PIC18F452 can act as either the master or slave on an I^2C network. If the PIC18F452 is configured as an I^2C master, it can act as a master-transmitter or master-receiver. Conversely, if the PIC18F452 is chosen to be a slave on the I^2C bus, its internal I^2C electronics can act in either slave-receiver or slave-transmitter mode. Remember one of my RS-232 "LAN" caveats and collisions? I^2C is a true multimaster bus that includes arbitration safeguards against data collisions, which prevents data corruption on the I^2C bus. Like RS-232, I^2C is an 8-bit bidirectional serial communications method. That's where the similarity ends. I^2C operates at a speed of 100 kbs in standard-mode, 400 kbs in fast-mode and up to 3.4 Mbps in high-speed mode. The only limitation as to how many devices can exist on a single I^2C bus is the total capacitance the devices place on the bus.

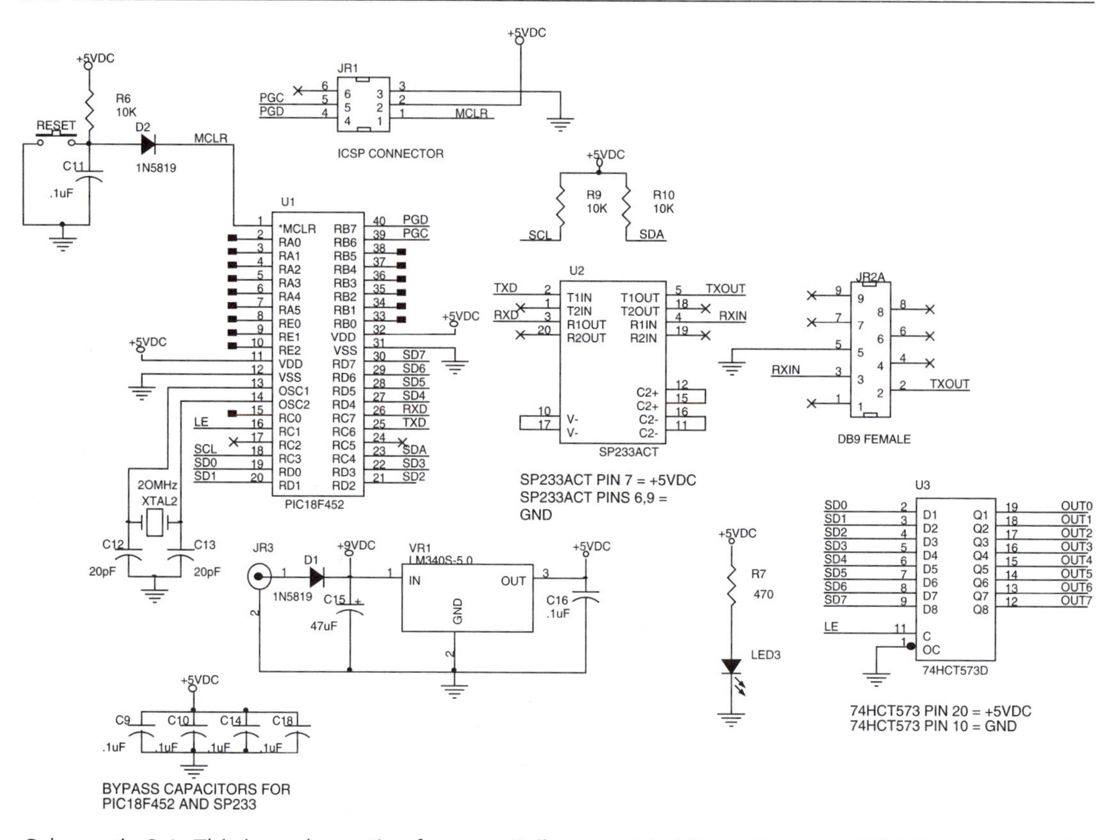

Schematic 6.1: This is a schematic of our partially assembled Easy Ethernet CS8900A. Resistors R9 and R10, plus the PIC18F452's internal I²C engine is all that's needed to effect an I²C network. Notice we've added a new component.

Advantages of using I²C are numerous, and there are a multitude of various I²C building blocks to choose from. By employing I²C in a design, we can eliminate much of the auxiliary support circuitry such as address decoders and standard logic gates needed for other communications methods.

In this chapter, we're going to use I²C to network our partially-assembled Easy Ethernet CS8900A to our partially-assembled Easy Ethernet AVR. Since each microcontroller in the network has on-chip I²C resources, we already have a solid basis for an I²C microcontroller network, but before we start slinging solder, let's take a course in I²C.

Why use I^2C?

For folks that make their living designing the neat gadgets we buy at department stores and over the Internet, putting out a product at the least possible cost is paramount. Chances are your television and stereo both contain an I^2C bus. Using I^2C is cheap because you don't have to do anything special to setup the physical communications link. Two wires or two traces are all that's needed for the physical I^2C-bus signal path. Although I^2C can operate at a very high speed, most of the time that's not a factor in the design. So, the serial nature of I^2C is well-suited for low-speed control type applications. I^2C also solves a majority of the design problems one would encounter when connecting dissimilar devices on a network. Slow devices must be able to talk to higher speed devices and vice versa, and everyone on the network must be able to speak the same language. Most importantly, somebody has to be the network boss and like the real world, there may be more than one boss on the bus. I^2C has an answer for all of these potential problems.

The I^2C bus

As you already know, I^2C is built around a two-wire serial bus, SDA (serial data) and SCL (serial clock). Each device on the I^2C bus is identified by a unique address. An I^2C device can be a microcontroller such as our PIC18F452, a memory device such as a standard I^2C EEPROM, or a special purpose device like an LED display driver. Some I^2C devices are capable of transmitting and receiving on the I^2C-bus while other I^2C devices may only be able to receive. In any case, a master-slave environment always exists on the I^2C bus. The I^2C master device always initiates an I^2C-bus data transfer and generates the clock signals to make the data transfer happen. The I^2C device that responds to the master's calling is considered the slave device.

Microcontrollers are normally defined as masters on a typical I^2C bus with other special-purpose I^2C devices acting as their slaves. In our application, even though all of the I^2C devices are microcontrollers capable of being an I^2C master, only one of the microcontrollers will be granted master status.

If more than one master exists on a single I^2C-bus, there will be conflict when one of the multiple masters attempts to transmit in unison with another peer master device on the I^2C bus. The I^2C specification solves this problem with a thing called arbitration. Arbitration is the process of allowing only one master to control the I^2C-bus at any time. Before I can really explain arbitration to you, there are some basic I^2C rules you need to know.

In an ideal world, if the master wanted to communicate with slave, the master would address the slave. The master is now in master-transmitter mode, and the slave is in slave-receiver mode. The master would clock-out data to the slave and terminate the data transfer after all of the desired bytes were transmitted.

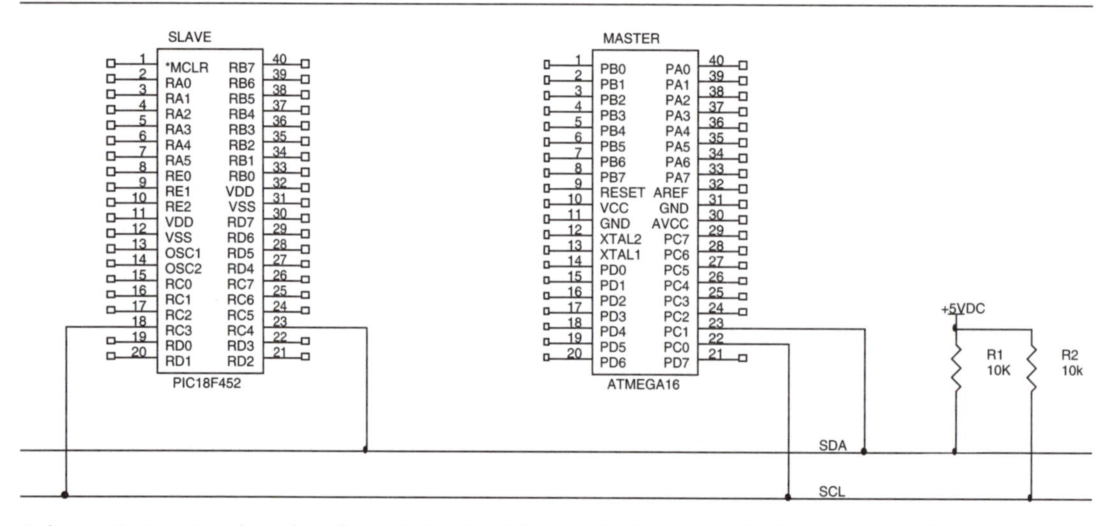

Schematic 6.2: For the sake of simplicity, I've left out all of the standard microcontroller connections to help us focus on the I²C bus. I flipped a coin to choose the master microcontroller.

On the other side of that, let's say that the master wanted to receive some data from the slave. Again, the master would clock-out an address aimed at the slave. Instead of assuming master-transmitter mode, this time the master would become the master-receiver with the slave acting as slave-transmitter. Data would be clocked-in by the master, which would terminate the transfer after receiving the desired bytes. For every bit of data moved, one clock pulse is generated and the data on the SDA line must be stable during the HIGH period of SCL. The logic level of the data line can only change when the SCL line is LOW.

Note that in either of the aforementioned cases, the master did all of the clocking and controlled the initiation and termination of the I²C session. The I²C master is always responsible for generating the clock on the I²C bus. For an I²C bus with multiple masters, each master generates its own specific clock. The only things that can alter a master's clock are a slower slave device holding down the clock line or another master I²C device during arbitration.

As you can see in both Schematic 6.2 and Schematic 6.3, the I²C bus SDA and SCL lines are pulled high by a pair of pull-up resistors. To participate on the I²C bus, an I²C device must present an open collector interface to the bidirectional SDA and SCL I²C-bus lines. This type of open collector interface performs a wired-AND function. As long as the 400 pF I²C-bus capacitance limit is not exceeded, any number of I²C devices can coexist on a single I²C bus.

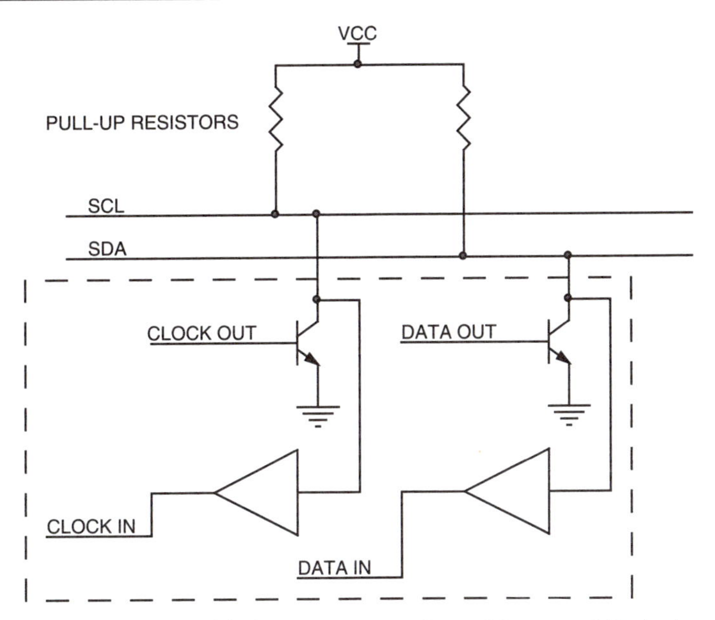

Schematic 6.3: This is a representation of how an I²C device connects to the I²C bus. Note that the type of transistors and associated circuitry would depend on the technology (CMOS, NMOS, bipolar) of the I²C device.

Figure 6.1: I²C itself is an abbreviation. So, why not abbreviate START and STOP with an S and a P.

The wired-AND configuration used in I²C could really cause lots of confusion on the bus if it were not for the strict protocol that makes up the logical side of the I²C bus. Remember the START and STOP bits you were exposed to in RS-232? Well, I²C has START and STOP bits too, but instead of bits they are technically known as I²C START and STOP conditions. The I²C START and STOP logic levels can be seen in Figure 6.1. The SCL line must be in a HIGH state for either a START or a STOP condition to occur. An I²C START condition is defined as a HIGH to LOW transition of the SDA line while the SCL line is HIGH. An I²C STOP condition occurs when the SDA line toggles from LOW to HIGH while the SCL line is HIGH. The I²C master always generates the *S* and *P* conditions. Once the I²C master initiates a START condition, the I²C bus is considered to be in a busy state.

I know what you're thinking. I²C has STOP and START bits like RS-232 does, and I²C transfers 8-bits of data in a data packet just like RS-232 does. That means that an I²C data packet is just like an RS-232 data packet with 8 bits of data sandwiched between a START and STOP bit. Not exactly…

It is true that I²C requires that the data be transferred in bytes. It is also true that I²C starts a transmission with a START condition and ends the transmission with a STOP condition. The difference between an RS-232 transmission and an I²C transmission is that an unrestricted number of data bytes can flow between an I²C START and STOP condition while only a single byte of information can be transferred between the START and STOP bits of an RS-232 data packet. Another major difference in I²C and RS-232 is that the data is transferred most significant bit first in an I²C data packet instead of least significant bit first as it is in RS-232. Regardless of the order in which the bits are transmitted, the real enabler for I²C multibyte transfers is the I²C acknowledge bit. Every byte that flows on the I²C bus must be followed by an acknowledge bit. Since the acknowledge bit is very important for I²C communications, let's get a better understanding of how it works.

I²C ACKS and NAKS

The acknowledge bit (ACK) rides on the master-generated clock pulse train. During an acknowledge, the transmitting device releases the SDA line and uses the wired-AND functionality of the I²C bus to pull the SDA line to a HIGH state. The I²C master generates an acknowledge clock pulse and during the acknowledge bit time (HIGH SCL), the receiving I²C device must pull the SDA line down to a LOW state for the time that SCL is in the acknowledge clock pulse HIGH state. Standard I²C protocol expects the receiving I²C device to acknowledge every byte that is received.

There may be times when the slave can't acknowledge the master. For instance, the slave is busy taking analog readings and "can't come to the I²C phone." In this case, the slave leaves the SDA line in a HIGH state. The I²C master senses this negative acknowledge (NAK) and can choose to either end the transaction with a STOP condition or begin a new transfer by issuing a repeated START condition. The repeated START condition allows the current I²C-bus master to keep control of the I²C-bus to issue another START bit instead of relinquishing the bus and attempting to recapture it to issue another START condition.

What if the slave "answers the I²C phone" in slave-receiver mode, but later gets called by a process that doesn't allow the slave to receive any more bytes? When the slave can't continue, it allows the SDA line to go HIGH during the acknowledge bit time, which in-turn sends a NAK to the I²C master. At this point, the I²C master can either abort the transfer or attempt a restart.

A NAK condition isn't always a bad thing. When the I²C master is in master-receiver mode, it signals the end of the data transfer from the slave-transmitter by generating a NAK on the last byte it clocked out of the slave-transmitter. The slave-transmitter senses the NAK and releases the SDA line so the I²C master can either generate a STOP condition or a repeated START condition. Logical examples of ACKs and NAKs are depicted in Figure 6.2.

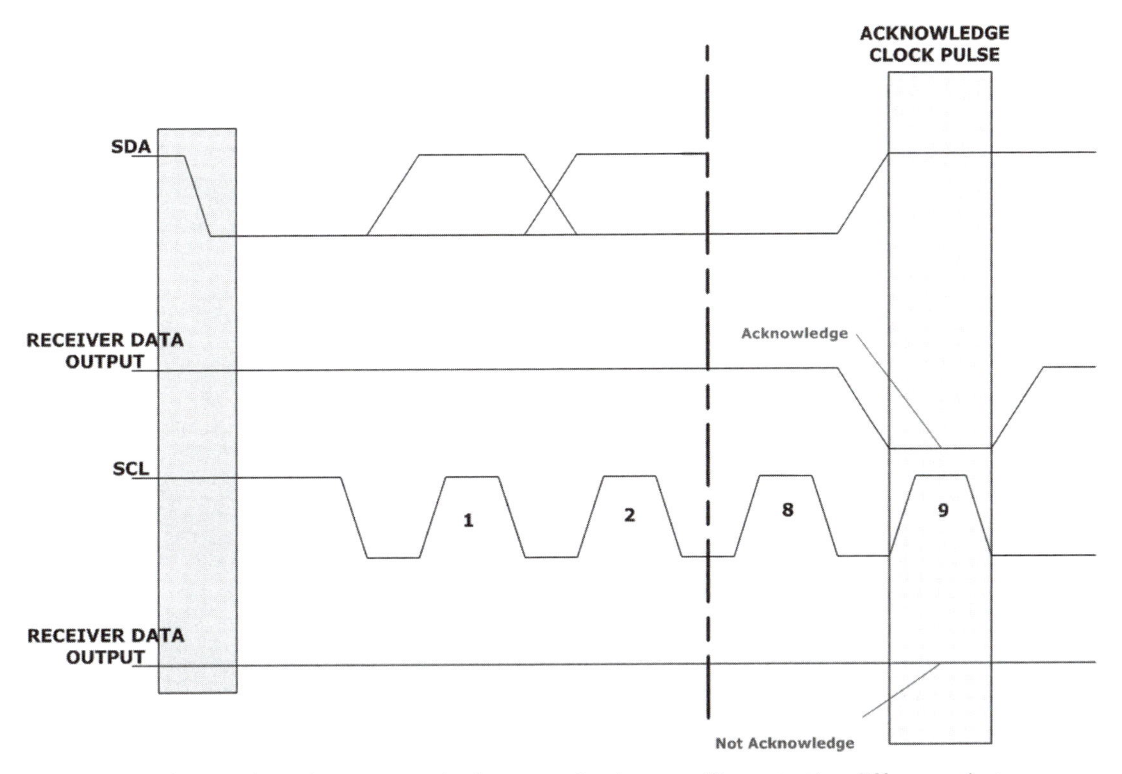

Figure 6.2: The receiver data output is shown twice here to illustrate the difference between an ACK and a NAK.

More on Arbitration and Clock Synchronization

Now that you're up to your ankles in I²C theory, let's talk a bit more about arbitration. I²C depends heavily on accurate clocking from each master on the I²C bus, and the wired-AND-based I²C bus connections have a hand in the clock synchronization process as well.

Data on the I²C bus is only valid when the SCL line is in the HIGH portion of a clock pulse. Let's use our example I²C bus with two microcontrollers attached as shown in Schematic 6.2. If each microcontroller can clock the I²C bus at a specific speed, that means that the internal master I²C engine of each microcontroller on the I²C bus has a means of counting to effect the elapsed times needed to swing the I²C bus HIGH and LOW at a specific rate.

The AVR being the master of the I²C bus wants to communicate with the PIC slave. The AVR generates the clock on the SCL line and sends a byte of data. The PIC acknowledges the data and then has to go off to service an external interrupt. If the AVR continues to try to communicate with PIC, the AVR will soon miss the acknowledgement it is expecting from the slave PIC, and the transmission would have to be aborted or restarted by the AVR. This is where the I²C-bus wired-AND logic comes into play to help avoid such a situation.

Think of the I²C bus as a simple AND gate. The truth table for a 2-input AND gate is shown graphically in Figure 6.3.

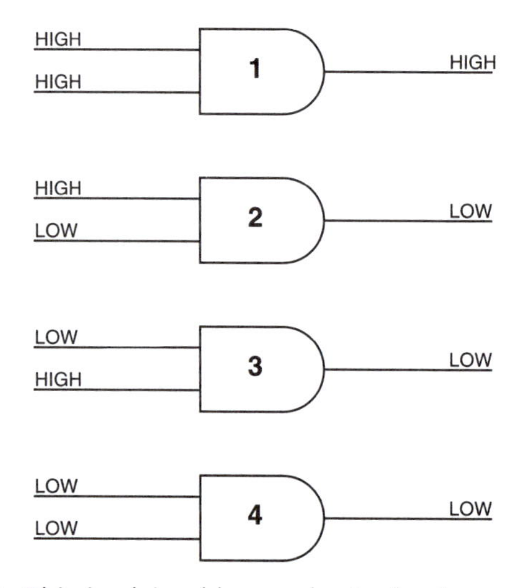

Figure 6.3: This is plain-old everyday logic. Any presence of a LOW on either of the inputs results in a LOW on the AND gate output.

Now, in Figure 6.4 let's substitute the AVR's and the PIC's SCL line states for the inputs with the AND gate outputs representing the resultant state of the I²C bus SCL line.

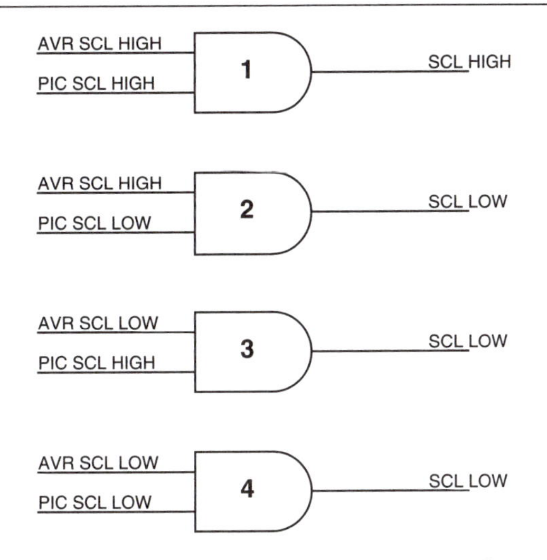

Figure 6.4: The I²C bus is a wired-AND configuration.

When the PIC is able to service the AVR's requests immediately, the PIC leaves the SCL line alone by driving its SCL interface HIGH. You can see this in states 1 and 3 of our substituted AND gate example in Figure 6.4. If the PIC needs more time to respond to AVR's requests, it can pull the SCL line down to a LOW state. The act of the PIC pulling down the SCL line is called *clock stretching*. As you can see in states 2 and 4, the AVR is unable to change the state of the SCL line when the PIC is holding the SCL line LOW. So, the AVR goes into a HIGH wait state and sits there until the PIC releases the SCL line and allows it to return to a HIGH state. The bottom line is that the SCL line will be held LOW by the I²C device with the longest LOW period. The I²C device with the shortest HIGH period determines how long the SCL line will remain in a HIGH state during clocking. This is how the I²C bus is synchronized.

It is possible for two or more I²C masters to initiate a start condition at the same time. When that occurs, the masters requesting the use of the I²C bus must utilize the I²C arbitration process. I²C arbitration is performed using the SDA line while the SCL line is at a HIGH level. Both the SCL and SDA lines are wired-AND configurations. So, we can apply the same logic to the arbitration process as we did to the I²C bus clock synchronization.

We must assume that both the AVR and the PIC in Schematic 6.2 are masters on the I²C bus. Figure 6.5 shows us that when any master on the I²C bus takes the SDA line LOW, the other masters on the I²C bus are unable to drive the SDA line high. Thus, the I²C-bus arbitration loser is the master that attempts to transmit a HIGH, while another master is transmitting a LOW on the SDA line. The master transmitting a HIGH when the SDA line is LOW senses that the SDA line is not at the same level as it is transmitting and switches off its data output

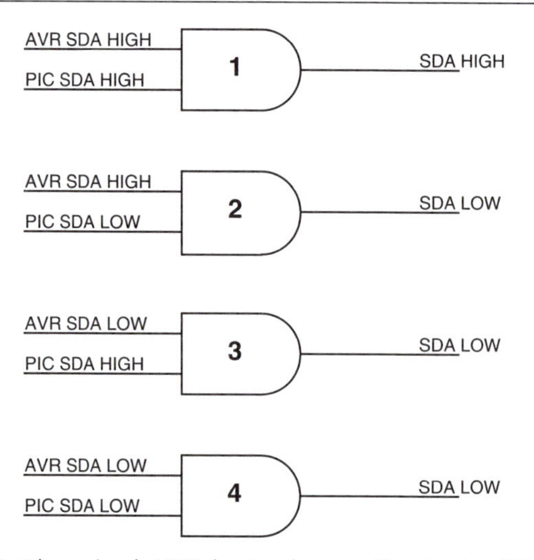

Figure 6.5: The wired-AND logic also applies to the I²C SDA line.

stage. The losing master applies a HIGH to the SDA line and reverts to slave mode if it is configured to perform the slave function. By presenting a HIGH to the SDA line, the losing master releases the SDA line to the winning master. Let's say the AVR is the winning master, and the PIC is the losing master. In Figure 6.5, states 1 and 3 define the state of the SDA line while the AVR was in charge of the I²C bus. If the PIC was declared the winner and the AVR the loser, states 1 and 2 would go into effect while the PIC was in control of the I²C bus.

Arbitration can be performed for a number of bits into the transaction. For instance, the masters may all be addressing the same slave in the same manner. In that case, the address bits from each master would be identical. The good news is that the winning master's address and data are the only valid items on the I²C bus and nothing in terms of address and data information is lost in the arbitration process.

Clock synchronization is always going on in the SCL domain while arbitration may be occurring at the SDA level. A slave I²C device can throttle the speed in which it accepts data bytes by dragging the SCL line LOW. In standard mode, any smart device on the I²C bus that can extend the LOW period of the clock can control the speed of other devices on the I²C bus because the device with the longest LOW period determines the top speed of every other master device on the I²C bus.

As long as we follow the rules and use a device with built-in I²C capability, I²C is dead-easy to implement. Before we write some I²C code to go along with our AVR and PIC RS-232 code modules, let's take a look at how data flows across an I²C bus.

I^{2}C Addressing

You already know that a START condition begins the I^{2}C data transfer process. Since multiple devices can coexist on the I^{2}C bus, there must be a way to differentiate them. This is done with I^{2}C addressing. I^{2}C devices can be addressed using a 7-bit or 10-bit format. I^{2}C 10-bit addressing isn't difficult to grasp once you understand 7-bit addressing. So, instead of trying to school you on 10-bit addressing, I'll concentrate on showing you how 7-bit addressing works as we'll only be using 7-bit addressing in our project.

The first byte sent on the I^{2}C bus after the start is usually an address byte. One exception involves sending a "general call" address following the start condition. The "general call" addresses everyone on the I^{2}C bus. Our project doesn't use the "general call." So, let's move on with picking apart the I^{2}C 7-bit address mechanism.

MSB							LSB
ADDR6	ADDR5	ADDR4	ADDR3	ADDR2	ADDR1	ADDR0	R/W

Figure 6.6: Think of this as subtracting 1 from the real I^{2}C address to write and adding 1 to the I^{2}C address to read. The ADDRX bits make up the actual slave address.

The seven *ADDRX* bits in the 7-bit address scheme shown in Figure 6.6 are taken from the first seven bits of the address byte that follows the start condition. Remember, in I^{2}C-land, the most significant bit is transmitted first. So, bits 7 through 1 of the address byte actually carry the I^{2}C address information. The least significant bit, bit 0, determines if the I^{2}C operation will be a read or write. A binary zero in bit 0 of the address byte tells the slave that the master will be writing data to the slave device. Conversely, a binary 1 in the LSB (least significant bit) position will allow the master to read information from the slave. Each device on the I^{2}C bus sees the address byte. Only the device that contains the match for the first seven bits of the address byte will ultimately respond to the I^{2}C master's call. If the I^{2}C operation is a write from the master, the slave device enters slave-receiver mode. An I^{2}C-bus read operation will put the addressed slave device into slave-transmitter mode. Let's write some I^{2}C code.

Some I^{2}C Firmware

Custom Computer Services PIC Compiler easily handles the I^{2}C master chores. Custom Computer Services C for PICs provides built-in code for the standard I^{2}C functions such as: i2c_start, i2c_read, i2c_write and i2c_stop. In this section, we're also going to be producing AVR I^{2}C code in parallel with the PIC C code using ICCAVR. The ImageCraft C compiler doesn't have built-in AVR I^{2}C functions but we can easily write our own. Reading and writing in I^{2}C master mode is straightforward. The real coding work comes in when exercising the slave side of these common I^{2}C functions.

You've already seen Schematic 6.1, which contains the PIC18F452 I^{2}C circuitry. Schematic 6.4 shows the AVR I^{2}C circuitry, which is very similar to the PIC I^{2}C circuitry.

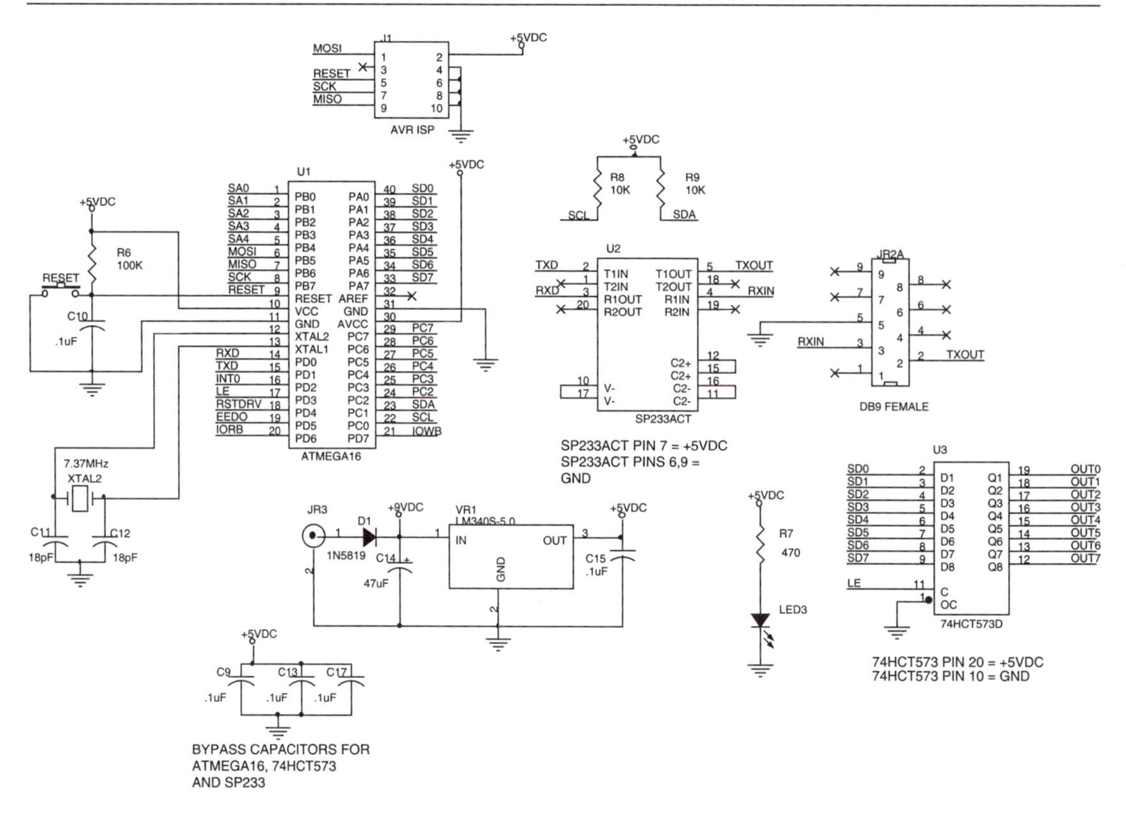

Schematic 6.4: This is the partially assembled Easy Ethernet AVR circuitry with an added 74HCT573D octal transparent latch. For both the PIC and the AVR, the only parts I've added that are really required are the I²C pull-up resistors. In some instances, the AVR doesn't require pull-up resistors as it can pull up the I²C port pins internally. You only need one set of pull-up resistors on the I²C bus.

The AVR Master I²C Code

Atmel's term for I²C is TWI (Two-Wire Interface). For I²C master operation, there are only four AVR registers we will be dealing with: the TWDR (Two-Wire Interface Data Register), the TWCR (Two-Wire Interface Control Register), the TWBR (Two-Wire Interface Bit Rate Register) and the TWSR (Two-Wire Interface Status Register). The TWBR is set and forget. So, we'll only be exercising the contents of three AVR I²C registers.

You can read datasheets as well as I can, so let's examine the AVR TWI subsystem as we write some code to drive it. To make this easier to digest, we want to write our AVR TWI code to look as much like our PIC I²C code as we can. So, I'll use the Custom Computer Services C Compiler nomenclature for I²C in the TWI AVR ICCAVR C source code.

The first thing we want to do is initialize the AVR's TWI module. The TWCR, which is used rather heavily, is shown in Figure 6.7.

TWCR

7	6	5	4	3	2	1	0
TWINT	TWEA	TWSTA	TWSTO	TWWC	TWEN	-	TWIE

Figure 6.7: You've already figured out that TW stands for Two-Wire Interface. Bits 7:4 are the busiest bits in this register.

Clearing the *TWEN* bit of the TWCR disables the AVR's TWI module, and stuffing 0x1E into the *TWBR* bit puts our I²C bus on the I²C SLOW train. There's a formula for calculating the I²C-bus bit rate in the datasheet, but there's an easier application to do the bit rate calculation included with ICCAVR (Figure 6.8).

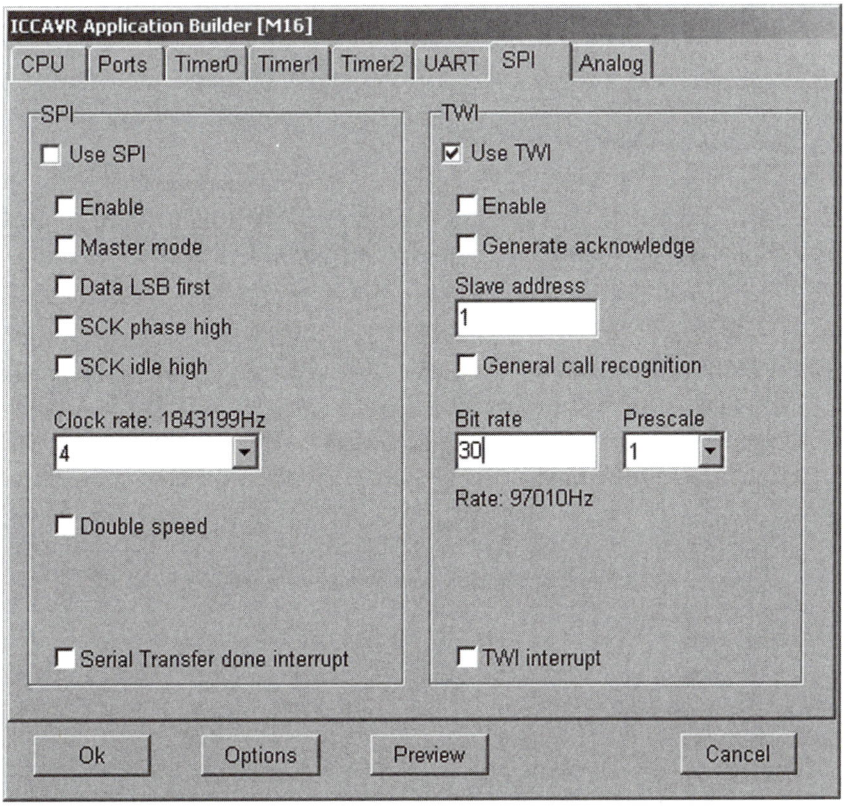

Figure 6.8: Hmmm...which method do you think I used to get the value for the I²C SLOW bit rate?

```
unsigned char flags;
//******************************************************************
//*   INITIALIZE THE TWI
//******************************************************************
void twi_init(void)
{
   flags = 0x00;
   TWCR= 0x00; //disable twi
   TWBR= 0x1E; //set bit rate
   TWSR= 0x00; //set prescale
   TWAR= 0x00; //set slave address
   TWCR= 0x04; //enable twi
}
```

Code Snippet 6.1: Since the AVR will be the master on the I²C bus, we'll leave the slave address at 0x00 for now.

Once the I²C bit rate is set, we can enable the AVR's TWI module. Our application will be simple enough to preclude the use of interrupts, and our AVR master will not be configured to also act as a slave. Therefore, the *TWIE* bit will remain clear for now. I've coded the TWI registers in Code Snippet 6.1 to reflect that. The *flags* variable is used to identify certain states of operation in our I²C code.

Like its PIC counterpart, our AVR I²C master will need some code to implement the basic elements of I²C that allow it to participate on an I²C bus. Since a START condition is the beginning of every I²C transfer, let's begin by writing the AVR I²C start routine. The Custom Computer Services C Compiler provides a built-in I²C start routine called *i2c_start*.

```
#define   START_i2c     0x08
//******************************************************************
//*   AVR i2c START
//******************************************************************
void i2c_start(void)
{
   TWCR = (1<<TWINT) | (1<<TWSTA) | (1<<TWEN);
   while (!(TWCR & (1<<TWINT)));
   if ((TWSR & 0xF8) != START_i2c)
      printf("i2c Start Error\r\n");
}
```

Code Snippet 6.2: Note that the 0xF8 masks out the prescale bits in the TWSR. The status codes specified in the AVR datasheet do not include the prescale bit values.

Writing a 1 to the *TWINT* bit of the TWCR clears the *TWINT* bit. Everything revolves around the state of the *TWINT* bit, as when it is set the TWI has finished an operation and is waiting for the application to respond. Normally an interrupt is generated every time the

TWINT goes from a low to high state. Since we're not using I²C interrupts, we must poll the *TWINT* bit after we reset it and look for it to return to a high state.

An I²C START is issued when the *TWINT*, *TWSTA* and *TWEN* bits are set. When the *TWINT* bit returns to a set state, the I²C START has completed. A successful I²C START condition is signaled by 0x08 in the TWSR. I've added some diagnostic *printf* code to flag an I²C START condition error.

We must also be able to stop the I²C transfer. That is done within the Custom Computer Services C Compiler with a built-in *i2c_stop* function. Guess what we will call our AVR stop function? Our AVR stop code is shown in Code Snippet 6.3

```
//*******************************************************************
//*   AVR i2c STOP
//*******************************************************************
void i2c_stop(void)
{
   TWCR = (1<<TWINT)|(1<<TWEN)  |  (1<<TWSTO);
}
```

Code Snippet 6.3: In slave mode, the STOP condition can be used to recover from an error condition by forcing the slave to release the SCL and SDA lines.

A STOP condition is generated by setting *TWINT*, *TWEN* and *TWSTO*. The *TWSTO* bit is automatically cleared once the STOP condition has executed on the I²C bus.

Once a START condition is generated, the next thing that happens in a normal I²C data transfer is the transmission of the slave address and mode bit. The slave address and mode bit are transmitted using an I²C write command. We'll name our AVR code in Code Snippet 6.4 after the Custom Computer Services C Compiler I²C function called *i2c_write.*

```
#define addrflag      0x01 //00000001
#define clr_modeSLA   flags &= ~addrflag
#define set_modeSLA   flags |= addrflag
#define MODE_SLA      (flags & addrflag)

#define modeMRflag    0x02 //00000010
#define clr_modeMR    flags &= ~modeMRflag
#define set_modeMR    flags |= modeMRflag
#define MODE_MR       (flags & modeMRflag)

#define modeMTflag    0x04 //00000100
#define clr_modeMT    flags &= ~modeMTflag
#define set_modeMT    flags |= modeMTflag
#define MODE_MT       (flags & modeMTflag)
```

```
//*********************************************************************
//*   MASTER TRANSMITTER MODE STATUS CODES
//*********************************************************************
#define  MT_SLA_ACK  0x18  //Master Transmitter Slave Addr ACK
#define  MT_DATA_ACK 0x28  //Master Transmitter Data ACK
//*********************************************************************
//*   MASTER RECEIVER MODE STATUS CODES
//*********************************************************************
#define  MR_SLA_ACK  0x40  //Master Receiver Slave Addr ACK

//*********************************************************************
//*   AVR i2c WRITE
//*********************************************************************
void i2c_write(unsigned char datum)
{
   TWDR = datum;
   TWCR = (1<<TWINT)|(1<<TWEN);
   while (!(TWCR & (1<<TWINT)));
   if(MODE_SLA && MODE_MT)
   {
      if ((TWSR & 0xF8) != MT_DATA_ACK)
      printf("i2c Data Transfer Error MT Mode %x\r\n",(TWSR & 0xF8));
      else
      {
         clr_modeSLA;
         clr_modeMT;
      }
   }
   else if (MODE_SLA && MODE_MR)
   {
      if ((TWSR & 0xF8) != MR_DATA_ACK)
      printf("i2c Data Transfer Error MR Mode %x\r\n",(TWSR & 0xF8));
      else
      {
         clr_modeSLA;
         clr_modeMR;
      }
   }
   else
   {
      if ((TWSR & 0xF8) == MT_SLA_ACK)
      {
         set_modeMT;
         set_modeSLA;
      }
```

```
        else if ((TWSR & 0xF8) == MR_SLA_ACK)
        {
           set_modeMR;
           set_modeSLA;
        }
        else
        {
           printf("i2c Start Error %x\r\n",(TWSR & 0xF8));
           clr_modeSLA;
           clr_modeMR;
           clr_modeMT;
        }
    }
}
```

Code Snippet 6.4: Everything in this snippet flows on status codes.

Before initiating the I²C transmission, the slave address and mode bit are loaded into the TWDR. Toggling the *TWINT* bit in the TWCR kicks off the slave address and mode bit write process. The *TWEN* bit is set to ensure that the AVR's I²C interface is activated.

When the slave address and mode bit write has completed without error, status codes of 0x18 (*MT_SLA_ACK*) or 0x40 (*MR_SLA_ACK*) will appear within the TWSR. If the mode bit is set, an I²C slave read operation will be performed and flags will be set to denote this state (*MODE_SLA* and *MODE_MR* for a read operation/*MODE_SLA* and *MODE_MT* for a write operation).

If the mode is set for the AVR to become a Master Transmitter (*MODE_MT*), the next I²C operation will perform the writing of the data. Our application will only send one byte per transmission, and again we will call upon the services of the AVR *i2c_write* function we just wrote. This time the slave address and mode bit are replaced by the actual data we want to send to the slave. At this point, the AVR is considered a Master Transmitter and the slave is in slave-receiver mode. Our AVR I²C code has set the *MODE_SLA* and *MODE_MT* flags indicating that the AVR is in Master Transmitter mode and that the slave has been successfully addressed. A clearing of the *TWINT* bit sends the data onto the I²C bus. If everything goes as planned, the TWSR will contain 0x28, which says that the slave acknowledged the data transfer. The AVR Master Transmitter then issues a STOP condition to end the I²C session.

The AVR I²C Master-Receiver Mode Code

There will be times with the AVR master must retrieve some information from the PIC slave. That's when we deploy the AVR *i2c_read* function in Code Snippet 6.5.

```
//*************************************************************
//*   MASTER RECEIVER MODE STATUS CODES
//*************************************************************
#define  MR_DATA_ACK 0x50  //Master Receiver Data ACK
#define  MR_DATA_NAK 0x58  //Master Receiver Data NAK

#define  ACK_i2c     0x01
#define  NAK_i2c     0x00
//*************************************************************
//*   AVR i2c READ
//*************************************************************
unsigned char i2c_read(unsigned char acknak)
{
   if(acknak == ACK_i2c)
   {
      TWCR = 0xC4;
      while (!(TWCR & (1<<TWINT)));
      if ((TWSR & 0xF8) != MR_DATA_ACK)
      printf("i2c Data Transfer Error MR Mode %x\r\n",(TWSR & 0xF8));
   }
   else  //acknak == NAK_i2c
   {
      TWCR = 0x84;
      while (!(TWCR & (1<<TWINT)));
      if ((TWSR & 0xF8) != MR_DATA_NAK)
      printf("i2c Data Transfer Error MR Mode %x\r\n",(TWSR & 0xF8));

      clr_modeSLA;
      clr_modeMR;
   }
   return(TWDR);
}
```

Code Snippet 6.5: The important thing to do here is to always send a NAK when reading the last byte from the slave.

Figure 6.9 lays out the bit pattern written to the TWCR after the START condition and slave addressing has successfully completed. The AVR is in master-receiver mode, and the slave is in slave-transmitter mode when the AVR *i2c_read* function is entered.

TWCR

7	6	5	4	3	2	1	0
TWINT	TWEA	TWSTA	TWSTO	TWWC	TWEN	–	TWIE
1	1	0	0	0	1	0	0

Figure 6.9: The TWEA (TWI Enable Acknowledge Bit) is a "don't care" bit until we enter master-receiver mode.

Notice that we purposely set the *TWEA* bit, which we have been ignoring up until this time. Setting the *TWEA* bit generates an ACK on the I²C bus when a data byte is received by the AVR master receiver. When things go right, the TWSR will hold the value of the *MR_DATA_ACK* (0x50) after each byte received by the AVR in Master Receiver mode. Our I²C application is setup to read four bytes from the slave device.

The last byte we receive from the slave transmitter must be NAKed. That's where the *TWEA* bit in Figure 6.10 gets the other 7.5 minutes of its 15 minutes of fame. By writing a 0 (zero) to the *TWEA* bit, a NAK is generated, which results in termination of the I²C read session between the master receiver and the slave transmitter. The TWSR will contain a 0x58 (*MR_DATA*_NAK) if all goes well with the NAK operation.

TWCR

7	6	5	4	3	2	1	0
TWINT	TWEA	TWSTA	TWSTO	TWWC	TWEN	–	TWIE
1	0	0	0	0	1	0	0

Figure 6.10: Writing a 0 to the TWEA *bit temporarily disconnects the AVR from the I²C bus.*

I have a project in mind. Let's combine our AVR RS-232 skills with our newfound AVR I²C skills to transfer data between the partially assembled Easy Ethernet AVR and the Easy Ethernet CS8900A boards. Before we put the whole of the AVR code together, let's write some PIC I²C slave code first.

The PIC I²C Slave-Transmitter Mode Code

To implement I²C on the Microchip PIC, there are only three PIC registers we need to be concerned with: SSPCON, SSPSTAT and SSPBUF. SSPCON is used to determine whether or not a collision has occurred (WCOL) and to ensure we are not stretching the clock when we shouldn't be (CKP = 1). Clock-stretching is legal for an I²C slave device when it can't respond in a timely manner. SSPSTAT gives us the status of the data transfer, while SSPBUF is the register that actually transfers the data to and from the I²C bus.

The PIC's MSSP (Master Synchronous Serial Port) does several other things for us including double-buffering our received I²C data using the SSPSR/SSPBUF register combination, providing a holding register for the slave address and generating I²C interrupts on START and STOP conditions. Double buffering is the act of holding or collecting data in an input or output buffer while operating on a totally separate input or output buffer. In short, double buffering allows data to be assembled for transmission while previously accumulated data is being transmitted. Receive double-buffering occurs when the microcontroller is working on pulling previously received data from an input buffer, while yet another input buffer is taking in new data and holding it until the microcontroller can start processing it.

As simple as the I²C concept is, if you're not careful, you can get your I²C code wrapped around the axel. To make I²C coding more manageable, the I²C transmission and reception process can be broken down into five states. Everything that's normal in I²C begins with a START condition. The START condition must be detected (S = 1) no matter what, and nothing begins until a valid START condition is sensed. Once we have detected a valid

START bit, we can use the other bits inside the SSPSTAT register to determine which state the I²C transaction is currently in. We used the TWSR for this in the AVR I²C code. The MSSP issues an interrupt on every byte transfer. This allows us to write I²C code, such as the code presented in Code Snippet 6.6, using the five states to take advantage of the MSSP module's interrupt generation.

```
//*********************************************************************
//*    SLAVE RAM DEFINITIONS
//*********************************************************************
int1  update_latch;
int8  index,digit;
int8  numbers[] = {0,1,2,3,4,};
//*********************************************************************
//*    I2C SLAVE RECEIVE
//*********************************************************************
#INT_SSP
   ssp_interrupt ()
{
//#bit SMP =    SSPSTAT.7
//#bit CKE =    SSPSTAT.6
//#bit D_A =    SSPSTAT.5
//#bit P =  SSPSTAT.4
//#bit S =  SSPSTAT.3
//#bit R_W =    SSPSTAT.2
//#bit UA = SSPSTAT.1
//#bit BF = SSPSTAT.0

   int8 dummy;
//-----------------------------------------------------
// The I2C code below checks for 5 states:
//-----------------------------------------------------
// State 1: I2C write operation, last byte was an address byte.
//
// SSPSTAT bits: S = 1, D_A = 0, R_W = 0, BF = 1
//
// State 2: I2C write operation, last byte was a data byte.
//
// SSPSTAT bits: S = 1, D_A = 1, R_W = 0, BF = 1
//
// State 3: I2C read operation, last byte was an address byte.
//
// SSPSTAT bits: S = 1, D_A = 0, R_W = 1, BF = 0
//
// State 4: I2C read operation, last byte was a data byte.
//
// SSPSTAT bits: S = 1, D_A = 1, R_W = 1, BF = 0
```

```
//
// State 5: Slave I2C logic reset by NACK from master.
//
// SSPSTAT bits: S = 1, D_A = 1, R_W = 0, BF = 0
//
//----------------------------------------------------

//State 1
   if(S && !D_A && !R_W && BF )
      dummy = SSPBUF;
//State 2
   else if(S && D_A && !R_W && BF )
      {
         digit = SSPBUF;
         update_latch = TRUE;
      }
//State 3
   else if(S && !D_A && R_W && !BF )
      {
         index = 0x00;
         while(BF);
         do{
         WCOL = 0;
         SSPBUF = numbers[index];
         }while(WCOL);
         ++index;
         CKP = 1;
      }
//State 4
   else if(S && D_A && R_W && !BF )
      {
         while(BF);
         do{
         WCOL = 0;
         SSPBUF = numbers[index];
         }while(WCOL);
         if(++index > 0x04)
         index = 0x00;
         CKP = 1;
      }
//State 5
   else if(S && D_A && !R_W && !BF )
      index = 0;
}
```

Code Snippet 6.6: The update_latch *variable and* numbers[] *array will be used by in our AVR-to-PIC grand I²C ball.*

The I²C SLAVE RECEIVE routine is the PIC18F452 I²C interrupt handler code that responds to every interrupt issued by the PIC18F452's microcontroller's MSSP module. I've moved the bit definitions of the SSPSTAT register into the routine's air space for clarity.

Notice that in each of the five defined states that *S = 1* is common. The bit *S* is defined as the third bit of the SSPSTAT register. If a valid START condition is detected, this bit will be set.

The slave address byte immediately follows the START bit. Since the slave microcontroller's MSSP will always generate an interrupt if the incoming address byte matches the slave's internally stored address (in SPPADD), the matching address byte just received triggers our first interrupt and its subsequent response. The MSSP module will also automatically issue an acknowledge (ACK) pulse upon detecting an address match.

The *D_A* bit signals if the last byte received was data or address. In this case, we know that a START bit was generated and was indeed followed by a 7-bit address. Therefore, *D_A* is cleared to zero indicating the last byte received was an address byte.

The R/W bit of the address is cleared for a write operation and set for a read operation. The *R_W* bit of the SSPSTAT registers reflects the level of the R/W bit in the address byte. Note that if the operation is a write operation, the BF (Buffer Full) bit is always set indicating data is in the buffer. The State 1 code runs following the reception of the address byte. The address byte is read and discarded as the slave MSSP module has already digested the address byte's contents. The act of reading SSPBUF also clears the BF bit. If the BF bit is not cleared at this point, the next incoming byte would cause an overflow condition. Let's follow the entire state-by-state chain of events involved with sending some data from the AVR master I²C microcontroller to the PIC slave I²C microcontroller.

Suppose that the AVR master I²C microcontroller needs to send a message via I²C to the PIC I²C slave microcontroller that tells the slave microcontroller to write 0x55 to its onboard 74HCT573 latch. The basic AVR TWI code would consist of what you see in Code Snippet 6.7.

```
i2c_start();
i2c_write(0x18);
i2c_write(0x55);
i2c_stop();
```

Code Snippet 6.7: The Easy Ethernet CS8900A's I²C address is 0x18.

After initiating a START condition, the master microcontroller clocks out the slave microcontroller's I²C address, hexadecimal 18 (0x18). The code *i2c_write(unsigned char datum)* indicates an I²C write operation has been requested as the R/W bit in the I²C address byte is cleared. At this point in time, every slave microcontroller on the I²C bus is listening on the I²C link looking to match its address against the incoming address byte. Our PIC I²C slave microcontroller compares the incoming address with the address stored in its SSPADD register and detects a match. The slave's BF bit is set, an ACK (acknowledge) pulse is generated by the slave microcontroller's MSSP hardware and an SSP interrupt is generated. The PIC I²C slave microcontroller enters the I²C SSP interrupt routine and using the SSPSTAT bits determines that the I²C transaction is in State 1, which tells us that the last byte received was an address byte. The BF bit is set, which means the contents of the SSPSR register have been transferred to the SSPBUF register. To avoid an overflow condition, the PIC's SSPBUF register must be read even though we don't have any further use for the address data.

It's the slave microcontroller's duty to translate the incoming I²C datastream.

```
#define  le_pin   PORTC,1

#define  latchdata   bit_set(le_pin);  \
   delay_us(1);        \
   bit_clear(le_pin);
//******************************************************************
//*   SLAVE MAIN
//******************************************************************
   do{
      {
   if(update_latch)
   {
      output_d(digit);
         latchdata;
      update_latch = FALSE;
   }
      }
      }while(1);
}
```

Code Snippet 6.8: Now you know what the update_latch *variable you saw in Code Snippet 6.6 is for.*

The data that was sent from the I²C master that is to be output to the slave's 74HCT573 latch was collected into the *digit* variable in the PIC's I²C interrupt handler routine. In the same stroke, the PIC I²C interrupt handler updated the *update_latch* flag to TRUE.

The code in Code Snippet 6.8 is the main routine that runs continuously inside the Easy Ethernet CS8900A's PIC18F452. The PIC I²C slave's code picks up the state of the *update_latch* variable. If the *update_latch* variable is TRUE, the data within the *digit* variable is output to the 74HCT573 latch by the *latchdata* macro and the *update_latch* variable is cleared to a FALSE condition. Each time a value is received by the slave via the I²C bus, it is transferred to the latch.

If the master microcontroller wants data from the slave microcontroller, State 3 starts things off and the slave microcontroller is coaxed into slave-transmitter mode while the master microcontroller becomes a master-receiver. In Code Snippet 6.9, the master microcontroller initiates a START condition and follows it with a "read" address byte. Since the R/W bit is the least significant bit in the address byte, the write address is simply the base address incremented by 1 (0x19 in our case). Incrementing the address byte has the effect of setting the R_W bit inside the I^{2}C address byte. In this mode the master microcontroller, not the slave microcontroller, generates the I^{2}C *ACKs* and *NAKs* on the I^{2}C bus.

```
#define  ACK_i2c       0x01
#define  NAK_i2c       0x00

   i2c_start();
   i2c_write(0x19);
   for(x=0;x<3;++x)
   {
      datum = i2c_read(ACK_i2c);
      printf("datum = 0x%x\r\n",datum);
   }
   datum = i2c_read(NAK_i2c);
   i2c_stop();
   printf("datum = 0x%x\r\n",datum);
```

Code Snippet 6.9: No worries...we read every byte except the last within the for *loop.*

Things on the I^{2}C bus are a bit busier when a master is reading from a slave. We already know that the slave microcontroller has four bytes of information the master can access stored in the *numbers[]* array. Let's use the AVR and the I^{2}C bus to retrieve the four bytes from the slave's *numbers[]* array and print them out to a master Tera Term Pro session.

The slave microcontroller must be ready to send the first byte of data after the ACK following the address byte. The State 3 code attempts to load the SSPBUF with that first byte of data while looking out to make sure the SSPBUF is clear and ready for the byte to be loaded. In our code, the first byte of the array *numbers[]* (0x00) is loaded and sent following the reception of the address byte. The *index* variable is incremented to point to the next element of the *numbers[]* array. Setting the CKP (SCK release control) bit assures that the slave microcontroller is not holding the clock line low, and thus "stretching" the clock.

The master microcontroller is coded to collect a total of four bytes. Since the last byte read was not the address byte, we can move on to State 4 in the PIC interrupt handler code. The remainder of the four bytes of data required by the master microcontroller are clocked out of the slave-transmitter microcontroller in State 4. To halt the I^{2}C read operation, the master generates a NAK after the last byte is read. The *NAK_i2c* in the *i2c_read(NAK_i2c)* tells the AVR I^{2}C read function to send the NAK. That brings us to State 5 and the end of the I^{2}C read operation.

The AVR-to-PIC I²C Communications Ball

Let's put everything we've written for RS-232 and I²C for the AVR together with everything we've written for RS-232 and I²C for the PIC and move some data. The source code PIC slave application and the AVR master I²C application is contained within Code Snippet 6.10 and Code Snippet 6.11, respectively.

```
////////////////////////////////////////////////////////////////////////////
// PIC I2C SLAVE DRIVER
// EASY ETHERNET CS8900A BOARD
// Author: Fred Eady
// Version: 1.0
// Date: 08/25/03
// Description: I2C SLAVE FUNCTION WITH 74HCT573 CODE
////////////////////////////////////////////////////////////////////////////
#include <18F452.h>
#include <f452.h>
#device ICD=TRUE
#fuses
DEBUG,HS,NOWRT,NOWDT,NOPUT,NOPROTECT,NOBROWNOUT,NOLVP,NOCPD,NOEBTR
#id 0x0812

#use fast_io(A)
#use fast_io(B)
#use fast_io(C)
#use fast_io(D)
#use fast_io(E)

#define  esc   0x1B

//*****************************************************************
//*   I2C SLAVE ADDRESS
//*****************************************************************
// LANE ADDRESS IS UPPER NIBBLE
#define i2c_addr 0x18
//*****************************************************************
//*   RS232 AND I2C DEFINITIONS
//*****************************************************************
#use delay(clock=20000000)
#use i2c(Slave,Slow,sda=PIN_C4,scl=PIN_C3,force_hw,address=i2c_addr)
#use rs232(baud=9600,parity=N,xmit=PIN_C6,rcv=PIN_C7)
//*****************************************************************
//*   SLAVE FUNCTION PROTOTYPES
//*****************************************************************
void cls(void);
```

```
//*******************************************************************
//*   SLAVE RAM DEFINITIONS
//*******************************************************************
int1  update_latch;
int8  index,digit;
int8  numbers[] = {0,1,2,3,4,};

#define  le_pin   PORTC,1

#define  latchdata   bit_set(le_pin);  \
   delay_us(1);       \
      bit_clear(le_pin);

//*******************************************************************
//*   I2C SLAVE RECEIVE
//*******************************************************************
#INT_SSP
   ssp_interrupt ()
{
//#bit SMP =   SSPSTAT.7
//#bit CKE =   SSPSTAT.6
//#bit D_A =   SSPSTAT.5
//#bit P =  SSPSTAT.4
//#bit S =  SSPSTAT.3
//#bit R_W =   SSPSTAT.2
//#bit UA = SSPSTAT.1
//#bit BF = SSPSTAT.0

   int8 dummy;
//;---------------------------------------------------
//;   The I2C code below checks for 5 states:
//;---------------------------------------------------
//;   State 1: I2C write operation, last byte was an address byte.
//;
//;   SSPSTAT bits: S = 1, D_A = 0, R_W = 0, BF = 1
//;
//;   State 2: I2C write operation, last byte was a data byte.
//;
//;   SSPSTAT bits: S = 1, D_A = 1, R_W = 0, BF = 1
//;
//;   State 3: I2C read operation, last byte was an address byte.
//;
//;   SSPSTAT bits: S = 1, D_A = 0, R_W = 1, BF = 0
//;
//;   State 4: I2C read operation, last byte was a data byte.
//;
```

```
//;    SSPSTAT bits: S = 1, D_A = 1, R_W = 1, BF = 0
//;
//;    State 5: Slave I2C logic reset by NACK from master.
//;
//;    SSPSTAT bits: S = 1, D_A = 1, R_W = 0, BF = 0
//;
//;-----------------------------------------------------

//State 1
   if(S && !D_A && !R_W && BF )
      dummy = SSPBUF;
//State 2
   else if(S && D_A && !R_W && BF )
      {
         digit = SSPBUF;
         update_latch = TRUE;
      }
//State 3
   else if(S && !D_A && R_W && !BF )
      {
         index = 0x00;
         while(BF);
         do{
         WCOL = 0;
         SSPBUF = numbers[index];
         }while(WCOL);
         ++index;
         CKP = 1;
      }
//State 4
   else if(S && D_A && R_W && !BF )
      {
         while(BF);
         do{
         WCOL = 0;
         SSPBUF = numbers[index];
         }while(WCOL);
         if(++index > 0x04)
         index = 0x00;
         CKP = 1;
      }
//State 5
   else if(S && D_A && !R_W && !BF )
      index = 0;
}
```

```
void main() {

   int8 x;
   SET_TRIS_A(0b11111111);
   SET_TRIS_B(0b11111111);
   SET_TRIS_C(0b11111101);
   SET_TRIS_D(0b00000000);
   ADCON1 = 0x06;        //00000110 all ports set for digital
   ADCON0 = 0;
   update_latch = FALSE;
//******************************************************************
//*   INITIALIZE COMMON VARIABLES
//******************************************************************
   SSPSTAT = 0x80;
   SSPCON2 = 0x00;

//******************************************************************
//*   ENABLE SLAVE INTERRUPTS
//******************************************************************
   enable_interrupts(INT_SSP);
   enable_interrupts(GLOBAL);
//******************************************************************
//*   SLAVE MAIN
//******************************************************************
   do{
      {
         if(update_latch)
         {
         output_d(digit);
      latchdata;
         update_latch = FALSE;
         }
      }
   }while(1);
}
```

Code Snippet 6.10: Don't worry; I've included the code on the CDROM so you won't have to burn up your fingers typing code.

You already have a good handle on the innerworkings of the PIC I²C slave code in Code Snippet 6.10. However, I've thrown in the kitchen sink in the AVR master code coming up in Code Snippet 6.11. So, I'll break it up and discuss the code parts as they are encountered. Consider the rest of the code in this chapter as part of Code Snippet 6.11.

```
///////////////////////////////////////////////////////////////////////////
// AVR I2C MASTER DRIVER
// EASY ETHERNET AVR BOARD
// Author: Fred Eady
// Version: 1.0
// Date: 08/26/03
// Description: RS232 FUNCTIONS AND I2C MASTER FUNCTIONS
///////////////////////////////////////////////////////////////////////////

#include <iom16v.h>
#include <stdio.h>
#include <macros.h>

#pragma interrupt_handler USART_RX_interrupt:iv_USART_RX
#pragma interrupt_handler USART_TX_interrupt:iv_USART_UDRE
```

Code Snippet 6.11a: There's nothing here you can't talk about intelligently.

It looks like we're going to include some interrupt driven RS-232 on the AVR side. The *#pragma* statements in Code Snippet 6.11a are a dead giveaway. The confirmation of an RS-232 resurrection is confirmed in Code Snippet 6.11b.

```
//*******************************************************************
//*   FUNCTION PROTOTYPES
//*******************************************************************
int recvchar( void );
int sendchar( int );
unsigned char CharInQueue(void);
void init_USART(unsigned int baud);

void twi_init(void);
void i2c_start(void);
void i2c_write(unsigned char datum);
unsigned char i2c_read(unsigned char acknak);
void i2c_stop(void);
```

Code Snippet 6.11b: These declarations are a preview of what's to come.

The code in Code Snippet 6.11c should look familiar as well. All of the USART-related code is contained in this snippet.

```
//*****************************************************************
//*   BAUD RATE NUMBERS FOR UBRR
//*****************************************************************
#define   b9600  47     // 7.3728MHz clock
#define   b19200 23
#define   b38400 11
#define   b57600 7

#define USART_RX_BUFFER_SIZE  16       /* 1,2,4,8,16,32,64,128 or 256
bytes */
#define USART_RX_BUFFER_MASK ( USART_RX_BUFFER_SIZE - 1 )
//#if ( USART_RX_BUFFER_SIZE & USART_RX_BUFFER_MASK )
//#error RX buffer size is not a power of 2
//#endif
#define USART_TX_BUFFER_SIZE  128      /* 1,2,4,8,16,32,64,128 or 256
bytes */
#define USART_TX_BUFFER_MASK ( USART_TX_BUFFER_SIZE - 1 )
//#if ( USART_TX_BUFFER_SIZE & USART_TX_BUFFER_MASK )
//#error TX buffer size is not a power of 2
//#endif
//*****************************************************************
//*   AVR RAM Definitions
//*****************************************************************
unsigned char
USART_RxBuf[USART_RX_BUFFER_SIZE],USART_TxBuf[USART_TX_BUFFER_SIZE];
unsigned char USART_TxHead,USART_TxTail,USART_RxHead,USART_RxTail;
unsigned char flags,datum,byteout,cntr;
//*****************************************************************
//*   Init USART Function
//*****************************************************************
void init_USART(unsigned int baud)
{
   UCSRB = 0x00; //disable while setting baud rate
   UCSRA = 0x00;
   UCSRC = 0x86;
   UBRRL = baud; //set baud rate lo
   UBRRH = 0x00; //set baud rate hi
   UCSRB = 0x98;
}
//*****************************************************************
//*   USART Receive Interrupt Handler
//*****************************************************************
```

```
void USART_RX_interrupt(void)
{
   unsigned char data;
   unsigned char tmphead;

   data = UDR;                      /* read the received data */
                                    /* calculate buffer index */
   tmphead = ( USART_RxHead + 1 ) & USART_RX_BUFFER_MASK;
   USART_RxHead = tmphead;          /* store new index */

   if ( tmphead == USART_RxTail )
   {
      /* ERROR! Receive buffer overflow */
   }

  USART_RxBuf[tmphead] = data;    /* store received data in buffer */
}
//********************************************************************
//*   USART Receive Character Function
//********************************************************************
int recvchar( void )
{
   unsigned char tmptail;
                                    /* wait for incoming data */
   while ( USART_RxHead == USART_RxTail );
                                    /* calculate buffer index */
   tmptail = ( USART_RxTail + 1 ) & USART_RX_BUFFER_MASK;
   USART_RxTail = tmptail;          /* store new index */

   return USART_RxBuf[tmptail];   /* return data */
}
//********************************************************************
//*   USART Transmit Interrupt Handler
//********************************************************************
//interrupt [iv_USART_UDRE]
void USART_TX_interrupt(void)
{
   unsigned char tmptail;
                                 /* check if all data is transmitted */
   if ( USART_TxHead != USART_TxTail )
   {
                                 /* calculate buffer index */
    tmptail = ( USART_TxTail + 1 ) & USART_TX_BUFFER_MASK;
    USART_TxTail = tmptail;         /* store new index */

    UDR = USART_TxBuf[tmptail];   /* start transmission */
```

```
    }
    else
    {
        UCSRB &= ~(1<<UDRIE);          /* disable UDRE interrupt */
    }
}
//*********************************************************************
//*   USART Transmit Character Function
//*********************************************************************
int sendchar( int data )
{
    unsigned char tmphead;
                                      /* calculate buffer index */
    tmphead = ( USART_TxHead + 1 ) & USART_TX_BUFFER_MASK;
                                      /* wait for free space in buffer */
    while ( tmphead == USART_TxTail );
                                      /* store data in buffer */
    USART_TxBuf[tmphead] = (unsigned char)data;
    USART_TxHead = tmphead;           /* store new index */

    UCSRB |= (1<<UDRIE);              /* enable UDRE interrupt */

    return data;
}
//*********************************************************************
//*   USART Character Waiting Function
//*********************************************************************
unsigned char CharInQueue(void)
{
    return(USART_RxHead != USART_RxTail);
}
```

Code Snippet 6.11c: We've already examined this code down to the bit level using emulators and in-circuit debuggers.

The code in Code Snippet 6.11d is the full complement of AVR I²C routines we cloned to match the built-in I²C functions provided by the Custom Computer Services C Compiler.

```
#define  addrflag     0x01 //00000001
#define  clr_modeSLA  flags &= ~addrflag
#define  set_modeSLA  flags |= addrflag
#define  MODE_SLA     (flags & addrflag)
```

```
#define  modeMRflag  0x02 //00000010
#define  clr_modeMR  flags &= ~modeMRflag
#define  set_modeMR  flags |= modeMRflag
#define  MODE_MR     (flags & modeMRflag)

#define  modeMTflag  0x04 //00000100
#define  clr_modeMT  flags &= ~modeMTflag
#define  set_modeMT  flags |= modeMTflag
#define  MODE_MT     (flags & modeMTflag)

#define  hexflagbit  0x08 //00001000
#define  clr_hex     flags &= ~hexflagbit
#define  set_hex     flags |= hexflagbit
#define  hexflag     (flags & hexflagbit)

#define  iorwport    PORTD
#define  LE_pin      0x08 //PORTD3 00001000
#define  set_le_pin  iorwport |= LE_pin
#define  clr_le_pin  iorwport &= ~LE_pin

#define latchdata    set_le_pin;     \
                     delay_us(1);      \
                     clr_le_pin;

#define  START_i2c   0x08
#define  ACK_i2c     0x01
#define  NAK_i2c     0x00
//*****************************************************************
//*   MASTER TRANSMITTER MODE STATUS CODES
//*****************************************************************
#define  MT_SLA_ACK   0x18  //Master Transmitter Slave Addr ACK

#define  MT_DATA_ACK 0x28  //Master Transmitter Data ACK
//*****************************************************************
//*   MASTER RECEIVER MODE STATUS CODES
//*****************************************************************
#define  MR_SLA_ACK   0x40  //Master Receiver Slave Addr ACK
#define  MR_DATA_ACK 0x50  //Master Receiver Data ACK
#define  MR_DATA_NAK 0x58  //Master Receiver Data NAK
//*****************************************************************
//*   INITIALIZE THE TWI
//*****************************************************************
void twi_init(void)
{
   flags = 0x00;
   TWCR= 0x00; //disable twi
```

```
   TWBR= 0x1E; //set bit rate
   TWSR= 0x00; //set prescale
   TWAR= 0x00; //set slave address
   TWCR= 0x04; //enable twi
}
//*********************************************************************
//*   AVR i2c START
//*********************************************************************
void i2c_start(void)
{
   TWCR = (1<<TWINT) | (1<<TWSTA) | (1<<TWEN);
   while (!(TWCR & (1<<TWINT)));
   if ((TWSR & 0xF8) != START_i2c)
      printf("i2c Start Error\r\n");
}
//*********************************************************************
//*   AVR i2c WRITE
//*********************************************************************
void i2c_write(unsigned char datum)
{
   TWDR = datum;
   TWCR = (1<<TWINT)|(1<<TWEN);
   while (!(TWCR & (1<<TWINT)));
   if(MODE_SLA && MODE_MT)
   {
      if ((TWSR & 0xF8) != MT_DATA_ACK)
      printf("i2c Data Transfer Error MT Mode %x\r\n",(TWSR & 0xF8));
   else
      {
         clr_modeSLA;
         clr_modeMT;
      }
   }
   else if (MODE_SLA && MODE_MR)
   {
      if ((TWSR & 0xF8) != MR_DATA_ACK)
      printf("i2c Data Transfer Error MR Mode %x\r\n",(TWSR & 0xF8));
   else
      {
         clr_modeSLA;
         clr_modeMR;
      }
   }
   else
   {
```

```
        if ((TWSR & 0xF8) == MT_SLA_ACK)
        {
            set_modeMT;
            set_modeSLA;
        }
    else if ((TWSR & 0xF8) == MR_SLA_ACK)
        {
            set_modeMR;
            set_modeSLA;
        }
    else
        {
            printf("i2c Start Error %x\r\n",(TWSR & 0xF8));
            clr_modeSLA;
            clr_modeMR;
            clr_modeMT;
        }
    }
}
//******************************************************************
//*   AVR i2c READ
//******************************************************************
unsigned char i2c_read(unsigned char acknak)
{
    if(acknak == ACK_i2c)
    {
        TWCR = 0xC4;
        while (!(TWCR & (1<<TWINT)));
        if ((TWSR & 0xF8) != MR_DATA_ACK)
        printf("i2c Data Transfer Error MR Mode %x\r\n",(TWSR & 0xF8));
    }
    else
    {
        TWCR = 0x84;
        while (!(TWCR & (1<<TWINT)));
        if ((TWSR & 0xF8) != MR_DATA_NAK)
            printf("i2c Data Transfer Error MR Mode %x\r\n",(TWSR & 0xF8));

        clr_modeSLA;
        clr_modeMR;
    }
    return(TWDR);
}
```

```
//*******************************************************************
//*   AVR i2c STOP
//*******************************************************************
void i2c_stop(void)
{
   TWCR = (1<<TWINT)|(1<<TWEN) | (1<<TWSTO);
}
```

Code Snippet 6.11d: Nothing to it so far. You haven't seen anything new unless you "chapter hopped" to this point.

Here's where all of our RS-232 and I²C work comes to fruition. I attached an MPLAB ICD 2 to the Easy Ethernet CS8900A and an AVR JTAG ICE to the Easy Ethernet AVR. The PIC slave code will run under control of MPLAB and the MPLAB ICD 2, and the AVR master code will run on the Easy Ethernet AVR under control of the AVR JTAG ICE and AVR Studio.

I also connected the PIC's I²C interface (SDA, SCL and ground) to the AVR's TWI. The Easy Ethernet CS8900A has an I²C "port," while the Easy Ethernet AVR's TWI is bundled in with the AVR's PORTC pins. The RS-232 communications will be handled by the AVR I²C master, and I've attached the Easy Ethernet AVR's serial port to a personal computer Tera Term Pro serial session. All of the in-circuit debuggers are attached to a single personal computer, and Tera Term Pro, MPLAB and AVR Studio are running on that same personal computer. I attached the Microchip MPLAB ICD 2 using USB, and the Atmel AVR JTAG ICE is communicating with AVR Studio using the COM1 serial port. The Easy Ethernet AVR's serial port is attached to the personal computer's COM2 serial port, which is under the control of Tera Term Pro.

OK...here's how it all works!

The slave Easy Ethernet CS8900A is started and is listening on the I²C bus. Once the Easy Ethernet AVR master's USART and TWI are initialized, the Easy Ethernet throws up the "Networking with Microcontrollers is dead easy..." banner in the Tera Term Pro window and waits for a character to be received by the Easy Ethernet AVR's serial port.

If the incoming character is a '*' (0x2A), the *hexflag* flag bit is set and the byte counter variable *cntr* is cleared. The '*' sets up the Easy Ethernet AVR to take the next two ASCII bytes following the '*' from the Easy Ethernet AVR's serial port and convert them into a single hexadecimal digit. Once the hexadecimal digit is assembled, the hex digit is sent via I²C to the slave, Easy Ethernet CS8900A, where it is latched out to the Easy Ethernet CS8900A's 74HCT573 latch. The Easy Ethernet AVR sends a message to the Tera Term Pro session informing you what was sent over the I²C bus.

Entering a '$' symbol from the Tera Term Pro session puts the Easy Ethernet AVR into master-receiver mode, and the four bytes stored in the slave's *number[]* array are read into the AVR's memory and displayed in the Tera Term Pro session.

If you don't enter a '*' or a '$' character, everything you type is echoed back to the Tera Term Pro session. I've got "film at 11" of the Tera Term Pro session in Figure 6.11.

```
//******************************************************************
//*   MAIN MAIN MAIN MAIN MAIN MAIN MAIN MAIN MAIN MAIN MAIN MAIN
//******************************************************************
void C_task main(void)
{
   unsigned char x;
   CLI(); //disable all interrupts
   PORTA = 0xFF;
   DDRA = 0x00;
   PORTB = 0xFF;
   DDRB = 0x00;
   PORTD = 0xFF;
   DDRD = 0x00;

   for(x=0;x<USART_RX_BUFFER_SIZE;++x)
      USART_RxBuf[x] = 'R';
   for(x=0;x<USART_TX_BUFFER_SIZE;++x)
      USART_TxBuf[x] = 'T';

   USART_RxTail = 0x00;
   USART_RxHead = 0x00;
   USART_TxTail = 0x00;
   USART_TxHead = 0x00;

   MCUCR = 0x00; //disable sleep modes
   GICR = 0x00; //set interrupt vectors at start of flash
   TIMSK = 0x00; //disable timer interrupt sources
   init_USART(47);
   twi_init();
   SEI(); //re-enable interrupts
   printf("Networking with Microcontrollers is dead easy...\r\n");

   while(1){
   ++cntr;
      while(!(CharInQueue()));
         datum = recvchar();
      if(hexflag)
   {
         if(datum >= '0' && datum <= '9')
      datum -= 0x30;
         else if(datum >= 'A' && datum <= 'F')
            datum -= 0x37;
         else if(datum >= 'a' && datum <= 'f')
```

```
            datum -= 0x67;
   else
   {
            cntr = 0x00;
         clr_hex;
   }

   if(cntr == 1)
      byteout = datum << 4;
         if(cntr == 2)
   {
      byteout |= datum & 0x0F;
            i2c_start();
               i2c_write(0x18);
               i2c_write(byteout);
               i2c_stop();
         clr_hex;
               printf("\r\nByte Sent Via i2c = 0x%x\r\n",byteout);
      }
   }
   if(datum == '*')
   {
      set_hex;
   cntr=0;
   }
   else if(datum == '$')
   {
      printf("\r\n");
      i2c_start();
      i2c_write(0x19);
      for(x=0;x<3;++x)
      {
         datum = i2c_read(1);
         printf("datum = 0x%x\r\n",datum);
      }
      datum = i2c_read(0);
      i2c_stop();
      printf("datum = 0x%x\r\n",datum);
   }
   else
      sendchar(datum);

   }
}
```

Code Snippet 6.11e: This little application shows just how easy it is to move data between multiple devices using RS-232 and I²C.

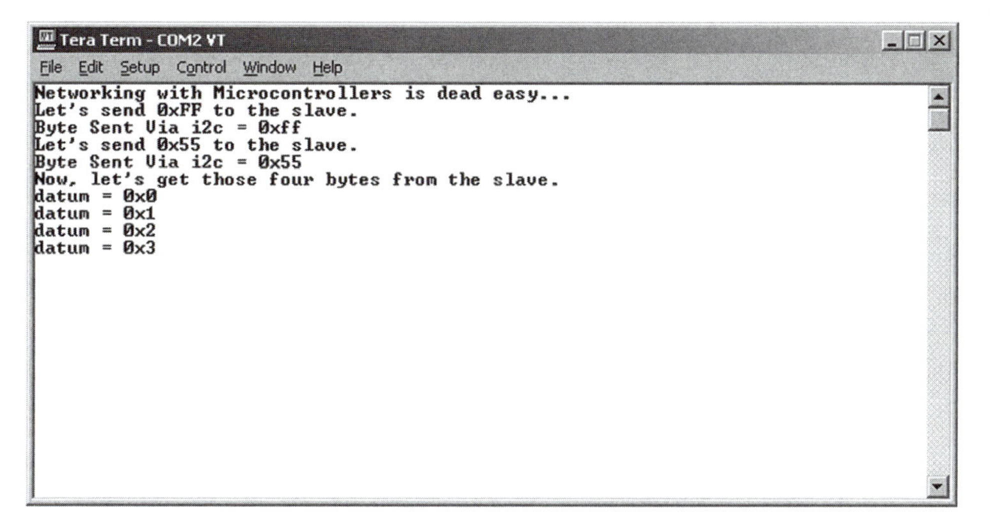

Figure 6.11: Everything you see in this shot was entered via the Tera Term Pro session into the Easy Ethernet AVR's serial port.

You've succeeded in building the RS-232 and I²C hardware for both a PIC and an AVR microcontroller. Along the way, you've also written some pretty nifty code to drive that hardware.

It's time to fill those empty component areas on the Easy Ethernet CS8900A and Easy Ethernet AVR, and in the process we'll spawn yet another device, the Easy Ethernet W.

CHAPTER 7

Ethernet

As Mario Andretti would say, "If everything seems under control, you're just not going fast enough."

You've conquered RS-232 and I^2C. Now it's time to go really fast. Let's put on our fire suits and drive the Ethernet.

What is Ethernet?

Ethernet is a faster, more powerful yet easy-to-implement alternative to RS-232 and I^2C. If you're into designing and building microcontroller-based devices that interface serially to the outside world using RS-232, I'm going to show you how to add an Ethernet interface to your microcontroller-based device. You will find that you can do the same things you do with a serial port faster and better with Ethernet. You will also find that you can do some things with the Ethernet interface you can't do with the RS-232 hardware.

The proof to the Ethernet versus RS-232 pudding is the inclusion of Ethernet interfaces as standard equipment on new personal computers along side the "old standard" serial ports. Think about this: In the "olden" days, the only way to get on the Internet was using a serial port and a dial-up modem connected to the POTS (Plain Old Telephone System). These days, if you really want to zip around on the Internet you're doing it via an Ethernet interface on your personal computer connected to a high-speed Internet service provider. And, by the way, Ethernet is the most widely used LAN technology in the world today.

All of this Ethernet stuff started in late 1972. I had graduated from high school, and Bob Metcalfe and some of his Xerox PARC pals were working on the first experimental Ethernet system. The idea at the time was to interconnect the Xerox Alto, a personal workstation with a graphical user interface. Bob and his buddies used the experimental Ethernet to link Xerox Alto computers to one another. In the heat of the moment, the experimental ether was also employed as a communications link to servers and laser printers. The original signal clock for the experimental Ethernet interface was derived from the Alto's system clock. The resultant data transmission rate on the experimental Ethernet, called the Alto Aloha Network, ended up being 2.94 Mbps. Later, Bob changed the name to Ethernet. His intention was to let everyone know that this new communications method could support any computer, not just the Xerox Alto. The name "Ethernet" was chosen because of the way the data was transmitted and received via a cable or through the "ether." In days of old, luminiferous ether was believed to propagate electromagnetic waves through space. So, just like this luminiferous substance was thought to carry electromagnetic waves, Ethernet could be thought of to do the same for computer-oriented data. Thus, Bob dubbed it Ethernet.

Ethernet is a multifaceted communications method that comes in a variety of "flavors." In this book, we'll cover a single flavor of 10-MB Ethernet that is based on the IEEE 802.3 standard with Ethernet engine ICs offered by two popular vendors:

- CS8900CQ – Cirrus Logic
- RTL8019AS – Realtek

The CS8900A-CQ will be the first Ethernet engine we will talk about. It's easy to confuse the CS8900A-CQ registers and their functionality. For the sake of clarity, I've assembled all of the definitions and code that sets up the prerequisite parameters needed to put the Easy Ethernet CS8900A board on a LAN. I've also provided register layouts so you can easily reference and assimilate the functionality of the code with the technical drawl of the CS8900A-CQ datasheet.

On the CD-ROM, I've included all of the datasheets and application notes that I used when developing the firmware and hardware for the Ethernet devices you're about to read about and possibly build for yourself. Although I'll be using plenty of visual aids to reinforce the concepts, I suggest having the datasheets and application notes handy to reference as you're reading about the Ethernet hardware and following the flow of the source code.

The CS8900A-CQ

The CS8900A-CQ is a full-function Ethernet IC capable of singlehandedly encoding and decoding standard Ethernet frames. Originally, the CS8900A-CQ was designed to operate on elaborate ISA-bus, personal-computer-based Ethernet adapters. However, the CS8900A-CQ lends itself well to smaller 8-bit microcontrollers as well. Unlike some of the cheaper NE2000 clone Ethernet ICs, the CS8900A-CQ comes equipped with plenty of on-chip RAM, and an internal analog module that includes integral 10Base-T transmit and receive filters. If you don't have a clue as to what NE2000 is, don't worry, as we'll walk that trail later. For now, all you really need to know is that the CS8900A-CQ is not natively NE2000 register compatible.

The 4 Kbytes of on-chip RAM contained within the CS8900A-CQ IC eliminate the need for a separate external static RAM IC, and since much of the necessary information needed to build an Ethernet session can be stored in the controlling microcontroller's program or data memory, there's really no need to add an external EEPROM either. In fact, the CS8900A-CQ datasheet explicitly says that when the CS8900A-CQ is used in 8-bit mode, an external EEPROM cannot be supported. Whether or not to use an EEPROM is one decision we won't have to make in our hardware design.

For the CS8900A-CQ, all of the analog circuitry necessary to process Ethernet frames is located on-chip with a single 4.99K 1% precision resistor being all it takes to awaken the CS8900A-CQ's analog functionality. The CS8900A-CQ contains everything necessary to assemble, disassemble and propagate Ethernet packets. Even with the sophisticated internal analog circuitry, the CS8900A-CQ is designed to function properly using a relatively inexpensive double-sided printed circuit board. That feature makes the CS8900A-CQ a prime candidate for homebrew Ethernet projects. My early CS8900A-CQ projects used a discreet isolation transformer and a separate RJ-45 jack to interface the CS8900A-CQ to the LAN segment. I've since then discovered an integrated interface magnetics package that houses the transmit and receive isolation transformers, the decoupling capacitors and the activity LEDs. I'll use the integrated magnetics packages for the projects presented in this book as using them makes the CS8900A-CQ easier to design into a small microcontroller environment.

Before you can feed meaningful data into the CS8900A-CQ for transmission onto the LAN or receive the same, the CS8900A-CQ must be powered up, reset and configured for Ethernet packet reception and transmission. The configuration parameters are written into the CS8900A-CQ's internal Control and Configuration Registers using routines drawn from the controlling microcontroller's firmware. The CS8900A-CQ Control and Configuration Registers hold information that determines things like the CS8900A-CQ memory base address, the Ethernet physical address (MAC address) and what types of Ethernet packets to receive.

CS8900A-CQ Reset Overview

The CS8900A-CQ's internal circuitry and operating registers are reset by taking the CS8900A-CQ RESET pin to a high logic level for more than 400 nS. In addition, if the power to the CS8900A-CQ's VDD pins falls below 2.5 volts, the CS8900A-CQ will enter a reset mode exiting only when the power supply voltage has risen above 2.5 volts and the CS8900A-CQ's oscillator has stabilized. A third and purely logical way of resetting the CS8900A-CQ is to set the RESET bit via a routine from a controlling microcontroller's firmware. In the case of any of the three reset methods, once a CS8900A-CQ reset is initiated, the CS8900A-CQ needs at least 10 ms to recalibrate its internal analog circuitry and initialize its internal registers. A CS8900A-CQ bit called INITD is set when the internal calibration and initialization has completed.

CS8900A-CQ Media Interface Overview

The CS8900A-CQ supports more than one media interface, and the desired media interface must also be selected during the configuration process. We'll be using 10Base-T over twisted pair, exclusively. The CS8900A-CQ has the ability to automatically detect the physical interface (10Base-T or AUI or Attachment Unit Interface), but we will override the auto-detect function and instruct the CS8900A-CQ to only operate using a 10Base-T interface. Using 10Base-T allows us to interface to our CS8900A-CQ-based device using ordinary Category 5 cabling that can be purchased most anywhere personal computer cables are sold.

CS8900A-CQ Transmit Process Overview

The transmission of an Ethernet packet using a CS8900A-CQ can be broken down into two distinct phases. Suppose that a microcontroller wants to transmit data to another station, or host, on the LAN. To begin Phase 1, the sending microcontroller must first issue a CS8900A-CQ transmit command and after permission is given by the CS8900A-CQ, load the Ethernet frame into the CS8900A-CQ's transmit buffer. The CS8900A-CQ transmit command tells the CS8900A-CQ that the controlling microcontroller has asked for a frame to be transmitted that it is holding in its buffer. In addition, the CS8900A-CQ transmit command issued by the controlling microcontroller tells the CS8900A-CQ when to start transmitting the data it receives from the controlling microcontroller. The bytes that follow the CS8900A-CQ transmit command tell the CS8900A-CQ how much of its internal buffer space will be needed to hold the frame to be transmitted. The frame passed from the controlling microcontroller normally includes destination and source address information, the type of data the frame is carrying, the data itself, and a checksum to help ensure the integrity of the data within the Ethernet packet.

When sufficient buffer memory is available within the CS8900A-CQ, the controlling microcontroller writes the frame into the CS8900A-CQ transmit buffer memory. Phase 1 of the transmit process ends here, and Phase 2 of the transmit process (and the CS8900A-CQ magic) begins.

In Phase 2, the CS8900A-CQ converts the frame data fed to it from the microcontroller into an Ethernet packet and puts the packet out onto the network. As soon as all of the designated data has been loaded into the CS8900A-CQ transmit buffer, the CS8900A-CQ generates a preamble, which consists of alternating binary "1's" and binary "0's" followed by a start of frame delimiter (SFD), which has a unique bit pattern. Then the rest of that frame stuff I mentioned earlier is transferred into the CS8900A-CQ transmit queue. If instructed to do so, the CS8900A-CQ is smart enough to "pad" the frame if the total length of the frame is less than the minimum 64 bytes. The last act the CS8900A-CQ performs before shooting the bits out onto the communications medium is to calculate and add the FCS value to the end of the outgoing bit stream.

The preamble is generated to allow the receiving Ethernet devices to sync-up their receiver circuits. An Ethernet device can afford to lose preamble bits, but the idea is not to lose any important data bits. Once an Ethernet device has adjusted to the preamble signal, the receiving Ethernet device regards any bits that follow the SFD as frame data. Figure 7.1 is a graphic representation of the types of Ethernet packets the CS8900A-CQ will handle in projects presented in this book.

CS8900A-CQ Receive Process Overview

Receiving an Ethernet packet with the CS8900A-CQ is also done in two phases. In Phase 1, the CS8900A-CQ's analog circuitry pulls an Ethernet packet into a portion of its 4-kilobyte memory area that it has reserved as the receive buffer. The incoming encoded data is decoded by a Manchester ENDEC (Encoder/Decoder) and converted for use by the CS8900A-CQ's 802.3 MAC Engine. The incoming preamble and start of frame delimiter bits are discarded by the CS8900A-CQ converting the incoming packet to an incoming frame. The incoming frame is then processed by the CS8900A-CQ's address filter. The CS8900A-CQ address filter is programmable, which allows various addressing schemes to be employed. If the incoming

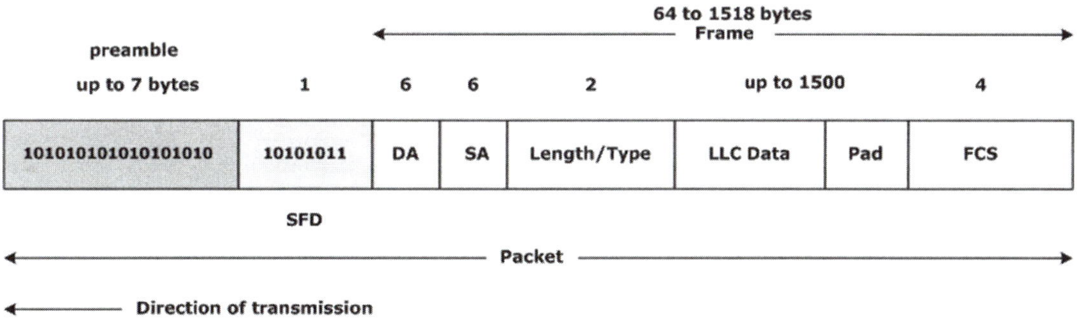

64 to 1518 bytes
Frame

preamble up to 7 bytes	1	6	6	2	up to 1500		4
10101010101010101010	10101011	DA	SA	Length/Type	LLC Data	Pad	FCS
	SFD						

Packet

Direction of transmission

Figure 7.1: The minimum of 64 bytes inside the Ethernet frame includes the 4 bytes of the checksum (FCS or Frame Check Sequence). The FCS is analogous to a CRC (Cyclic Redundancy Check) value.

frame's destination address (DA) passes the CS8900A-CQ address filter's test, the rest of the incoming frame information is stored in the CS8900A-CQ's on-chip memory. The integrity of the stored frame is checked using the CRC (FCS) bytes, and if everything is OK, the CS8900A-CQ can signal the controlling microcontroller that its receive buffer contains a valid frame that is ready to be transferred and processed. The microcontroller then initiates Phase 2 and transfers the data from the CS8900A-CQ receive buffer memory into its buffer for processing.

During Phase 2 of the receive process, the controlling microcontroller transfers the data from the CS8900A-CQ receive buffer via an I/O portal. The portal that allows the controlling microcontroller to access the CS8900A-CQ's internal registers and data is called *PacketPage*.

CS8900A-CQ External Storage Overview

CS8900A-CQ configuration data can be loaded using firmware routines executed by the controlling microcontroller or from an external EEPROM. In our case, there's no need to store any configuration data in an external EEPROM, as we can integrate all of our configuration parameters into our operating code. Once the proper configuration values have been stored in the CS8900A-CQ's register banks, we can then turn the CS8900A-CQ loose on a LAN (Local Area Network) or the Internet.

Although we won't be incorporating an external EEPROM into our design, the CS8900A-CQ always checks for the presence of an external EEPROM after each reset. If the EEDI (EEDataIn) pin is at a high logic level, the presence of an external EEPROM is assumed and the CS8900A-CQ automatically loads configuration data that is stored in the EEPROM. So, in our firmware, we must make sure the EEDI is low after a reset. When the CS8900A-CQ reads a low logic level on the EEDI pin, the CS8900A-CQ enters a default configuration mode, which sets the base address to 0x0300 and initializes the CS8900A-CQ's internal registers to predetermined values. Besides, the EEPROM is not a valid option when running the CS8900A-CQ in 8-bit mode.

CS8900A-CQ Status Indicators

Most commercial Ethernet cards have LED indicators to provide a visual status of what's happening on the Ethernet communications medium. These LED indicators are optional, as the CS8900A-CQ and any other Ethernet engine IC will operate just fine without them. However, we will incorporate them into our design simply because we can. Having the blinking LEDs also helps in troubleshooting when things just aren't working right. The CS8900A-CQ has two indicator LED drivers and one logic level output pin: the LANLED, the LINKLED and HC1. The LANLED blinks whenever a frame is transmitted or received by the CS8900A-CQ. It also winks when a collision occurs. A valid 10Base-T link pulse excites the LINKLED. Very little additional circuitry (two LED current limiting resistors) is needed to use the CS8900A-CQ's LED indicators.

In addition to driving LEDs, the LANLED and LINKLED pins can also be programmed to simply output a logic level just as the HC1 pin does. When the CS8900A-CQ is instructed to put the LED driver pins in logic output mode, the logic output pins are controlled by setting and clearing bits associated with the CS8900A-CQ logic output pins. The logic levels on these output pins can then be used to trigger other circuitry or indicate status to an external device.

The CS8900A-CQ MAC Engine

The magic of Ethernet connectivity that emanates from the CS8900A-CQ is partially made possible by the CS8900A-CQ's Ethernet Media Access Control, or MAC, engine. The CS8900A-CQ's MAC engine is responsible for Ethernet frame transmission and reception, which includes collision detection, preamble generation and detection and CRC generation. The MAC is also the entity that pads the outgoing frames when they're discovered to be too short.

You've probably heard the term "802.3" thrown about when folks that claim they know what they're talking about are discussing Ethernet LANs. The "802.3" they're speaking of refers to an IEEE Ethernet standard upon which the CS8900A-CQ MAC and many other Ethernet IC MAC engines are built.

As you can see in Figure 7.2, the CS8900A-CQ MAC engine sits between the CS8900A-CQ's internal bus and the CS8900A-CQ's on-chip Manchester encoder/decoder, or ENDEC. The ENDEC encodes (EN) and decodes (DEC) the bits passed to it by the 10Base-T interface. The other side of the 10Base-T interface is where the action is as far as the CS8900A-CQ is concerned. In a lab environment like mine, the 10Base-T interface may draw data from a small LAN (Local Area Network) environment. The other side of the 10Base-T interface could just as well be connected to a full-blown Internet router or tied directly to another Ethernet device using a crossover cable. The bottom line is that the CS8900A-CQ 802.3 MAC engine automatically assembles and disassembles Ethernet packets passing between the CS8900A-CQ's internal bus and the ENDEC.

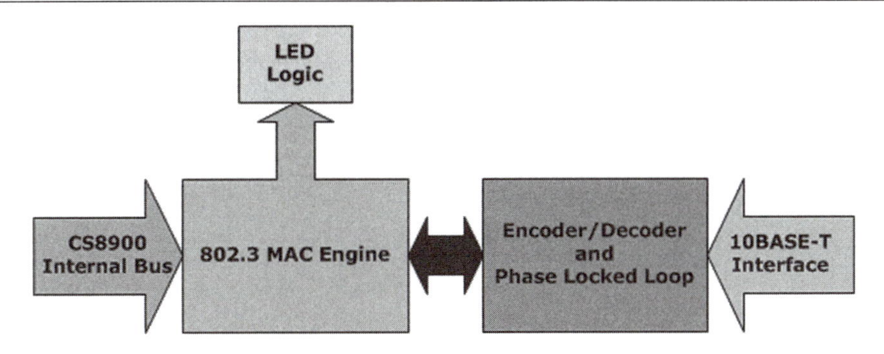

MAC Interface

Figure 7.2: The primary function of the MAC are frame encapsulation/ decapsulation, error detection/handling and media access management.

The CS8900A-CQ's 802.3 MAC engine is just like the engine in your car. Unless you're a bona fide mechanic, you can't fix your car's motor when it breaks or soup it up to make it go faster. All you know about your car's engine is how to put gas in the tank, how to start it and how to stop it. When the car's engine is running, you press the accelerator pedal to give the engine gas and make the car move or hit the brake pedal to make the car stop. The same thought applies to what you need to know about the CS8900A-CQ's MAC engine. In this case, the car is analogous to an Ethernet packet and gas is data. All you need to know about the CS8900A-CQ MAC engine is how to put gas in it, start it, drive it and stop it. The Ethernet packet will move accordingly.

We already know that an Ethernet packet transmission originating from a CS8900A-CQ begins with the 802.3 MAC engine generating and transmitting a 7-byte preamble. We also know that the preamble consists of alternating binary '1's and '0's (10101010) and is used to allow other Ethernet devices listening on the wire to sync-up for the upcoming data bits. We're also aware that immediately following the preamble is the SFD, which is coded in binary as "10101011" and as far as the Ethernet devices receiving the bits are concerned, everything following the SFD had better be valid frame data. At this point in the transmission process, the CS8900A-CQ's 802.3 MAC engine has done most of its job by providing the wake-up call for the devices listening on the Ethernet network the CS8900A-CQ's transmission is addressing.

The controlling microcontroller or whatever is in charge of the data must provide the destination address (DA), the source address (SA), the length or type of transmission and the data. If programmed to do so, the CS8900A-CQ MAC engine performs the padding and generates the FCS checksum after the controlling microcontroller has filled in the frame components it is responsible for. The responsibilities of the CS8900A-CQ MAC engine and the controlling microcontroller are clarified in Figure 7.3.

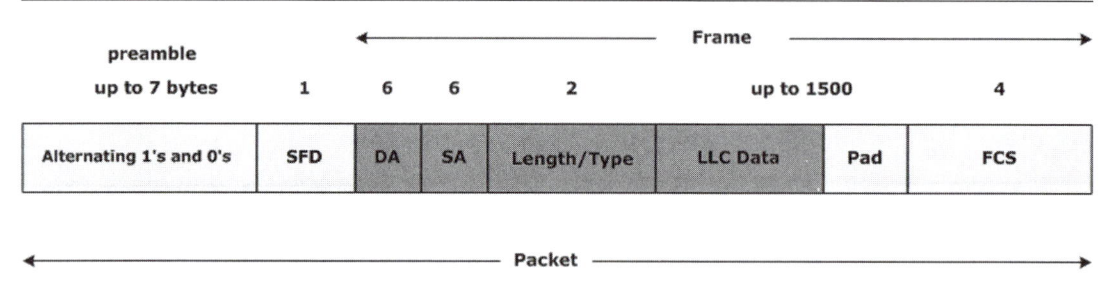

Figure 7.3: The CS8900A-CQ MAC engine supplies the contents of the fields in yellow (white area). The red fields (gray area) represent physical addressing, logical addressing and data information. If the protocol carried inside the LLC Data area requires any special logical addressing schemes, the special logical addressing data is woven into the initial bytes of the data contained in the Type and LLC Data fields.

The Manchester decoder provides a stream of NRZ (Non-Return to Zero) data to the CS8900A-CQ's MAC engine. The best way to explain Manchester and NRZ encoding is to show them to you side by side as I've done in Figure 7.4.

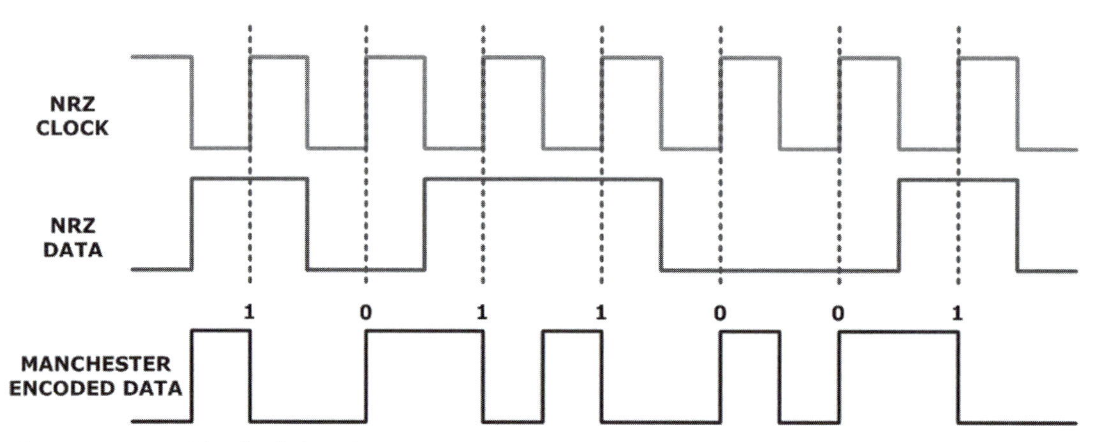

Figure 7.4: I could talk all day about this and still not make any sense. It's a pretty simple concept now that you can see it, huh? Manchester encoding is also sometimes called Biphase Coding because each of the Manchester encoded bits can be defined as a positive or negative 90-degree phase transition.

NRZ data is just that, Non-Return to Zero data. Simply stated, NRZ does not encode the data. An NRZ logical '1' is represented by positive voltage, and an NRZ logical '0' is represented by zero voltage. There are no in-between voltages to specify logic levels in NRZ encoding. Put in another way, NRZ data doesn't have a resting voltage. The voltage level of the data is either positive or zero, with positive being the voltage level that represents a logical '1', and zero being the voltage level that represents a logical '0'.

Bit timing for NRZ data is provided by an NRZ clock. As you can see in Figure 7.4, the logic level of the NRZ data is valid at every rising edge of the NRZ clock. I've enhanced the rising edges of the NRZ clock with vertical dotted lines in Figure 4. In the Manchester encoding example I've drawn for us here, a logic '1' is a high-to-low transition at the center of the bit time. The center of the bit time occurs at the rising edge of the NRZ clock. Conversely, a Manchester encoded logic '0' is a low-to-high transition in the center of the bit time. In our example, the bit boundaries are denoted by falling edges of the NRZ clock. Notice that the Manchester encoded data may or may not transition on a bit boundary but it always transitions at the center of each NRZ bit. Simply stated, each NRZ bit is converted to a Manchester bit using a transition. Figure 7.5 is what the Ethernet preamble looks like in both the NRZ and Manchester domains.

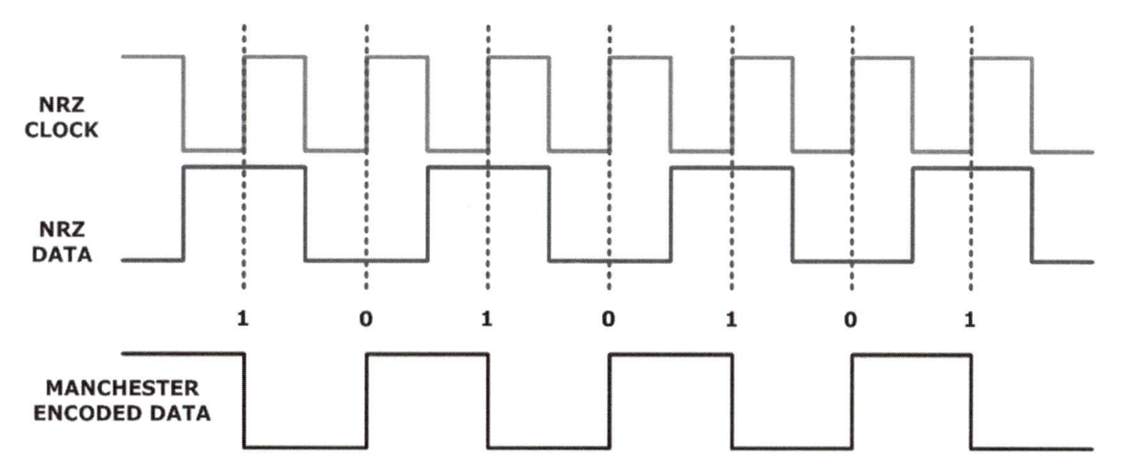

Figure 7.5: What do you think? That "101010" NRZ-bit pattern looks like a square wave to me and so does the Manchester bit pattern. So, using some simple math and knowing the CS8900A-CQ clocks the LAN at 10 Mbps, I figure a bit time is equal to .1µS bit for a 10-Mbps LAN speed (1/10,000,000). It takes 2 bit times (.2µS) to complete a full cycle (also known as the period). Just a little more math (frequency = 1/period) tells me the preamble square wave has a frequency of 5 MHz.

The inherent nature of Manchester encoding ensures that there are enough of the center-of-the-bit transitions to allow the CS8900A-CQ receiver's Phase-Locked Loop circuitry to extract the clock signal from the incoming Manchester-encoded bit stream and correctly decode the incoming Manchester-encoded bits to NRZ data.

Assuming the receiver has synced on the preamble signal, the MAC engine checks for a valid SFD and if the SFD checks out, the DA is read to determine if the address matches what the CS8900A-CQ's address filter wants to see. Once the DA is determined to be "ours," the rest of the incoming frame data is loaded into the CS8900A-CQ buffer. The CS8900A-CQ can be programmed to ignore the incoming FCS bytes or load the FCS bytes and have them checked for validity by the MAC engine.

There are lots of CS8900A-CQ buttons and knobs we can twist and turn using firmware that determine how the incoming and outgoing data is handled. The goal here is to get some data moving on a LAN using the services of a microcontroller and the CS8900A-CQ. So, instead of trying to describe every little nuance of every feature in detail, I'll concentrate on what we need to get the job done. You've had enough theory to last a while, and you've already built the power supply and microcontroller circuitry (we used it in the RS-232 and I²C chapters). So, let's finish designing and building the CS8900A-CQ Ethernet hardware. By the time you've finished reading the text and assembling your own CS8900A-CQ-based Ethernet device, you'll see just how easy working with the CS8900A-CQ is.

Easy Ethernet CS8900A Hardware

Because of the unique PacketPage I/O port access scheme, we can use the CS8900A-CQ with most any of the commonly available 8-bit or 16-bit microcontrollers. To that end, I've chosen the Microchip PIC16F877 as the primary microcontroller for this project.

The PIC16F877 Microcontroller

The PIC16F877 is an easy-to-understand part with ample and easily accessible resources. Being flash based and available in the standard 40-pin DIP package, the PIC16F877 is easy to incorporate into designs, and is capable of being programmed in-circuit using the Microchip ICSP (In-Circuit Serial Programming) programming algorithm. The PIC16F877 microcontroller can be programmed hundreds of times, and the Easy Ethernet CS8900A's onboard ICSP circuitry allows you to program the PIC16F877 without removing it from its socket. Another advantage to using the PIC16F877 is that we can simply drop in a Microchip PIC18F452 and gain additional computing resources as well as speed without having to significantly change the PIC16F877 source code.

I've chosen to design the Easy Ethernet CS8900A using the Microchip flash-based microcontrollers because Microchip's line of flash microcontrollers are inexpensive and easily obtained. In addition, the flash-based PICs don't require the support hardware a standard-windowed PIC needs. For instance, using flash devices eliminates the need for an ultraviolet EPROM eraser. And, since flash parts can be programmed and reprogrammed in-circuit using ICSP, the development and debug cycle time is reduced significantly as fewer microcontroller parts are needed in the development cycle. Using flash-based microcontrollers means that there is no need to rotate a number of the same type of microcontrollers through the ultraviolet eraser to save time between runs while you're debugging your code.

When it comes to on-chip resources, the PIC16F877 is the largest part in the PIC16F87X crew. The PIC16F877 can operate with a 20 MHz clock, which gives an instruction cycle time of 200 nS. There are 8K words of program flash and 368 bytes of SRAM or data memory crammed inside a PIC16F877. Should our design require it, there is also a block of 256 bytes of EEPROM (Electrically Erasable Programmable Read Only Memory) data memory available for storing constants or whatever else we decide is important to keep in on-chip memory. The PIC16F877 microcontroller's EEPROM is nonvolatile memory and retains any information we place there even after the power is removed from the part.

Having ample microcontroller I/O is very important in a networking design like the Easy Ethernet CS8900A. Not only do we need enough I/O to perform tasks like monitoring a voltage or turning an external device on or off, there have to be some I/O pins dedicated to the CS8900A-CQ. For instance, the Easy Ethernet CS8900A microcontroller Ethernet design requires at least 16 dedicated microcontroller I/O pins. The PIC16F877 has 33 I/O lines we can put to work, which leaves some I/O for things that microcontrollers do best…control.

The Microchip PIC18F452

The PIC18F452 is pin for pin compatible with the PIC16F877. That's where most of the familiarity stops. The PIC18F452 comes with 16K on on-chip program memory backed-up by 1.5K of RAM. The PIC16F877 is equipped with 368 bytes of SRAM. The PIC18F452's abundance of RAM makes the PIC18F452 a viable candidate for Ethernet LAN applications like our Easy Ethernet CS8900A. In addition to the increased internal memory area, the PIC18F452 can run twice as fast as the PIC16F877 at 40 MHz. All of the PIC16F877 communications peripherals operate in the same manner on the PIC18F452.

The CS8900A-CQ Ethernet Engine

To facilitate the movement of data back and forth between the PIC16F877 microcontroller and the CS8900A-CQ, the CS8900A-CQ employs a unique I/O port scheme that is called PacketPage. The CS8900A-CQ's PacketPage architecture is supported by 4 Kbytes of CS8900A-CQ on-chip memory called PacketPage memory. CS8900A-CQ PacketPage memory is used for both CS8900A-CQ internal registers and for the storage of Ethernet transmit and receive frame data. PacketPage technology consists of a set of eight 16-bit I/O ports that are mapped into the CS8900A-CQ I/O space. The PacketPage ports are located between offsets 0x0000 and 0x000F and allow the PIC16F877 to access the CS8900A-CQ's internal registers and select portions of the CS8900A-CQ's 4-Kbyte chunk of on-chip memory.

To access the PacketPage I/O ports, the Ethernet adapter base address is added to the offsets laid out in Table 7.1. In our design, the base adapter address, or I/O Base Address, is hard-wired for 0x0300. The "adapter" referenced in the base adapter address, in this case, is actually the entire complement of the Easy Ethernet CS8900A circuitry. Setting the CS8900A-CQ base address is done by forcing (hard wiring) the external CS8900A-CQ address lines SA4 through SA15 into a permanent 0x0300 pattern, with only the lower 4 bits capable of being altered by the Easy Ethernet CS8900A's PIC16F877 microcontroller.

Since there will be no external EEPROM in our Easy Ethernet CS8900A design, the CS8900A-CQ internal I/O Base Address register defaults to 0x0300 on power up. The I/O Base Address is kept at PacketPage Address 0x20 in the format shown in Figure 7.6.

PacketPage Port Layout

Offset	Type	Description
0000h	Read/Write	Receive/Transmit Data (Port 0)
0002h	Read/Write	Receive/Transmit Data (Port 1)
0004h	Write-only	TxCMD (Transmit Command)
0006h	Write-only	TxLength (Transmit Length)
0008h	Read-only	Interrupt Status Queue
000Ah	Read/Write	PacketPage Pointer
000Ch	Read/Write	PacketPage Data (Port 0)
000Eh	Read/Write	PacketPage Data (Port 1)

Table 7.1: Even though the PIC16F877 is an 8-bit microcontroller, the CS8900A-CQ requires that data be transferred to and from the PacketPage portal using 16-bit transfers.

Address 0x21	Address 0x20
Most significant byte of I/O Base Address	Least significant byte of I/O Base Address

Figure 7.6: The binary reset value of the I/O Base Address is 0000 0011 0000 0000.

The combination of the CS8900A-CQ default I/O Base Address register contents and external hardwiring of the CS8900A-CQ address lines results in our PacketPage I/O port addresses being defined between offsets 0x0300 to 0x030F. All of the necessary configuration data will be supplied by the PIC16F877 microcontroller, and its firmware and the need for a dedicated boot EEPROM is eliminated.

Powering the CS8900A-CQ

Some of the CS8900A-CQ power pins are designated strictly for the powering of the analog circuitry found within the CS8900A-CQ. To properly bias these internal analog modules, R1 (a 4.99K 1% resistor), is installed as close as possible to the CS8900A-CQ RES pin (93) and a neighboring ground pin (94). Resistor R2 (4.75K 1%) is used to pull up the CS8900A-CQ SLEEP line at pin 77. Although the CS8900A-CQ SLEEP input has its own internal weak pull-up, it is recommended to add the external pull-up resistor. The active low CS8900A-CQ SLEEP pin enables the Hardware Suspend and Hardware Standby sleep modes. The Easy Ethernet CS8900A design never activates the CS8900A-CQ sleep modes.

The CS8900A-CQ Ethernet Magnetics

The receive magnetics (isolation transformers) are integrated into a single unit with the transmit magnetics and the indicator LEDs. The receiver magnetics have a primary-to-secondary turns ratio of 1:1. Thus, for optimal transfer characteristics, the termination resistor value follows the communications cable's impedance. R5, a 100 Ohm 1% part, provides receive-side impedance matching for 100 Ohm Category 5 communications cable.

The transmit side impedance matching is performed by resistors R3 and R4, with some help from C1 and the windings ratio of the isolation transformer. On the transmit side, the ratio of primary to secondary windings is 1:1.414. If you're wondering what's inside the can, a schematic representation of the NU1S114-XXX can be seen in Schematic 7.1. Lots of soldering is avoided by using the NU1S114-XXX in the Easy Ethernet CS8900A design. Without the NU1S114-XXX, the isolation transformer, the RJ-45 jack and the indicator LEDs are all be separate components that would have to be soldered onto the Easy Ethernet CS8900A printed circuit board separately.

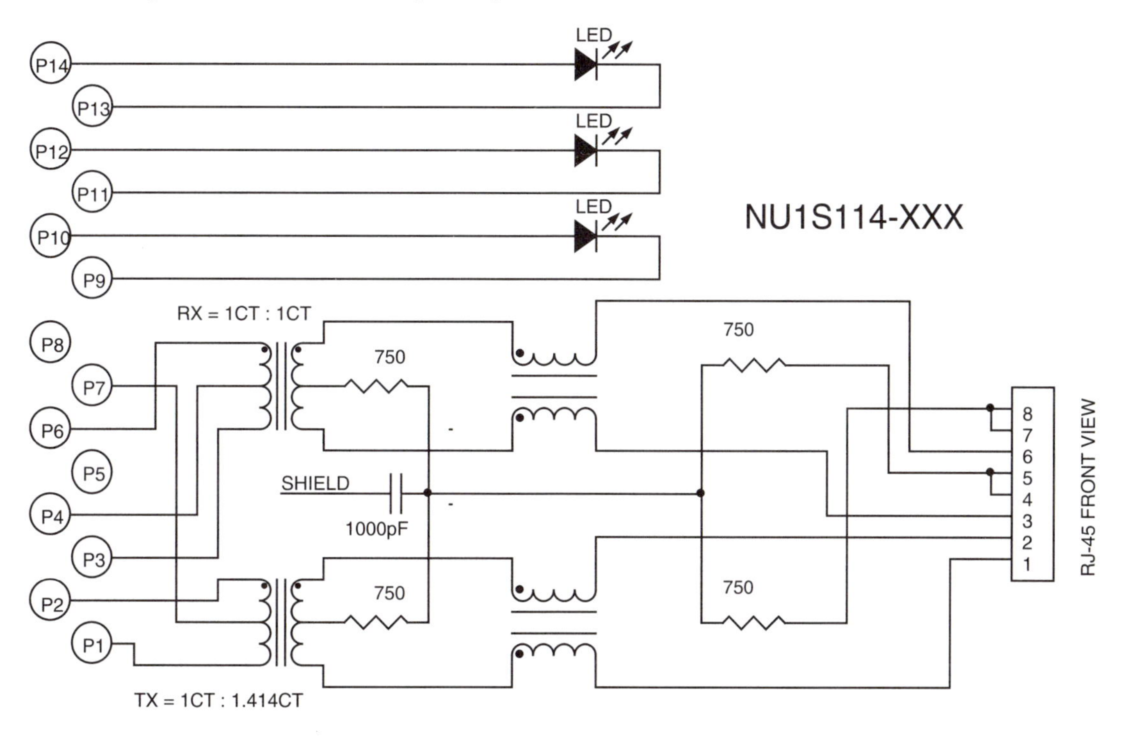

Schematic 7.1: The XXX in the magnetics assembly part number designates the color of the in-can LED indicators. The Easy Ethernet CS8900A uses a NU1S114-434, which sports green and yellow LED indicators.

The Link Good LED indicator on the NU1S114-434 is driven by the LINKLED pin, (pin 99), of the CS8900A-CQ. The HCE0 bit of the Self Control Register is set to force the pin low when valid link pulses are detected on the communications segment. The PIC16F877 microcontroller controls the logic level of this pin if the HCE0 bit is clear. Link pulses are generated by transmitters on the Ethernet segment if no Ethernet packet activity is detected. Transmitted link pulses are 1-bit wide positive pulses that are generated by an Ethernet transmitter every 16 ms. At the completion of an EOF (End of Frame) sequence, a 16 ms timer kicks-off. If no other packets appear on the Ethernet segment before the 16 ms EOF timer times out, a link pulse is generated.

Pin 100 of the CS8900A-CQ is called the LANLED pin and is the driving force behind the transmit/receive/collision indicator LED, which is also an integral part of the Ethernet magnetics assembly. The LANLED pin is driven low for 6ms whenever a collision, transmit or receive operation occurs. Unlike the LINKLED, the LANLED pin doesn't do dual duty as a microcontroller-controlled I/O pin.

I didn't incorporate the services of the third status output pin, HC1, into the design of the Easy Ethernet CS8900A. To use the HC1 output, a PacketPage I/O transaction must be performed by the PIC16F877 microcontroller. Since the Easy Ethernet CS8900A's microcontroller is driving the state of the HC1 line anyway, it's much easier and faster to use a PIC16F877 microcontroller I/O line instead of the HC1 line. On the other side of that, the HC1 is a "free" output line in that it doesn't take a physical I/O pin from the microcontroller attached to the CS8900A-CQ.

The schematic view of the CS8900A-CQ portion of our design (Schematic 7.2), shows you that a minimal number of external components are needed to support the CS8900A-CQ running in 8-bit mode. Easy Ethernet indeed.

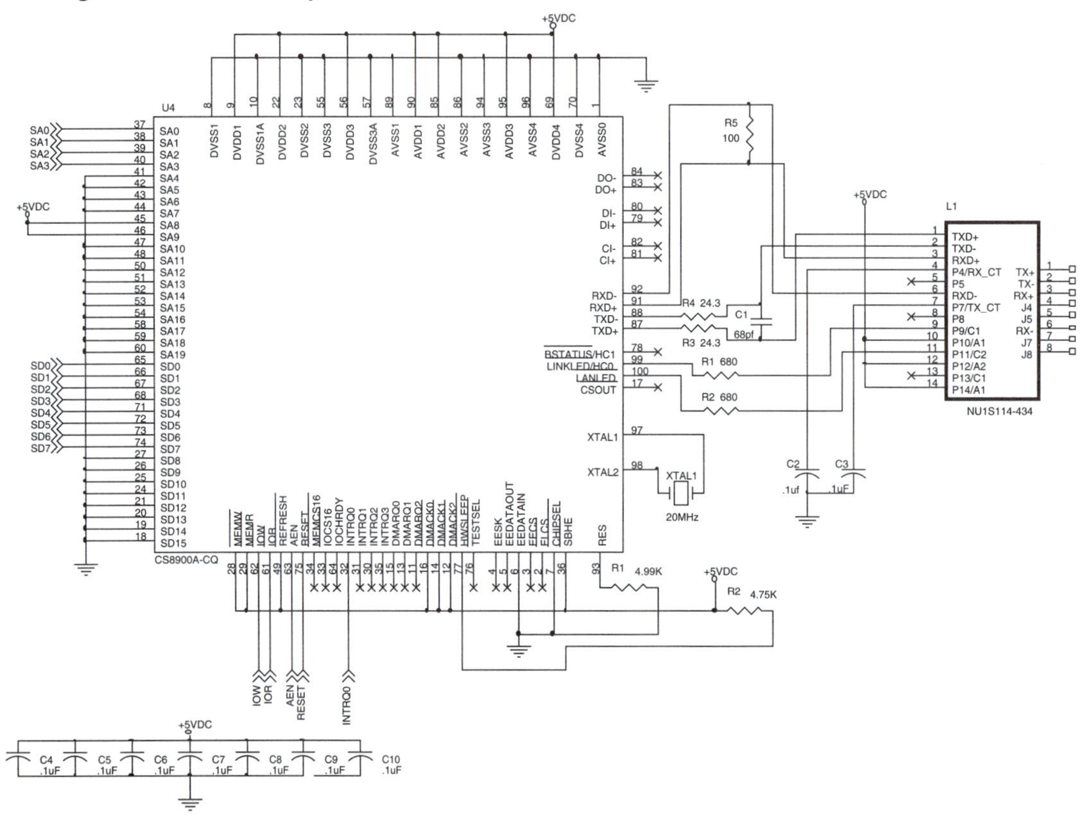

Schematic 7.2: The SA0:SA15 0x0300 base address pattern is apparent in the schematic view of the CS8900A-CQ Ethernet module. By the way, you'll see a colon used a lot in this text between address and data line descriptions. Just read the colon as the word "through." For instance, SA0:SA15 translates as: "SA0 through SA15."

Designing in the Easy Ethernet CS8900A's PIC16F877 Microcontroller

There are a total of twenty CS8900A-CQ address lines (SA0:SA19), but only the lower sixteen (SA0:SA15), are used for I/O read and write operations. To navigate the entire PacketPage port structure, it is only necessary to manipulate the least significant nibble (SA0:SA3) of the 0x0300 address configuration. Thus, we only need four of the microcontroller's I/O lines to serve as the PacketPage address lines. In this design, I've assigned pins RB0:RB3 of the PIC16F877 microcontroller to address line duty.

In addition to the four CS8900A-CQ address lines provided by PORTB, the PIC16F877 microcontroller will provide 8 bits of CS8900A-CQ bidirectional data bus from its PORTD I/O pins. In a preliminary design, I used the PIC16F877's PORTC as the data bus. I went ahead with writing the PIC16F877 microcontroller firmware with the PORTC data bus configuration and everything worked as expected. After deciding that I was wasting a perfectly good RS-232 port and I²C interface, I decided to move the data bus to PORTD of the PIC16F877.

An interesting thing happened during the initial phases of testing the CS8900A-CQ/PIC16F877 microcontroller union. I noticed that I could run without pull-up resistors on PORTC or PORTD when I used a "real" PIC16F877 microcontroller. When I emulated the PIC16F877 microcontroller with the Microchip ICE 2000, I found that I had to install the PORTD pull-up resistors for proper operation. Resistors are cheap and easy to install. So, I kept the PORTD pull-up resistors in the final design. This bug kept me scratching my head for at least a week.

To access the CS8900A-CQ's PacketPage memory, we must employ the services of the CS8900A-CQ AEN, IOR and IOW pins in concert with the CS8900A-CQ data and address busses that are tied to the PIC16F877 microcontroller's PORTB and PORTD. The PIC16F877 microcontroller's I/O pin RB4 has been tapped to handle the CS8900A-CQ AEN pin, and two bits of the PIC16F877's PORTE I/O port, RE0 and RE1, will control the logic levels of the CS8900A-CQ IOR and IOW lines, respectively. The CS8900A-CQ AEN input serves as a pseudo-chip select in our design. Asserting IOR with a valid address puts the contents of the selected CS8900A-CQ 16-bit I/O register on the data bus. Conversely, a valid address coupled with the active low IOW signal writes the contents of the data bus into the selected CS8900A-CQ 16-bit I/O register.

Even though the PIC16F877 microcontroller can use the CS8900A-CQ's PacketPage architecture to see into the CS8900A-CQ, we still must have a way to sense that a good packet is sitting in the CS8900A-CQ receive queue. Since the CS8900A-CQ cannot reliably support interrupts in 8-bit mode, the status of incoming packets is obtained by having the Easy Ethernet CS8900A's onboard PIC16F877 poll the CS8900A-CQ RxOK bit inside the CS8900A-CQ Receiver Event Register. If the CS8900A-CQ's RxOK bit is set, the length of the valid received frame is loaded into PacketPage Address 0x0402.

Even though the CS8900A-CQ can be reset using microcontroller control using bit 6 of the CS8900A-CQ Self Control Register, I've also elected to externally control the CS8900A-CQ RESET line using the remaining PORTE I/O pin, RE2.

My PIC16F877 pin assignments intentionally leave the entire PIC PORTA I/O bank open. With the PIC16F877 I/O pin assignments I have made, PORTA can now be used as general purpose I/O, A/D (analog-to-digital) inputs or a combination of both A/D and I/O. Since I went out of my way to preserve the I^2C and RS-232 ports, I've provided hardware and component pads on the Easy Ethernet CS8900A printed circuit board for the optional use of I^2C and RS-232 by way of PORTC of the PIC16F877 microcontroller. Otherwise, most of the I/O pins of PORTC are available as only one other PORTC pin is assigned as the latch enable signal for the 74HCT573 transparent latch. Speaking of latches, since there is not a full 8-bit output port available, I decided to add a 74HCT573 transparent latch to PORTD. Using a pin from PORTC (RC1) as the 74HCT573 latch enable enables an 8-bit output port that is multiplexed with the CS8900A-CQ data bus. The inclusion of the 74HCT573 provides 6 bits of analog or digital input or general purpose I/O via PORTA and 8 bits of digital output complements of the 74HCT573.

The ICSP (In-Circuit Serial Programming) Interface

The ICSP header is wired in the Microchip suggested standard manner and can be used with the Microchip MPLAB® ICD, MPLAB ICD 2 or PRO MATE® II. A 6-pin RJ-11 connection is the ICSP connection method used by the MPLAB ICD and MPLAB ICD 2. The PRO MATE II uses a more elaborate 15-pin ICSP interface. If you have access to a PRO MATE II and the PRO MATE II ICSP module, a simple 15-pin to RJ-11 adapter cable can be easily fabricated to allow the PRO MATE II's ICSP module to service the Easy Ethernet CS8900A's ICSP interface. The RJ-11 interface cable comes as standard equipment when you purchase either of the ICD modules.

No matter what type of PIC programmer you use on the PIC16F877 microcontroller, PORTB pins RB6 and RB7 are used to receive program clock and program data, respectively. The ICSP algorithm uses the signals driving RB6 and RB7 to program the PIC16F877's internal flash memory. The PIC programmer supplies the programming voltage (usually +13 VDC) through the MCLR line. At +13 VDC, the programming voltage applied to the MCLR pin is high enough to possibly destroy other components on the Easy Ethernet CS8900A. So, blocking diode D2 is used to isolate the potentially damaging programming voltage from the rest of the Easy Ethernet CS8900A's circuitry.

The PIC16F877 microcontroller section of the Easy Ethernet CS8900A is shown schematically in Schematic 7.3. All of the Easy Ethernet CS8900A IC components are powered with +5 VDC. The LM340S-based power supply circuit gets its power from a standard 9V-center positive wall wart. Diode D1 protects against accidental polarity reversal at the power supply input. The capacitors surrounding the LM340S voltage regulator provide filtering to provide a clean +5 VDC output voltage for the Easy Ethernet CS8900A.

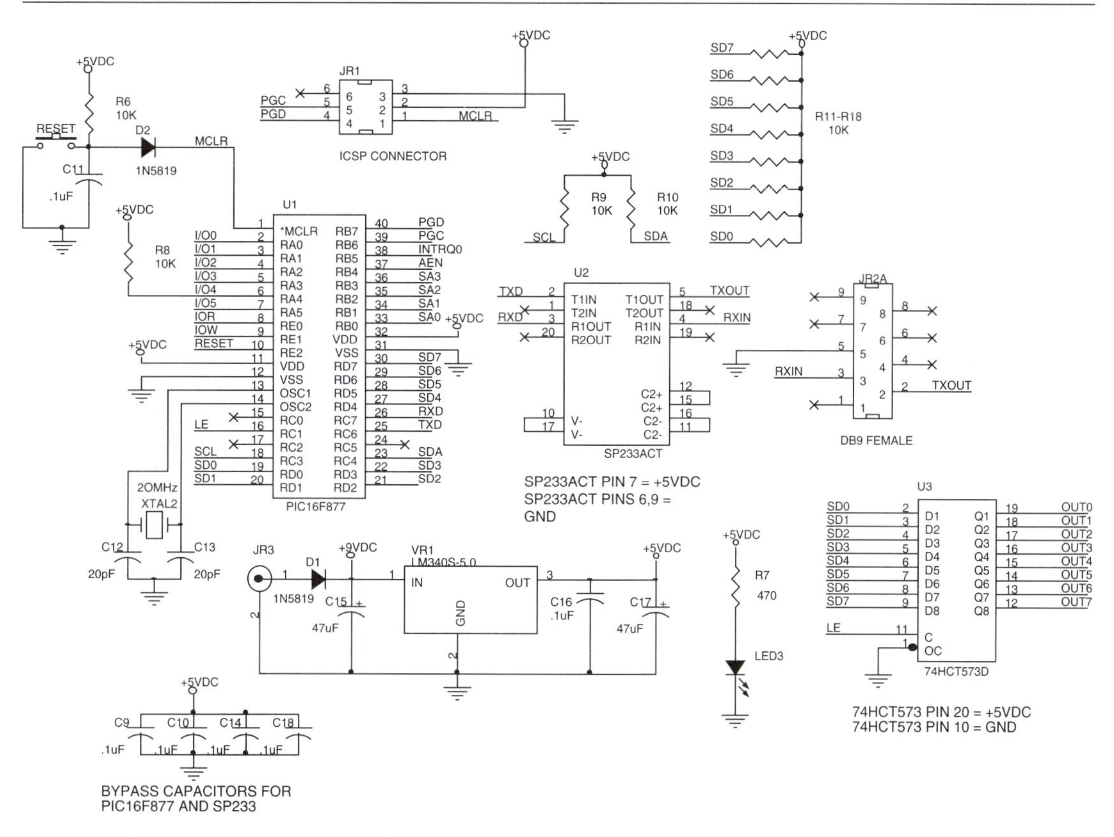

Schematic 7.3: There's no rocket science here. The Easy Ethernet CS8900A's PIC16F877 microcontroller module is just as simple as the CS8900A-CQ Ethernet engine module. Don't let the simplicity of the physical circuitry fool you. Despite the simple nature of the hardware, the Easy Ethernet CS8900A is a very powerful embedded device.

Pin RA4 of the PIC16F877 microcontroller's PORTA is an open collector pin. If you decide to use RA4 as an I/O pin, the open collector must be pulled to VCC for proper operation of the pin. Pads for a pull-up resistor, (R8), are designed into the printed circuit board layout just in case you require the use of this pin.

I²C is activated by turning on the PIC16F877's MSSP module's functionality via firmware and by installing SCL and SDA pull-up resistors R9 and R10. The I²C interface code is not included as part of the Easy Ethernet CS8900A source code. However, you can use the source code found in the I²C section of this book to implement I²C on the Easy Ethernet CS8900A.

A standard implementation of the SP233ACT results in a two-component (the SP233ACT and its power supply bypass capacitor) serial interface for the Easy Ethernet CS8900A. The SP233ACT is tied directly to the PIC16F877's USART pins RC6 and RC7. I used the Easy

Ethernet CS8900A serial interface to send test and debug messages to a terminal emulator program that was running on a personal computer. The C printf statement drives the messages I placed in various areas of the Easy Ethernet CS8900A source code. There is no RS-232 interrupt code in the current Easy Ethernet CS8900A firmware. Should you need to add a serious serial interface, the interrupt-driven RS-232 I/O code described in the RS-232 section of this book will work just fine with the Easy Ethernet CS8900A RS-232 hardware.

The 74HCT573 inputs sit on the data bus along with the CS8900A-CQ data bus pins SD0:SD7. The contents of the CS8900A-CQ data bus pins are only visible to the data bus when the AEN pin is low along with either the IOR pin or IOW pin being low as well. A write to PORTD with the CS8900A-CQ AEN, IOR and IOW lines high allows a pulse from RC1 (LE or Latch Enable) to latch the contents of PORTD into the 74HCT573. The output pins of the 74HCT573 latch are always active and the incoming latched data is always represented on the 74HCT573 output pins.

All of the componentry you see in the Easy Ethernet CS8900A schematics in Figures 7.2 and 7.3 fits on a 2.7 x 4.3 inch double-sided printed circuit board. The printed circuit board layout for the Easy Ethernet CS8900A is included on the CDROM that accompanies this book. The production version of the Easy Ethernet CS8900A is displayed in Photo 7.1.

Photo 7.1: We assembled most of the Easy Ethernet CS8900A in the RS-232 and I²C chapters.

Developing the Easy Ethernet CS8900A Firmware

Now that the hardware design is nailed down, the next step in the design of the Easy Ethernet CS8900A is to produce a group of C routines that will enable the Easy Ethernet CS8900A to speak and listen on a LAN or WAN (Wide Area Network) using some of the standard Internet protocols.

In this section, we will examine the firmware that will ultimately be programmed into the Easy Ethernet CS8900A's PIC16F877 microcontroller. There's a fair amount of C source code involved with activating the Easy Ethernet CS8900A. So, to make the job of understanding the Easy Ethernet CS8900A source code easier, the Easy Ethernet CS8900A firmware will be broken down into smaller and more manageable modules:

- The PIC16F877 microcontroller setup module
- The function prototypes
- The global variable, constant and global array definitions
- The macros
- The PacketPage Register Set definitions
- The Ethernet frame layout and definitions
- The core firmware functions
- The protocol functions (ICMP, UDP, ARP, IP, TCP)

We'll examine the code within each module line by line. I'll break the source code down into chewable snippets and then explain each line of the snippet. All hexadecimal values in the Easy Ethernet CS8900A source code are denoted with a leading "0x." There are no binary representations of numbers in the actual code that is used as a variable or argument, but you may see a binary equivalent of a hexadecimal number displayed in the comments area of the Easy Ethernet CS8900A code for clarity.

There will be times in this text that all of the code in a module will not be shown. Complete or not I'll identify the code by preceding the snippet with the banner from the module the piece of code belongs to. That should make it a bit easier to run down a line of code I'm talking about in the overall Easy Ethernet CS8900A source code listing.

If a set of definitions relates to the code snippet, I'll pull those from their actual locations in the source code and include them with the code snippet. I'll put the associated definition statements right above the code module's banner. Putting the definitions in the same area as the code snippet eliminates you from having to search through the full listing of source code to relate the definition to the code snippet. The full source code listing does have a purpose as once all of the snippets of a module have been analyzed, you can then reference the full version of the Easy Ethernet CS8900A C source code to correlate what you've learned about that specific module with the rest of the Easy Ethernet CS8900A code. A full version of the Easy Ethernet CS8900A C source code is supplied on the CD-ROM.

Here's an example:

```
int8 aux_data[16];          //tcp application received data area
int8 const telnet_banner[] = "\r\nEDTP Telnet Server>";
int8  packet[96];           //50 bytes of UDP data available
int16 tcpdatalen_out;
//*******************************************************************
//*      Application Code
//*   Your application code goes here.
//*   Following a * this module writes the hex value that follows
//*   the * to the 74HCT573 latch..
//*   Use Telnet to connect.
//*   Example:  *55 writes 01010101 to the 74HCT573 latch
//*******************************************************************
   int8 i,j;
   if (aux_data[0] == 0x0D)
      {
         j = sizeof(telnet_banner);
            for(i=0;i<j;++i)
          packet[TCP_data+i] = telnet_banner[i];
         tcpdatalen_out = j;
      }
```

Code Snippet Example: I'll put some additional comments here.

The four global variable definition statements that are involved with the Code Snippet Example are placed before the banner identifying the location of the code snippet in the full listing. If any local variables are associated with the code snippet, I'll place them just above the code snippet and below the banner (int8 i,j;). Having everything that relates to the code snippet in a standardized location keeps you from having to search the full listing for the code snippet's parent code module and supporting variables.

The Easy Ethernet CS8900A firmware was written using the Custom Computer Services C Compiler. The Custom Computer Services C Compiler is totally tilted towards the PIC microcontroller; therefore, some of the built-in functions may not be available in other C compiler packages. The Custom Computer Services C Compiler built-in functions are added as a convenience for PIC C programmers and can easily be ported to native C code for use with other microcontroller C compilers.

To enhance your understanding of how the Easy Ethernet CS8900A firmware actually works, I'll include personal computer screen shots of actual MPLAB ICE 2000 PIC16F877 emulation sessions. The MPLAB ICE 2000 screen shots sometimes show a bit too much information, some of which is not related to the idea we're pursuing at the time. So, where it is appropriate, I will extract data from an MPLAB ICE 2000 PIC16F877 emulation session and show it to you in text format. You'll see me do this most often when displaying the data within a PIC16F877 hexadecimal memory dump.

During our discussions of the Easy Ethernet CS8900A code, I'll also employ the services of Network Associate's Sniffer. The Sniffer personal computer screen shots provide a detailed look at the contents of an Ethernet frame. I used the Sniffer to verify the Easy Ethernet CS8900A firmware as it was developed. One good example of the usefulness of the Sniffer during the code development process was the Sniffer's ability to certify the validity of the checksum routines I had written. The Sniffer was also instrumental in the debugging process. While I was in the process of porting the Easy Ethernet CS8900A code from the original assembler version I had written, the Easy Ethernet CS8900A seemed to be dead on the network. An inspection of the frames I trapped with the Sniffer showed that the Easy Ethernet CS8900A was answering ARP requests but doing nothing else. Using that information, I looked at a memory dump I captured with the MPLAB ICE 2000 and eventually found a bug in the new C code that was not transferring frames from the CS8900A-CQ's frame buffer to the PIC16F877's buffer correctly. For those of you that might be interested in how this was done in assembler, I've also included the full version of the MPLAB assembler source code on the CD-ROM.

One more debugging tactic I employed was the use of messages sent from the Easy Ethernet CS8900A's RS-232 port. You'll see commented out (preceded by //) printf statements announcing the start or completion of an event in parts of the source code. You may uncomment these if they help you better understand what and where things are happening in the flow of the Easy Ethernet CS8900A code.

Setting up the PIC16F877 Microcontroller

One of the advantages of using the Custom Computer Services C Compiler is the abundance of built-in functionality aimed towards the resources inside the PIC microcontroller. The Custom Computer Services C Compiler allows the use of a "wizard" to preconfigure some of the internal resources of the selected PIC. If you're an old-hand at using PICs, you can also write the C declarations in yourself to tailor the PIC manually.

```
#include <16f877.h>
#device *=16
#include <f877.h>
#use delay(clock=20000000)
#fuses HS,WRT,NOWDT,NOPUT,NOPROTECT,NOBROWNOUT,NOLVP,NOCPD
//#use i2c(Master,Slow,sda=PIN_C4,scl=PIN_C3,force_hw)
#use rs232(baud=57600, xmit=PIN_C6,rcv=PIN_C7)
#id 0x0802

#use fast_io(A)
#use fast_io(B)
#use fast_io(C)
#use fast_io(D)
#use fast_io(E)
```

Code Snippet 7.1: The PIC16F877's I²C can be activated by simply removing the comment characters (//) from the beginning of the #use I²C statement.

The snippet of code you see in Code Snippet 7.1 was partially generated with the Custom Computer Services C Compiler project wizard. The #include <16f877.h> statement brings in definitions that refer to the physical architecture of the PIC16F877 microcontroller from a file that is included with Custom Computer Services C Compiler. The PIC microcontroller include files that come with the Custom Computer Services C Compiler normally contain information that relates a human name to a number or Boolean function that can be used to control the actions of a PIC16F877 internal resource. For instance, to turn off all analog inputs, a 0x86 must be written to the PIC16F877's ADCON1 register. The 16f877.h file includes a definition called NO_ANALOGS. Code Snippet 7.2 is a look into the Easy Ethernet CS8900A's main C function that shows the NO_ANALOGS definition found in the include file, and it shows what was coded into the Easy Ethernet CS8900A source code when I told the Custom Computer Services C Compiler project wizard to disable the analog inputs and configure them as standard I/O pins.

```
#define NO_ANALOGS                0x86          // None
//**********************************************************************
//*   Absolute Start Point
//**********************************************************************
void main() {
   int16 scratch16;

   setup_adc_ports(NO_ANALOGS);
```

Code Snippet 7.2: The setup_adc_ports function is one of many built-in Custom Computer Services C Compiler functions. The A/D setup statement was generated entirely by the Custom Computer Services C Compiler project wizard.

The f877.h include file is a homegrown include file that I assembled by using parts of the original Microchip MPLAB assembler PIC16F877 include file that comes with MPLAB. The f877.h include file augments the Custom Computer Services C Compiler 16f877.h include file and contains some PIC16F877 definitions not found in the Custom Computer Services C Compiler include file that I needed for the Easy Ethernet CS8900A.

Yet another feature of the Custom Computer Services C Compiler is its ability to relieve the C programmer from having to keep up with what's where in PIC memory. The device #device *=16 turns on full 16-bit PIC RAM pointers. Note that as you travel through the Easy Ethernet CS8900A code, I never specify a memory location to hold a value.

There are many built-in functions that rely on the microcontroller clock speed. My favorite built-in Custom Computer Services C Compiler functions are the delay_ms, delay_us and delay_cycles functions. I absolutely hate writing timing and delay routines because if you want your timing to be precise you are compelled to count the microcontroller instruction cycles inside of each and every timing loop instruction. With the Custom Computer Services C Compiler, once the compiler is informed as to what the clock speed is with the #use delay(clock=20000000) statement, all of the delay routines I just mentioned are automatically calculated using the PIC microcontroller cycle time for a 20 MHz microcontroller clock. For instance, to delay for 1 ms, the C statement is delay_ms(1). Want to delay for exactly 1 second? Then all that you have to enter is: delay_ms(1000) (1000 ms = 1 second). If you're familiar with PIC assembler, you know that an NOP is a do nothing instruction that kills time for one PIC microcontroller instruction cycle. An NOP in Custom Computer Services C Compiler lingo is delay_cycles(1).

The baud rate register value for the PIC serial port is also calculated using the microcontroller clock speed as a reference. The C statement #use rs232(baud=57600, xmit=PIN_C6,rcv=PIN_C7) sets the USART up for asynchronous operation with a speed of 57600 bps. In addition, the #use rs232 statement allocates the serial port's transmit and receive pins. In the Easy Ethernet CS8900A design, the serial port pins match the USART transmit and receive pins. Using the Custom Computer Services C Compiler, it is possible to allocate almost any pin for serial port duty even if the PIC being used doesn't have a USART.

The I²C statement that activates the PIC's I²C engine (#use i2c(Master,Slow,sda=PIN_C4,scl=PIN_C3,force_hw) is commented out as there is no I²C code incorporated into this version of the Easy Ethernet CS8900A firmware. Like the #use rs232 statement, the #use i2c statement also uses the microcontroller's clock speed to time the I²C data stream.

Most of the PIC16F877's I/O pins are bidirectional and that is true for any other PIC microcontroller. For each of a PIC's bidirectional I/O pins, a hardware mechanism is in place that allows the PIC programmer to determine which direction the PIC I/O pin will operate in, input or output. If you choose to ignore the PIC's I/O direction mechanism and code accordingly (#use standard_I/O(X) where X is a port A, B, C, and so forth), the Custom Computer Services C Compiler will automatically adjust the pin your C statement is addressing for input or output operation depending on the C statement you are using. I've chosen to pay attention to the PIC I/O pin direction mechanism. However, I've also chosen to turn the PIC I/O pin direction wheel myself with the #use fast_I/O(X) statements. I've written the macros dataport_in and dataport_out to instruct the PIC16F877 as to how to program the direction of a particular I/O port's pins.

When I first began to write PIC code for the public, I used to enter a date code into the PIC's ID words. I used this date code to determine which revision of firmware was in the customer's PIC and I still use that identification method today. The #id 0x0802 means that I touched the Easy Ethernet CS8900A code last on August 2nd.

Carving up the PIC16F877's Memory Resources

Function Prototypes

In C, all of the identifiers including functions and variables must be declared before they are used. That's what the function prototype area is all about. Using a function prototype allows the C compiler to check for the correct number and type of arguments within each function that will be called in the course of the program flow.

```
//******************************************************************
//*   FUNCTION PROTOTYPES
//******************************************************************
void application_code();
void tcp();
void assemble_ack();
void get_frame();
void setipaddrs();
void cksum();
void echo_packet();
void send_tcp_packet();
void arp();
void icmp();
void udp();
```

Code Snippet 7.3: There's not much to say right now about what you see in this code snippet. However, you will see each of the functions again.

The function prototypes in Code Snippet 7.3 give you a feel for what is to come. To keep things as simple as possible, I purposely don't have any functions that return values. I chose to allow each function to stand alone as much as possible, and use global variables if data had to flow from one function to another.

Defining the Variables

If you study the original assembler version of this code, you'll see that every variable had to be assigned to a particular memory location in the PIC16F877 memory map. Yes, I could have used each memory location for more than one variable, but that would have made the assembler code hard to read, hard to follow and hard to maintain. Using C, all I have to do is allocate each variable I want to use by name and by type. Bits are identified as *int1*. Bytes are tagged as *int8*, and word and double words are identified as *int16* and *int32*, respectively. The byte, word and double word identifiers can also be used to build arrays. For BASIC programmers out there, an array in C is exactly like an array in BASIC as far as the human programmer is concerned. As you can see in Code Snippet 7.4, I've tried to name the variables so they will have some meaning when you see them in their code modules.

```
//******************************************************************
//*   PIC16F87X Global Variable Definitions
//******************************************************************
int8 ppoffsetH,ppoffsetL,cntr,byteout;
int8 aux_data[16];           //received data area
int8 *addr;
int8 data_H,data_L;
int16 i,txlen,rxlen,chksum16,hdrlen,tcplen,tcpdatalen_in;
int16 tcpdatalen_out,ISN,portaddr,ip_packet_len;
```

```
int32 hdr_chksum,my_seqnum,client_seqnum,incoming_ack,expected_ack;
int1 synflag,finflag,hexflag;
```

Code Snippet 7.4: The "" is an indirection symbol. Adding the "*" to the variable *addr defines addr as a pointer.*

The variables defined in the code you see in Code Snippet 7.4 reside in the PIC16F877 RAM. It is also possible to place data into the flash program memory area by defining the data as constant.

```
//******************************************************************
//*    TELNET SERVER BANNER STATEMENT CONSTANT
//******************************************************************
int8 const telnet_banner[] = "\r\nEDTP Telnet Server>";

//******************************************************************
//*    Receive a Frame
//******************************************************************
void get_frame()

    printf("trashed\r\n");
```

Code Snippet 7.5: I exercised editorial privilege and threw in the printf tidbit. I've actually run out of PIC program memory because of having too many printf statements in a program.

Hex Dump 7.1 is a PIC16F877 program memory capture of the statements shown in Code Snippet 7.5. The telnet_banner constant array elements are stored as a lookup table in PIC16F877 program memory. The 0x34 preceding each ASCII character is a PIC assembler RETLW instruction, and that usually is a good indicator that the data is in a standard PIC lookup table format. If you pick through the dump you'll eek out "EDTP Telnet Server>."

I included the printf statement found in the get_frame function to show how printf data was stored in the PIC16F877's program memory area. Can you see "trashed" followed by a carriage return and line feed in Hex Dump 7.1?

```
Address                                                  ASCII

 0008   340D 340A 3445 3444 3454 3450 3420 3454 .4.4E4D4 T4P4 4T4
 0010   3465 346C 346E 3465 3474 3420 3453 3465 e4l4n4e4 t4 4S4e4
 0018   3472 3476 3465 3472 343E 3400 100A 108A r4v4e4r4 >4.4....
 0020   110A 0782 3474 3472 3461 3473 3468 3465 ....t4r4 a4s4h4e4
 0028   3464 340D 340A 3400 3000 1683 0088 30FA d4.4.4.4 .0.....0
```

Hex Dump 7.1: Carriage return and line feed (/r/n) begin the Telnet banner message. The 0x34 preceding each character is a PIC RETLW assembler instruction.

Without writing a single line of C, we already know that we'll need to establish an IP address and a hardware (MAC) address for the Easy Ethernet CS8900A. Any valid IP address will work, and your choice of IP addresses depends upon the network on which the Easy Ethernet CS8900A will participate. If you don't plan to have your Ethernet packets leave your home LAN, 192.168.XXX.XXX is a good choice as it is one of the addresses reserved for private networks.

```
//*******************************************************************
//*    IP ADDRESS DEFINITION
//*    YOU MAY CHANGE THIS TO ANY VALID IP ADDRESS
//*******************************************************************
int8 MYIP[4] = { 192,168,0,150 };
```

Code Snippet 7.6: The array MYIP contains the logical or protocol address, which isn't written in stone and can identify any number of hosts on a network.

The IP address of the Easy Ethernet CS8900A, 192.168.0.150, is specified in the IP ADDRESS DEFINITION area of the code. Hexadecimal notation could have been used in the definition, but it's not as easy to read and remember as dotted decimal notation. The Easy Ethernet CS8900A IP address is held in a byte array called MYIP. Placing the IP in an array within the operating firmware allows easy access to the IP address's individual components and makes changing the IP address simple.

Dotted decimal notation of the IP address is for human consumption. The PIC16F877 wants to see the IP address as a series of binary values as shown in the PIC16F877 RAM dump in Hex Dump 7.2.

```
Address   00 01 02 03 04 05 06 07 08 09 0A 0B 0C 0D 0E 0F        ASCII

  0010    00 00 00 00 00 00 00 00 90 00 00 00 00 00 00 00  ........ ........
  0020    01 12 02 AA FA 0A 00 00 00 00 00 00 00 00 00 00  ........ ........
  0030    00 00 00 00 5A 01 00 00 00 D3 4A 00 4A 00 4A 00  ....Z... ..J.J.J.
  0040    5B B3 00 00 15 00 00 00 01 00 06 00 98 1F 29 00  [....... ......).
  0050    9D 4C 07 00 1E 00 07 00 1E 23 22 BB 1F 00 07 5F  .L...... .#"...._
  0060    1F 00 07 00 00 C0 A8 00 96 00 00 45 44 54 50 05  ........ ...EDTP.
```

Hex Dump 7.2: The Easy Ethernet CS8900A's IP address of 192.168.0.150 begins at offset 0x0065, while the Easy Ethernet CS8900A's MAC address (00EDTP) is found beginning at offset 0x0069.

The Easy Ethernet CS8900A's IP address is assigned according to the needs and requirements of the network the Easy Ethernet CS8900A will attach to. Other devices on the network will need to know the hardware address of the Easy Ethernet CS8900A to communicate with it. So, the next order of business is to assign a hardware or MAC (Media Access Control) address to the Easy Ethernet CS8900A. The hardware address is normally a purchased item that is regulated by the IEEE. If you plan to use the Easy Ethernet CS8900A in a commercial environment, you will need to purchase a unique hardware identifier. There's a blurb in the CS8900A-CQ datasheet that tells you who to contact to purchase an OUI, or Organizationally

Unique Identifier. The OUI forms a basis, that when mixed with data of your choice, becomes your equipment's assigned hardware address. Basically, you mix your OUI with each piece of hardware's serial number or such to create a unique hardware address for every piece of equipment branded with your OUI combination. The idea is to not have any identical NIC (Network Interface Card) hardware addresses. You can normally find this OUI, printed on a paper tag glued to most commercial Ethernet NICs. Issuing ARP –a in a command window of a network enabled Windows operating system will also show you the OUI if the NIC has registered itself in the ARP cache. Another common way to see the NIC's MAC address is to issue the IPCONFIG /ALL command. As we progress and view some network dumps, you'll notice that the personal computer Ethernet NIC I'm using has a common set of digits that identify it as an SMC NIC. I'm not allowing my Easy Ethernet CS8900A to interface to the Internet directly just yet, so a homebrewed MAC address of '00EDTP' is assigned to the Easy Ethernet CS8900A in the CS8900A-CQ driver firmware included with this book.

```
//******************************************************************
//*    HARDWARE (MAC) ADDRESS DEFINITION
//*    YOU MAY CHANGE THIS TO ANY VALID MAC ADDRESS
//******************************************************************
int8 MYMAC[6] = { 0,0,'E','D','T','P' };
```

Code Snippet 7.7: You can see the MYMAC array beginning at memory location 0x0069 in the PIC16F877 RAM hex dump shown in Hex Dump 7.2.

Once again, I've chosen a byte array to store the six bytes of the Easy Ethernet CS8900A's MAC address. However, unlike the IP address, the MAC address is stored within the CS8900A-CQ register set. The CS8900A-CQ doesn't store the MAC internally as you think it would, and having each MAC address component in its own individual array "container" will help make the process of placing the MAC address inside the CS8900A-CQ a bit easier.

The MAC address is stored in the CS8900A-CQ in the Individual Address (IA) register set beginning at PacketPage address 0x0158 as shown in Figure 7.7. The layout of the IA register set makes it necessary to place the most significant octet of the MAC address (MYMAC[0]) into the least significant octet of the IA register set, and so on. I used the Custom Computer Services C Compiler's built-in *make16* function and a 16-bit scratch register to put the MAC octets in the right order before loading them into the CS8900A-CQ's IA register set. The *make16* function takes two bytes and combines them into a 16-bit word (make16(0x12,0x34) = 0x1234).

INDIVIDUAL ADDRESS (IEEE ADDRESS)

	MYMAC[5]	MYMAC[4]	MYMAC[3]	MYMAC[2]	MYMAC[1]	MYMAC[0]
REGISTER	0x015D	0x015C	0x015B	0x015A	0x0159	0x0158
MAC DATA	P	T	D	E	0	0

Figure 7.7: There is really a great deal of order in this. The MAC address will be transmitted beginning with the octet (byte) at IA location 0x0158, and then 0x0159 will be transmitted, and so forth.

In a personal computer environment, the NIC's EEPROM would hold the MAC address and other information that on the Easy Ethernet CS8900A will be loaded from the bowels of the PIC16F877 microcontroller. Code Snippet 7.8 details the code that performs the loading of the CS8900A-CQ IA register set with the MAC address.

A temporary 16-bit memory location (scratch16) is used to assemble two octets of the MAC address into a single 16-bit word. The reasons for doing this include packing the MAC octets in the correct order, and the WPP function is written to accept only 16-bit arguments.

```
//*******************************************************************
//*   PacketPage Internal Register Definitions
//*******************************************************************
#define ppageIA   0x0158  //Individual Address
int16 scratch16;
//*******************************************************************
//*   Load the CS8900 IA
//*   INDIVIDUAL ADDRESS LAYOUT IN CS8900
//*******************************************************************
      scratch16 = make16(MYMAC[1],MYMAC[0]);
      WPP(ppageIA,scratch16);
      scratch16 = make16(MYMAC[3],MYMAC[2]);
      WPP(ppageIA+2,scratch16);
      scratch16 = make16(MYMAC[5],MYMAC[4]);
      WPP(ppageIA+4,scratch16);

//uncomment this code to see the MAC address as it has been entered
      //RPP(ppageIA);
      //printf("%x%x \r\n",data_H,data_L);
      //RPP(ppageIA+2);
      //printf("%x%x \r\n",data_H,data_L);
      //RPP(ppageIA+4);
      //printf("%x%x \r\n",data_H,data_L);
//end commented code
      //printf("CS8900A-CQ MAC Address LOADED.\r\n");
```

Code Snippet 7.8: You can use the extra debug code to verify that the MYMAC array elements were loaded into the CS8900A-CQ's IA register set in the correct order.

After ordering the IA register bytes, the WPP (Write PacketPage) function writes the newly created 16-bit MAC address fragment into the proper slots in the IA register set. As you will see throughout the CS8900A-CQ code, all CS8900A-CQ register accesses are word (16-bit) accesses.

The WPP function uses the Custom Computer Services C Compiler function "make8," which breaks down a 16-bit or 32-bit variable into a byte. You can choose which byte of the word/double word variable you want to keep. For instance, the first line of code inside the

WPP function breaks out and keeps the least significant byte of the 16-bit ppoffset value. The second line of WPP retains the upper byte of ppoffset. The function argument ppoffset in this case is the 16-bit address of an octet in the PacketPage IA register set. PPWrite uses the resulting high-order and low-order bytes sorted out by WPP to call the WpppL and WpppH macros that actually write the data into the designated CS8900A-CQ registers.

Note that in Code Snippet 7.9, the Custom Computer Services C Compiler make8 function is used to break the 16-bit arguments ppoffset and datum into octets (bytes) that are passed to the PPWrite function via global variables ppoffsetL, ppoffsetH, data_L and data_H.

```
int8  ppoffsetH,ppoffsetL;
int8 data_H,data_L;
//******************************************************************
//*   WPP (Write PacketPage)
//*   Writes Data (datum) at PacketPage Offset
//*   PPoffset = PacketPage Data Offset
//******************************************************************
void WPP(int16 ppoffset, int16 datum)
{
   ppoffsetL = make8(ppoffset,0);
   ppoffsetH = make8(ppoffset,1);
   data_L = make8(datum,0);
   data_H = make8(datum,1);
   PPWrite();
}
#define  pageport_Ptr  0x0A   //PacketPage Pointer
//******************************************************************
//*   PPWrite (PacketPage Write)
//*   Writes Data to ppoffsetH/L
//******************************************************************
void PPWrite()
{
   dataport_out;
   WpppL(pageport_Ptr,ppoffsetL);
   WpppH(pageport_Ptr,ppoffsetH);
   WpppL(pageport_Data0,data_L);
   WpppH(pageport_Data0,data_H);
}
```

Code Snippet 7.9: The arguments for the WPP function, ppoffsetL, ppoffsetH, data_L and data_H, are previously defined in the PIC16F877 global variable definitions in Code Snippet 7.4.

The PPWrite function makes sure that the PIC16F877 data bus I/O pins are configured as output pins before invoking a series of macros to write the data to the CS8900A-CQ's PacketPage memory. We haven't discussed the Easy Ethernet CS8900A macros yet but you can easily use substitution to move the PacketPage Addresses and data from the PPWrite function to the macros in Code Snippet 7.10.

```
//*******************************************************************
//*   WpppL (Write packetpage port Low)
//*   Writes low byte to specified PacketPage Port
//*   pp_port = PacketPage Port - datum = data to write
//*******************************************************************
#define  WpppL(pp_port,datum) writeaddrport(pp_port);          \
                              writedataport(datum);            \
                              clr_aen;                         \
                              clr_iow;                         \
                              delay_cycles(1);                 \
                              set_iow;                         \
                              set_aen;
//*******************************************************************
//*   WpppH (Write packetpage port High)
//*   Writes high byte to specified PacketPage Port
//*   pp_port = PacketPage Port - datum = data to write
//*******************************************************************
#define  WpppH(pp_port,datum) writeaddrport(pp_port+1);        \
                              writedataport(datum) ;           \
                              clr_aen;                         \
                              clr_iow;                         \
                              delay_cycles(1);                 \
                              set_iow;                         \
                              set_aen;
```

Code Snippet 7.10: You can already see how easy it is to follow the C source code when the abstract machine terms are manipulated as human language.

I inserted some extra "commented out" code (code preceded by //) that uses the Easy Ethernet CS8900A's onboard RS-232 circuitry to show you what is really inside the IA register set. Set your personal computer terminal emulator (HyperTerminal, Tera Term Pro, and so forth) for no flow control, 57600 bps, 8 data bits, no parity and 1 stop bit to get a human readable look at what was inserted into the IA register set.

The port address definition you see in Code Snippet 7.11 will be used by code in our TCP/IP module. It is an arbitrary address and there's nothing significant about it except that it is not in what is called the well-known port list. You can find MY_PORT_ADDRESS at memory locations 0x004C and 0x004D in the PIC16F877 RAM hex dump shown in Hex Dump 7.2.

```
//*******************************************************************
//* PORT ADDRESS DEFINITION
//*   YOU MAY CHANGE THIS TO ANY VALID PORT ADDRESS
//*******************************************************************
#define  MY_PORT_ADDRESS      0x1F98  // 8088 DECIMAL
```

Code Snippet 7.11: Depending upon how much you already know about TCP/IP, this definition may or may not mean much to you right now. If you're in the dark about it, I'll show you the light when we discuss TCP/IP.

The Easy Ethernet CS8900A Macros

There are lots of port I/O operations performed in the Easy Ethernet CS8900A firmware. Rather than try to remember which port is data and which port is address, I wrote macros with descriptive names that performed the desired I/O functions using the datasheet names and associated built-in Custom Computer Services C Compiler functions.

For the programmer, the idea behind creating macros is to ease the firmware design and coding process by providing easy-to-remember labels for often-used functions. Writing code rich with macros gives the user/analyst/project builder studying the source code listing an easier-to-read description of the flow of the firmware. Another big advantage to using macros is that if something needs to be changed like a port or a pin location and the information resides within a macro it need only be changed at one point in the program, which is inside the macro.

I like writing macros. I know that as I'm writing them they will in the end save me lots of physical typing and programming think time as I build code around them. I like to include as many microcontroller port and pin definitions as I can in the macro area as the macros depend heavily on these base definitions. The advantage to having the microcontroller port and pin definitions handy in the macro area is that a port or pin change in the code only requires changing the port or pin definition in one place, the macro area. Another advantage to listing port and pin connections in the macro area is that one could look at the pin descriptions and pinouts in the macro area and not have to consult a schematic for connection points in the actual circuitry. Knowing the microcontroller is a PIC16F877 and it is connected in a standard manner to a CS8900A-CQ, you can tell a great deal about the Easy Ethernet CS8900A circuitry using only what you see in Code Snippet 7.12.

Lots of times, folks that are recreating my projects ask for a "corrected" schematic when they can't get the project to work. I learned long ago that the schematic is only part of the documentation needed to assemble hardware. So, every time I design something that depends on a microcontroller, I always provide a list of the microcontroller port and pin definitions within the source code. Listing the hardware connections within the source code allows the user/builder to verify the schematic connections against the logical operations of the microcontroller firmware.

```
//DEFINITIONS FROM f877.h
#byte PORTA    =0x005
#byte PORTB    =0x006
#byte PORTC    =0x007
#byte PORTD    =0x008
#byte PORTE    =0x009
//DEFINITONS FROM 16f877.h
#define  PIN_B4   52
#define  PIN_B5   53
#define  PIN_C1   57
```

```
#define  PIN_E0   72
#define  PIN_E1   73
#define  PIN_E2   74
//******************************************************************
//    MACROS AND PORT/PIN DEFINITIONS
//******************************************************************
#define  dataport PORTD
#define  addrport PORTB
#define  cntlport PORTE

#define  IOR      PIN_E0
#define  IOW      PIN_E1
#define  RESET    PIN_E2
#define  AEN      PIN_B4
#define  LE       PIN_C1
```

Code Snippet 7.12: The names in the definitions are all homebrewed. The port and pin equates are drawn from the 16f877.h and f877.h include files.

The reason the PIC16F877 port definitions use a *#byte X = Y* statement instead of a #define statement is that the numbers being equated to the port names are the actual port locations in the PIC16F877 memory map. The PIN_XX definitions are native to the Custom Computer Services C Compiler. In my years of writing code and building microcontroller-based gadgets, I've found that a bit of sanity is kept by keeping the names of the ports and pins in the source code identical to their associated names in the schematics.

The port and pin definitions establish a base for the rest of the macros. To keep with the human readable idea, I take function names and equate them to more meaningful names. For instance, in Code Snippet 7.13 I've taken common built-in Custom Computer Services C Compiler functions and simply renamed them.

```
//******************************************************************
//    MACROS AND PORT/PIN DEFINITIONS
//******************************************************************
#define  writedataport(datum) output_d(datum);
#define  writeaddrport(datum) output_b(datum | 0xF0);
#define  readdataport         input_d()
#define  dataport_out         set_tris_d(0x00)
#define  dataport_in          set_tris_d(0xFF)
```

Code Snippet 7.13: Since the PIC16F877's PORTD has been defined previously as the data port, writedataport makes a bit more sense than output_d when you're trying to read through the Easy Ethernet CS8900A source code.

Sometimes a simple renamed macro definition like the ones in Code Snippet 7.13 can be structured to reap an additional benefit over the stock function it is emulating. Using an inclusive OR (|) I took the liberty of setting the upper nibble of the output data to 1111 in the *writeaddrport(datum)* macro as the upper four bits of PORTB should remain high when the address information is written to the lower nibble of the PIC16F877's PORTB. If you check this against the PIC16F877 module of the Easy Ethernet CS8900A schematic you'll find that AEN occupies the RB4 bit position and its output logic level should not be changed by the address information entered for output on PORTB. So, instead of having to remember to OR the PORTB data with 0xF0 every time data is written to PORTB, I simply included the rule in the macro definition.

If it's good for ports, it's usually just as good for pins. The aforementioned axiom of wisdom is proven in the pin macro set of Code Snippet 7.14. Every time data is transferred between the microcontroller and the CS8900A-CQ the IOR, IOW and AEN lines of the CS8900A-CQ are used. I used macros named for their functionality to clear and set these control lines.

```
//****************************************************************
//    MACROS AND PORT/PIN DEFINITIONS
//****************************************************************
#define  clr_reset    output_low(RESET)
#define  clr_ior      output_low(IOR)
#define  clr_iow      output_low(IOW)
#define  clr_aen      output_low(AEN)
#define  set_reset    output_high(RESET)
#define  set_ior      output_high(IOR)
#define  set_iow      output_high(IOW)
#define  set_aen      output_high(AEN)
#define  latchdata    output_high(LE);   \
                      delay_us(1);       \
                      output_low(LE);
```

Code Snippet 7.14: I was just reading a magazine answer to a question of which programming language to use to program small microcontrollers like the PIC. I was amazed to read the "expert" tell the reader that C was difficult to learn. It doesn't get any easier than this. And, if you don't believe that, compare the work it takes to write an assembler routine that only a single line of C source code performs. In many cases the C compiler writes better code behind the C than a human can using assembler.

Built-in functions and macros make life easy on the C ranch. I've left more than enough latch-enable time between the pin toggle statements in the latchdata macro you see at the bottom of Code Snippet 7.14.

Macros are also useful when an often-used function consists of multiple lines of code. For instance, the LE (Latch Enable) control line connecting the microcontroller to the 74HCT573 transparent latch must be toggled every time the latch is accessed. It's a trivial three lines of C source code but it's worth placing into a macro. Now instead of three lines of nondescript code, the word "latchdata" is all that's needed to manipulate the LE control line.

Remember the bit definitions in the global variable definitions (*int1, synflag, finflag, hexflag*)? Well, macros work for them too. There is a flag bit that needs to be set and cleared in the UDP application code. It's easy enough to set and clear the hexflag using C constructs, but it's better reading when you mean what you say. Check out the flag raising and lowering macro code in Code Snippet 7.15.

```
int1 hexflag;
//*******************************************************************
//    MACROS AND PORT/PIN DEFINITIONS
//*******************************************************************
#define  set_hex  hexflag=1
#define  clr_hex  hexflag=0
```

Code Snippet 7.15: Which one would you rather use?

Even with the superb built-in functions offered up by the Custom Computer Services C Compiler, sometimes you still have to roll your own. When we visit TCP/IP land you'll find that TCP/IP likes 32-bit numbers. 32-bit numbers are no problem that can't be handled with a C statement or C macro. Code Snippet 7.16 is a macro I put together to disassemble a 32-bit number the way TCP/IP wants it broken down.

```
#define set_packet32(d,s)  packet[d] = make8(s,3);     \
                           packet[d+1] = make8(s,2);   \
                           packet[d+2] = make8(s,1);   \
                           packet[d+3]= make8(s,0);
```

Code Snippet 7.16: This macro stuffs a 32-bit number into a byte array. You'll find this macro twiddling TCP sequence numbers in the TCP module of the Easy Ethernet CS8900A firmware.

The real work of reading and writing the CS8900A-CQ is done by a quartet of macros. The RpppL (Read PacketPage Port Low) and RpppH (Read PacketPage Port High) macros in Code Snippet 7.17 are very simple to read and understand when you know what the macros inside the RpppL and RpppH macros do. Even if you didn't know about the I/O pin clear and set macros, the language of the macros give their intent away.

```
//*****************************************************************
//*   RpppL (Read packetpage port Low)
//*   Reads low byte of specified PacketPage Port
//*   pp_port = PacketPage Port - dest = where to store data
//*****************************************************************
#define  RpppL(pp_port,dest)      writeaddrport(pp_port);     \
                                  clr_aen;                    \
                                  clr_ior;                    \
                                  delay_cycles(1);            \
                                  dest = readdataport;        \
                                  set_ior;                    \
                                  set_aen;

//*****************************************************************
//*   RpppH (read packetpage port High)
//*   Reads high byte of specified PacketPage Port
//*   pp_port = PacketPage Port - dest = where to store data
//*****************************************************************
#define RpppH(pp_port,dest)       writeaddrport(pp_port+1);   \
                                  clr_aen;                    \
                                  clr_ior;                    \
                                  delay_cycles(1);            \
                                  dest = readdataport;        \
                                  set_ior;                    \
                                  set_aen;
```

Code Snippet 7.17: I can't assume everyone reading this section knows what the backward slash (\) behind some of the macro lines is for. The backward slashes (\) are macro statement continuation symbols. If another macro statement needs to follow, the backward slash (\) is added to signify that to the C compiler. Note that single-line macros and the last line of the multilined macros do not require the continuation slash.

Having the base macros already in place makes writing other macros based on them very easy to do. In fact, the PacketPage write macros in Code Snippet 7.18 were generated from copies of the RpppL/RpppH macros you see in Code Snippet 7.17.

```
//*****************************************************************
//*   WpppL (Write packetpage port Low)
//*   Writes low byte to specified PacketPage Port
//*   pp_port = PacketPage Port - datum = data to write
//*****************************************************************
#define WpppL(pp_port,datum)      writeaddrport(pp_port);     \
                                  writedataport(datum);       \
                                  clr_aen;                    \
                                  clr_iow;                    \
                                  delay_cycles(1);            \
```

```
                                  set_iow;                       \
                                  set_aen;
//*****************************************************************
//*   WpppH (Write packetpage port High)
//*   Writes high byte to specified PacketPage Port
//*   pp_port = PacketPage Port - datum = data to write
//*****************************************************************
#define WpppH(pp_port,datum)      writeaddrport(pp_port+1);   \
                                  writedataport(datum) ;      \
                                  clr_aen;                    \
                                  clr_iow;                    \
                                  delay_cycles(1);            \
                                  set_iow;                    \
                                  set_aen;
```

Code Snippet 7.18: When a good code foundation is laid such as with our macro base, the further you get into writing the code the more it seems the code is writing itself.

It's never too late to add a macro. Many of the macros in the Easy Ethernet CS8900A firmware were added after the code was "finished" and working.

Defining the CS8900A-CQ PacketPage Register Set

You've already figured out that there are a bunch of CS8900A-CQ registers. The best way to keep up with them is to predefine them with names or labels that relate to the register's function or name. In the CS8900A-CQ driver firmware beginning with the PacketPage I/O Port Definitions, all of the internal CS8900A-CQ registers are listed and defined, whether we use them in this design or not.

You were briefly introduced to the CS8900A-CQ PacketPage registers at the beginning of our CS8900A-CQ discussion. In Code Snippet 7.19, I've mapped out the PacketPage port structure so it can be used in our Easy Ethernet CS8900A firmware.

```
//*****************************************************************
//*   PacketPage I/O Port Definitions
//*****************************************************************
#define  pageport_RxTxData0   0x00  //Receive/Transmit data Port 0
#define  pageport_RxTxData1   0x02  //Receive/Transmit data Port 1
#define  pageport_TxCmd       0x04  //Transmit Command
#define  pageport_TxLen       0x06  //Transmit Length
#define  pageport_ISQ         0x08  //Interrupt Status Queue
#define  pageport_Ptr         0x0A  //PacketPage Pointer
#define  pageport_Data0       0x0C  //PacketPage Data Port 0
#define  pageport_Data1       0x0E  //PacketPage Data Port 1
```

Code Snippet 7.19: There's nothing new here. Again, it's easier to remember a name associated with a function than remember a number associated with a function.

In the Easy Ethernet CS8900A source code labels beginning with pageport represent the 16 base PacketPage I/O ports. PacketPage internal registers that are accessed using the PacketPage I/O ports are prefixed by ppage. The ppage area is where the CS8900A-CQ initialization and operational registers are located. Also, you'll find the counters and status registers in the group of PacketPage registers you see listed in Code Snippet 7.20.

```
//******************************************************************
//*   PacketPage Internal Register Definitions
//******************************************************************
#define  ppageEISA       0x0000  //EISA Registration number of CS8900
#define  ppagePID        0x0002  //Product ID Number
#define  ppageBaseIO     0x0020  //I/O Base Address
#define  ppageINT        0x0022  //Interrupt number (0,1,2, or 3)
#define  ppageBaseMemory 0x002C  //20-bit Memory Base address register
#define  ppageRxCFG      0x0102  //Receiver Configuration
#define  ppageRxCTL      0x0104  //Receiver Control
#define  ppageTxCFG      0x0106  //Transmit Configuration
#define  ppageTxCmdRO    0x0108  //Transmit Command Read Only Status
#define  ppageBufCFG     0x010A  //Buffer Configuration
#define  ppageLineCTL    0x0112  //Line Control
#define  ppageSelfCTL    0x0114  //Self Control
#define  ppageBusCTL     0x0116  //Bus Control
#define  ppageTestCTL    0x0118  //Test Control
#define  ppageISQ        0x0120  //Interrupt status queue
#define  ppageRxEvent    0x0124  //Receiver Event
#define  ppageTxEvent    0x0128  //Transmitter Event
#define  ppageBufEvent   0x012C  //Buffer Event
#define  ppageRxMiss     0x0130  //Receiver Miss Counter
#define  ppageTxColl     0x0132  //Transmit Collision Counter
#define  ppageLineStatus 0x0134  //Line Status
#define  ppageSelfStatus 0x0136  //Self Status
#define  ppageBusStatus  0x0138  //Bus Status
#define  ppageTxCmd      0x0144  //Transmit Command Request
#define  ppageTxLength   0x0146  //Transmit Length
#define  ppageIA         0x0158  //Individual Address
#define  ppageRxStatus   0x0400  //Receive Status
#define  ppageRxLength   0x0402  //Receive Length
#define  ppageRxFrame    0x0404  //Receive Frame Offset
#define  ppageTxFrame    0x0A00  //Transmit Frame Offset
```

Code Snippet 7.20: Some of the PacketPage registers are "set and forget," while other PacketPage registers may report the status of a frame or report an error condition.

There are five PacketPage Event Registers and they are defined by their datasheet names in Code Snippet 7.21.

```
//*******************************************************************
//*   PacketPage Event Register Definitions
//*******************************************************************
#define  RXEVENT_REG         0x0004
#define  TXEVENT_REG         0x0008
#define  BUFEVENT_REG        0x000C
#define  RXMISS_REG          0x0010
#define  TXCOLL_REG          0x0012
```

Code Snippet 7.21: The register definitions in this code snippet are actually the register numbers.

PacketPage registers and register sets can be subdivided into six major categories:

- Bus Interface Registers
- Status and Control Registers
- Initiate Transmit Registers
- Address Filter Registers
- Receive Frame Location
- Transmit Frame Location

The six categories including all of their subsets are found within the CS8900A-CQ's 4 Kbyte PacketPage memory area.

CS8900A-CQ Bus Interface Registers

The Bus Interface Registers are primarily intended for interfacing the CS8900A-CQ to a personal computer ISA bus. Since there's no place for an ISA connector on the Easy Ethernet CS8900A, we won't be using much of the Bus Interface Register's functionality. The ppageBaseIO, which is part of the Bus Interface Register group, defaults to 0x0300 after a chip-wide reset, which is the Easy Ethernet CS8900A's preselected I/O base address. The advantage to using the default address is that we don't have to write any code to set anything up in the I/O base address department. You've already been introduced to the I/O Base Address register layout earlier in Figure 7.6.

Product Identification Code

I purposely didn't write any algorithms to obtain the hardware level of the Easy Ethernet CS8900A's CS8900A-CQ IC. We could have used the ppagePID register to determine which hardware level of CS8900A-CQ we soldered to our Easy Ethernet CS8900A printed circuit board, but in reality that is just wasted code and wasted time, as the data won't be of use to us. The information concerning the CS8900A-CQ we could obtain is contained in the Product Identification Code register as shown in Figure 7.8.

Product Identification Code

Address 0x0000	Address 0x0001	Address0x0002	Address 0x0003
First byte of EISA registration number for Crystal Semiconductor	Second byte of EISA registration number for Crystal Semiconductor	First 8 bits of Product ID number	Last 3 bits of the Product ID number

Figure 7.8: These are the first four bytes of the PacketPage. Once you've driven through this code a few times you'll know how to read and retrieve these bytes if you really want them.

The rest of the Bus Status Registers are geared for ISA operation or interrupt handling and we can simply ignore them. Let's move ahead and take a look at the next set of CS8900A-CQ registers called Status and Control Registers.

CS8900A-CQ Status and Control Registers

Instances of the Status and Control Registers can be found throughout the Easy Ethernet CS8900A code, as their job is to report the status of transmitted and received frames. Status and Control Register activity can also be found in situations where we need to know what's going on inside the CS8900A-CQ. The Status and Control Registers can be subdivided into two groups:

- Configuration and Control Registers
- Status and Event Registers

CS8900A-CQ Configuration and Control Registers

The Easy Ethernet CS8900A utilizes many of the Configuration and Control Registers early in the Easy Ethernet CS8900A code to set up various areas that deal with CS8900A-CQ receive-and-transmit functions. Configuration and Control registers also determine how Ethernet frames are transmitted and received.

The CS8900A-CQ Receiver Configuration Register

For easier identification, all of the Configuration and Control registers are odd numbered. Configuration and Control Register 1 is reserved. So, our discussion of Configuration and Control Registers begins with Register 3, the Receiver Configuration Register, which is shown graphically in Figure 7.9.

RECEIVER CONFIGURATION REGISTER

7	6	5	4	3	2	1	0
StreamE	Skip_1	000011					

F	E	D	C	B	A	9	8
	ExtradataiE	RuntiE	CRCerroriE	BufferCRC	AutoRxDMAE	RxDMAonly	RxOKie

Figure 7.9: In the case of the Easy Ethernet CS8900A, this is a set-and-forget register except when a frame needs to be trashed using the Skip_1 bit.

Operating the CS8900A-CQ in 8-bit mode really narrows the choices when it comes to which bits we can twiddle. In the Receiver Configuration Register we can count out any bit that ends with "iE." Any bit with DMA in its name including StreamE can be pushed off over side as well.

Only two of the bits in Figure 7.9 and Code Snippet 7.22 are of interest to our Easy Ethernet CS8900A application. The RXCFG_SKIP_BIT bit is used to skip over and never return to a frame in the CS8900A-CQ receive buffer. Hopefully, we won't be employing the services of the Receiver Configuration Register very often.

We also want to make sure that the BufferCRC bit is clear. Clearing the BufferCRC bit instructs the CS8900A-CQ not to load the incoming CRC bytes into the CS8900A-CQ receive buffer. The BufferCRC bit should be clear on CS8900A-CQ power up, but we'll write some code to assure that.

```
//******************************************************************
//*   PacketPage Receiver Configuration Bit Definitions
//******************************************************************
#define  RXCFG_NOBUF_CRC      0x0000
#define  RXCFG_SKIP_BIT       0x0006
#define  RXCFG_SKIP           0x0040
#define  RXCFG_RX_OK_IE       0x0100
#define  RXCFG_CRC_ERR_IE     0x1000
#define  RXCFG_RUNT_IE        0x2000
#define  RXCFG_X_DATA_IE      0x4000
```

Code Snippet 7.22: RXCFG_SKIP is a mask that could be used to manipulate the Skip_1 bit. If bit manipulation routines that require a bit number are required, RXCFG_SKIP_BIT would be the definition of choice. Clearing the Receiver Configuration Register is insurance to make certain that the BufferCRC bit is clear when we startup the CS8900A-CQ and go online.

The "iE" behind some of the bits in the Receiver Configuration Register stand for "interrupt enable." According to the CS8900A-CQ datasheet, any event that occurs and has a corresponding "iE" bit set will generate an interrupt. We can't use any of those bits and rely on getting valid data as interrupt use on the CS8900A-CQ is forbidden in 8-bit mode.

The original CS8900A-CQ assembler driver uses the interrupt mechanism of the CS8900A-CQ and the services of the RXCFG_RX_OK_IE bit. The current version of the Easy Ethernet CS8900A hardware makes an exception and includes the INTRQ0 physical connection between the PIC16F877 and the CS8900A-CQ (PIC16F877 pin RB5).

You'll find that the interrupts do indeed work on the CS8900A-CQ in 8-bit mode as long as you keep the Ethernet traffic very light. To that end, I've also included a C version of the "illegal" interrupt driven Easy Ethernet CS8900A firmware on the CD-ROM so you can see how CS8900A-CQ interrupts would work if we were "allowed" to use them.

The CS8900A-CQ Receiver Control Register

The next Configuration and Control Register in line is Register 5, the Receiver Control Register.

RECEIVER CONTROL REGISTER

7	6	5	4	3	2	1	0
PromiscuousA	IAHashA	000101					

F	E	D	C	B	A	9	8
	ExtradataA	RuntA	CRCerrorA	BroadcastA	IndividualA	MulticastA	RxOKA

Figure 7.10: This register must be really smart—Look at all of the good grades its bits have! Bit 7 is of particular interest because if it is set the CS8900A-CQ will accept a frame with any address.

Figure 7.10, the Receiver Control Register, is made up of nine Accept bits. Setting an Accept bit indicates that the CS8900A-CQ will accept that the type of frame the Accept bit represents. An accepted frame is one that is validated and placed in the CS8900A-CQ's on-chip memory. The first definition in Code Snippet 7.23 is a good indication of which Receiver Control Register bits we're interested in using in the Easy Ethernet CS8900A firmware.

```
//*******************************************************************
//*   PacketPage Receiver Control Register Bit Definitions
//*******************************************************************
#define RXCTL_SETUP      (RXCTL_RX_OK_A|RXCTL_IND_A|RXCTL_BCAST_A)
#define RXCTL_RX_OK_A    0x0100
#define RXCTL_MCAST_A    0x0200
#define RXCTL_IND_A      0x0400
#define RXCTL_BCAST_A    0x0800
#define RXCTL_CRC_ERR_A  0x1000
#define RXCTL_RUNT_A     0x2000
#define RXCTL_X_DATA_A   0x4000
```

Code Snippet 7.23: It's pretty obvious that the RXCTL_SETUP value will be loaded into the Receiver Control Register. The interesting thing about some of the choices is that they are undesirable in the everyday working world of networking. The idea behind having unlimited frame access is to be able to receive packets and frames no matter what kind of shape they're in. The uninhibited reception of packets is good for debugging networking designs and making useful tools like Sniffers.

The CS8900A-CQ Transmit Configuration Register

One look at the Transmit Configuration Register in Figure 7.11 and taking into account the number of bits followed by "iE" leads to one conclusion; there's nothing in this register we can use in 8-bit mode. Every bit in the Transmit Configuration Register is an interrupt enable bit.

TRANSMIT CONFIGURATION REGISTER

7	6	5	4	3	2	1	0
SQEerroriE	Loss-of-CRSiE	000111					

F	E	D	C	B	A	9	8
16colliE				AnycolliE	JabberiE	Out-of-windowiE	TxOKie

Figure 7.11: Since the Transmit Configuration Register is useless to us, this is a good place to point out that every register we've looked at so far is identified numerically in the first five bits of the low byte of the register. The Transmit Configuration Register is register number 7.

Even though the contents of the Transmit Configuration Register are null and void as far as we are concerned, it's bit pattern is still defined in the Easy Ethernet CS8900A source code as you see it in Code Snippet 7.24.

```
//*****************************************************************
//*   PacketPage Transmit Configuration Register Bit Definitions
//*****************************************************************
#define  TXCFG_LOSS_CRS_IE    0x0040
#define  TXCFG_SQE_ERR_IE     0x0080
#define  TXCFG_TX_OK_IE       0x0100
#define  TXCFG_OUT_WIN_IE     0x0200
#define  TXCFG_JABBER_IE      0x0400
#define  TXCFG_16_COLL_IE     0x8000
#define  TXCFG_ALL_IE         0x8FC0
```

Code Snippet 7.24: The current version of the Easy Ethernet CS8900A firmware doesn't use the CS8900A-CQ interrupts. However, you can get a taste of what it's like to use the TXCFG_TX_ALL_IE bit structure in the interrupt-enabled version of the Easy Ethernet CS8900A firmware included on the CD-ROM that is included with this book.

The CS8900A-CQ Transmit Command Status Register

Number 9, Number 9, Number 9…the Transmit Command Status Register is strange, but not in the sense of that Number 9 line from a popular psychedelic recording I grew up with. There are two Transmit Command Status Registers. Transmit Command Status Register 9 is read only and resides at PacketPage Address 0x0108. The other Transmit Command Status Register, which is also Register 9, is write only and is located at PacketPage Address 0x0144. The Transmit Command Status Register at PacketPage Address 0x0144 takes the transmit commands issued by the Easy Ethernet CS8900A firmware. If necessary, the contents of the Transmit Command Status Register are read from the Transmit Command Status Register at PacketPage Address 0x0108. Isn't that lovely?

The Transmit Command Status Register at PacketPage Address 0x0144 actually belongs to the Initiate Transmit Registers group. While we have the Transmit Command Status Register's attention, let's go ahead and describe our use of its bits in the Easy Ethernet CS8900A firmware.

TRANSMIT COMMAND STATUS REGISTER

7	6	5	4	3	2	1	0
TxStart		001001					

F	E	D	C	B	A	9	8
		TxPadDis	InhibitCRC			Onecoll	Force

Figure 7.12: The Transmit Command Status Register at PacketPage Address 0x0144 is touched every time a packet is transmitted.

The TxStart bits shown in the Transmit Command Status Register in Figure 7.12 are always set in the Easy Ethernet CS8900A code. Setting both TxStart bits tells the CS8900A-CQ to start the transmission of the frame after the entire frame is transferred to the CS8900A-CQ transmit queue.

The Transmit Command Status Register at PacketPage Address 0x0108 is mapped in the PacketPage register definitions but there is no bit group definition associated with it. The bit definitions in Code Snippet 7.25 belong to the Transmit Command Status Register at PacketPage Address 0x0144.

```
//********************************************************************
//*   PacketPage Transmit Command Register Bit Definitions
//********************************************************************
#define  TXCMD_AFTER_5      0x0000
#define  TXCMD_AFTER_381    0x0080
#define  TXCMD_AFTER_1021   0x0040
#define  TXCMD_AFTER_ALL    0x00C0
#define  TXCMD_FORCE        0x0100
#define  TXCMD_ONE_COLL     0x0200
#define  TXCMD_NO_CRC       0x1000
#define  TXCMD_NO_PAD       0x2000
```

Code Snippet 7.25: Since the Transmit Command Status Register at PacketPage Address 0x0144 is updated before each transmission, we could actually change the way each packet is transmitted on the fly. The Easy Ethernet CS8900A firmware uses a standard combination of the bits in this register for every transmission.

The CS8900A-CQ Buffer Configuration Register

The Buffer Configuration Register in Figure 7.13 is another of those "iE" registers.

BUFFER CONFIGURATION REGISTER

7	6	5	4	3	2	1	0
RxDMAiE	Swint	001011					

F	E	D	C	B	A	9	8
RxDestiE		Miss OvfloiE	TxCol OvfloiE	Rx128iE	RxMissiE	TxUnder runtiE	Rdy4TxiE

Figure 7.13: About the only things useful to us are the Register identification bits.

```
//*********************************************************************
//*   PacketPage Buffer Configuration Register Bit Definitions
//*********************************************************************
#define  BUFCFG_SW_INT      0x0040
#define  BUFCFG_RDY4TX_IE   0x0100
#define  BUFCFG_TX_UNDR_IE  0x0200
```

Code Snippet 7.26: There just isn't anything to talk about here.

The Buffer Configuration Register is not even used in our "illegal" Easy Ethernet CS8900A interrupt code, and the code in Code Snippet 7.26 is virtually useless. Enough said.

The CS8900A-CQ Line Control Register

The CS8900A-CQ Line Control Register (Figure 7.14) is of a bit more use than the Buffer Configuration Register. The Line Control Register bits enable the CS8900A-CQ transmitter and receiver and determine what type of interface is attached to the CS8900A-CQ. The LINECTL_10BASET definition in Code Snippet 7.27 for 10Base-T operation is what we'll use in the Easy Ethernet CS8900A firmware.

To set the Ethernet interface for 10Base-T requires both the AutoAUI/10BT and the AUIonly bits to be clear. That's not obvious, looking at the bit descriptions for bits 8 and 9 of the Line Control Register, and you must consult the CS8900A-CQ datasheet description for that information.

LINE CONTROL REGISTER

7	6	5	4	3	2	1	0
SerTxON	SerRxON	010011					

F	E	D	C	B	A	9	8
	LoRxSquelch	2partDefDis	PloarityDis	ModBackoffE		AutoAUI/10BT	AUIonly

Figure 7.14: An interesting feature of the CS8900A-CQ is its ability to extend its range by setting the LoRxSquelch bit. The LoRxSquelch bit lowers the 10Base-T receiver squelch thresholds and allows operation on "quiet" cables with lengths in excess of 100 meters.

```
//*********************************************************************
//*   PacketPage Line Control Bit Definitions
//*********************************************************************
#define  LINECTL_RX_ON_BIT     0x0006
#define  LINECTL_RX_ON         0x0040
#define  LINECTL_TX_ON_BIT     0x0007
#define  LINECTL_TX_ON         0x0080
#define  LINECTL_AUI_ONLY      0x0100
#define  LINECTL_10BASET       0x0000
```

Code Snippet 7.27: The CS8900A-CQ can be set to detect the interface (AUI or 10Base-T) automatically.

The CS8900A-CQ Self Control Register

The Self Control Register (Figure 7.15) has the potential of being one of the most interesting registers to work with. If you decide to manipulate the LED driver/logic level output pins, this is your register. The Self Control Register is also the holder of the CS8900A-CQ RESET bit and can control the CS8900A-CQ's power modes as well.

SELF CONTROL REGISTER

7	6	5	4	3	2	1	0
	RESET	010101					

F	E	D	C	B	A	9	8
HCB1	HCB0	HC1E	HC0E		HW Standby	HWSleepE	SW Suspend

Figure 7.15: I personally like the blinky LED indicators. However, if you're down to needing just one more output pin, you can steal one from the CS8900A-CQ using the bits in this register.

Used or not, the Self Control Register bits are included in the Easy Ethernet CS8900A firmware and are laid out as shown in Code Snippet 7.28.

```
//*****************************************************************
//*   PacketPage Self Control Register Bit Definitions
//*****************************************************************
#define  SELFCTL_RESET     0x0040
#define  SELFCTL_HC1E      0x2000
#define  SELFCTL_HCB1      0x8000
```

Code Snippet 7.28: I figured most of you would want the LED indicators, so I didn't "pen in" the LED's alternate bit definitions.

The CS8900A-CQ Bus Control Register

The Bus Control Register (Figure 7.16) is intended for use with ISA systems. The only bit of any interest to us is the "illegal" EnableIRQ bit that is used in the "illegal" interrupt versions of the Easy Ethernet CS8900A firmware.

BUS CONTROL REGISTER

7	6	5	4	3	2	1	0
	Reset RxDMA	010111					

F	E	D	C	B	A	9	8
EnableIRQ		RxDMAsize	IOCH RDYE	DABurst	MemoryE	UseSA	DMAextend

Figure 7.16: The EnableIRQ bit tells the CS8900A-CQ to generate an interrupt when an interrupt-tagged event occurs. This bit enables the "illegal" 8-bit mode code to control the logic level of the CS8900A-CQ's INTRQ0 pin.

```
//************************************************************************
//*   PacketPage Bus Control Bit Definitions
//************************************************************************
#define  BUSCTL_USE_SA         0x0200
#define  BUSCTL_MEM_MODE       0x0400
#define  BUSCTL_IOCHRDY        0x1000
#define  BUSCTL_INT_ENBL_BIT   0x0007
#define  BUSCTL_INT_ENBL       0x8000
```

Code Snippet 7.29: This is another register that you can see in action if you run the older Easy Ethernet CS8900A interrupt code.

The Bus Control Register bit definitions are holdovers from the earlier Easy Ethernet CS8900A code that used the INTRQ0 interrupt line. The BUSCTL_INT_ENBL_BIT in Code Snippet 7.29 is not used in the current spin of the Easy Ethernet CS8900A firmware.

The CS8900A-CQ Test Control Register

The Test Control Register (Figure 7.17) is primarily used to test the Manchester ENDEC by looping its output back into its input. There are also a couple of bits in the Test Control Register that deal with standard operation of the CS8900A-CQ. The Disable Backoff bit turns off the "wait after a collision" algorithms. If the Disable Backoff bit is set, the CS8900A-CQ will attempt to transmit on the time interval between Ethernet packets no matter what. The minimum inter packet gap interval is 9.6 µS.

TEST CONTROL REGISTER

7	6	5	4	3	2	1	0
DisableLT		011001					

F	E	D	C	B	A	9	8
	FDX			Disable Backoff	AUIloop	ENDEC loop	

Figure 7.17: The Easy Ethernet CS8900A firmware doesn't take advantage of the CS8900A-CQ's internal loopback features. However, the FDX bit is used.

The TESTCTL_FDX bit in Code Snippet 7.20 puts the CS8900A-CQ into full duplex mode. Full duplex mode allows transmission and reception to occur simultaneously. In most cases, you would need separate transmit and receive wiring to achieve full duplex mode. If you take a look at standard Category 5 twisted pair and how it's used in an Ethernet LAN, you'll notice that there is a transmit and receive pair of wires allocated within the four twisted pairs.

```
//*****************************************************************
//*   PacketPage Test Control Bit Definitions
//*****************************************************************
#define  TESTCTL_DIS_LT        0x0080
#define  TESTCTL_ENDEC_LP      0x0200
#define  TESTCTL_AUI_LOOP      0x0400
#define  TESTCTL_DIS_BKOFF     0x0800
#define  TESTCTL_FDX           0x4000
```

Code Snippet 7.30: Just enabling the loopback bits doesn't a loopback test make—one must interpret the loopback data with supporting software.

Let's move on and investigate the next major set of CS8900A-CQ operational registers.

CS8900A-CQ Status and Event Registers

The CS8900A-CQ Interrupt Status Queue

The CS8900A-CQ Status and Event Registers aren't as prolific as the CS8900A-CQ's Control and Configuration Registers. And, I shouldn't even be discussing the very first Status and Event Register, the Interrupt Status Queue (Figure 7.18), as interrupt operation in 8-bit mode is not recommended.

INTERRUPT STATUS QUEUE

7	6	5	4	3	2	1	0
RegContent		RegNum					

F	E	D	C	B	A	9	8
RegContent							

Figure 7.18: The Interrupt Status Queue maps the register number and contents of certain interrupt-enabled registers for use by the application.

If you're interested in how to use the Interrupt Status Queue, you can get plenty of Interrupt Status Queue experience from the down-level and "illegal" Easy Ethernet CS8900A interrupt-laden firmware.

The CS8900A-CQ Receiver Event Register

When we get into looking at the Easy Ethernet CS8900A source code, you'll find the Receiver Event Register (Figure 7.19) is instrumental in the Ethernet frame receive and transfer process. The Receiver Event Register is the register that is polled to check for valid incoming Ethernet frames held in the CS8900A-CQ's receive buffer.

RECEIVER EVENT REGISTER

7	6	5	4	3	2	1	0
Dribblebits	IAHash	000100					

F	E	D	C	B	A	9	8
	Extradata	Runt	CRCerror	Broadcast	Individual Adr	Hashed	RxOK

Figure 7.19: Using the CS8900A-CQ interrupts maps this register number and its contents into the Interrupt Status Queue when a receive event occurs. For polled operation, which is used in the current version of the Easy Ethernet CS8900A firmware, this register is continually read and its bits evaluated to determine if a frame is positioned in the CS8900A-CQ receive queue.

The Receiver Event Register is one of the event registers that is mapped to the Interrupt Service Queue when interrupts are enabled for the CS8900A-CQ. The difference in polling and interrupt modes is that when polled, the Receiver Event Register's contents aren't moved to the Interrupt Status Queue and the status of the receive event must be taken from the condition of the Receiver Event Register's bits. This will all make more sense when we discuss the Receiver Event Register's role in the Easy Ethernet CS8900A firmware.

The CS8900A-CQ Transmitter Event Register

The Transmitter Event Register (Figure 7.20) is a neat register if statistics are your thing. However, it's not used at all in the Easy Ethernet CS8900A firmware.

TRANSMITTER EVENT REGISTER

7	6	5	4	3	2	1	0
SQEerror	Loss-of-CRS	010011					

F	E	D	C	B	A	9	8
16coll	Number-of-Tx-collisions				Jabber	Out-of-window	TxOK

Figure 7.20: Some of the bits pertain to AUI (Loss-of-CRS and SQEerror). But, you can use the rest of the bits in this register to collect data about the link your Easy Ethernet CS8900A is operating on.

The idea behind this book is to give you working hardware and software. There may be some things you read in this text that will fuel an idea or tweak your curiosity, so I've included all of the CS8900A-CQ register definitions in the Easy Ethernet CS8900A source code instead of leaving out what I feel to be unimportant. Code Snippet 7.31 is another example of register definition code not used by the Easy Ethernet CS8900A but available for use by you.

```
//*******************************************************************
//*   PacketPage Transmit Event Register Bit Definitions
//*******************************************************************
#define  TXEVENT_TX_OK      0x0100
#define  TXEVENT_OUT_WIN    0x0200
#define  TXEVENT_JABBER     0x0400
#define  TXEVENT_16COLLS    0x1000
```

Code Snippet 7.31: The bit definitions for the Transmitter Event Register are in the Easy Ethernet CS8900A source code if you wish to use them.

The CS8900A-CQ Buffer Event Register

This is yet another CS8900A-CQ register that goes unused in the Easy Ethernet CS8900A firmware. The Buffer Event Register (Figure 7.21) provides the status of the CS8900A-CQ transmit and receive buffers.

BUFFER EVENT REGISTER

7	6	5	4	3	2	1	0
RxDMA frame	SWint	001100					

F	E	D	C	B	A	9	8
RxDest				Rx128	RxMiss	TxUnderrun	Rdy4Tx

Figure 7.21: The RxDest and Rx128 bits can be used in the preprocessing of incoming frames. By checking the status of the RxDest and Rx128 bits, the application can get a jump start on processing the incoming frame.

The Rdy4Tx bit is directly tied to an interrupt process. The CS8900A-CQ datasheet says the Rdy4Tx bit is very similar to the Rdy4TxNOW bit. So, I substituted the Rdy4Tx bit for the Rdy4TxNOW bit in the transmit code. It didn't work. So, unequivocally the Rdy4Tx bit cannot be used in 8-bit mode to signal a successful bid for transmission. In true CS8900A-CQ 8-bit mode (no interrupts enabled), the microcontroller must poll the Rdy4TxNOW bit before gaining access to the CS8900A-CQ's transmit buffer.

The CS8900A-CQ Receiver Miss Counter and Transmit Collision Counter Registers

Neither of these registers are used in the Easy Ethernet CS8900A firmware. In a nutshell, both registers are 10-bit counters that can be used to signal the networking application that something is really wrong with the way things are going on the transmit and/or receive side of the link. The PacketPage Addresses of both counters is defined for those of you that need to write code to employ this functionality.

The CS8900A-CQ Line Status Register

This is another register that can be used at your discretion, as nothing is coded for its use in the Easy Ethernet CS8900A firmware. The Line Status Register simply reports the status of the Ethernet physical interface.

The CS8900A-CQ Self Status Register

Just when you were thinking that the rest of the Status and Event Registers were absolutely good for nothing, a bit inside the Self Status Register (Figure 7.22) rises from the ashes. The INITD bit of the Self Status Register is used by the Easy Ethernet CS8900A firmware to check for a valid initialization of the CS8900A-CQ.

SELF STATUS REGISTER

7	6	5	4	3	2	1	0
INITD	3.3V Active	010110					

F	E	D	C	B	A	9	8
			EEsize	ELPresent	EEPROM OK	EEPROM present	SIBUSY

Figure 7.22: Most of the bits in this register have to do with an EEPROM, which isn't used in 8-bit mode and not included in the hardware design of the Easy Ethernet CS8900A.

The Easy Ethernet CS8900A code does a bit test operation to check the INITD bit. So, the SELFSTAT_INIT_DONE_BIT definition in Code Snippet 7.28 is used. The SELFSTAT_INIT_DONE definition can be used as a mask to determine if the INITD bit is set.

```
//*****************************************************************
//*   PacketPage Self Status Bit Definitions
//*****************************************************************
#define  SELFSTAT_INIT_DONE_BIT  0x0007
#define  SELFSTAT_INIT_DONE      0x0080
#define  SELFSTAT_SI_BUSY        0x0100
#define  SELFSTAT_EEP_PRES       0x0200
#define  SELFSTAT_EEP_OK         0x0400
#define  SELFSTAT_EL_PRES        0x0800
```

Code Snippet 7.28: Note that the definitions following the SELFSTAT_INIT_DONE_BIT definition are all mask values.

The CS8900A-CQ Bus Status Register

Only two bits of information exist in the Bus Status Register (Figure 7.23) and both relate to transmission. The Rdy4TxNOW bit tells the PIC16F877 that the CS8900A-CQ is ready to accept the transfer of a frame from the PIC16F877 for transmission.

BUS STATUS REGISTER

7	6	5	4	3	2	1	0
TxBidErr		011000					

F	E	D	C	B	A	9	8
							Rdy4TxNOW

Figure 7.23: The maximum number of bytes you can stuff into an Ethernet frame is 1518. Now, consider that the last 4 bytes of a frame are the CRC. That's 1518 – 4, or 1514 bytes of stuff you can cram into a frame. The TxBidErr bit will set when you try to exceed the maximum Ethernet frame size. If you decide to modify the maximum frame size in the Easy Ethernet CS8900A firmware, be sure to write some code to check this bit if you think you'll get close to the max frame size.

The BUSSTA_RDY4TXNOW_BIT definition represents bit 0 of the upper byte of the mask definition BUSSTA_RDY4TXNOW in Code Snippet 7.29.

```
//*******************************************************************
//*   PacketPage Bus Status Bit Definitions
//*******************************************************************
#define  BUSSTA_TX_BID_ERR    0x0080
#define  BUSSTA_RDY4TXNOW_BIT 0x0000
#define  BUSSTA_RDY4TXNOW     0x0100
```

Code Snippet 7.29: As you will see later, a bit test operation is performed on the Bus Status Register's high byte to check for the OK to transfer a frame to the CS8900A-CQ transmit queue.

CS8900A-CQ Address Filter Registers

The CS8900A-CQ comes equipped with a destination address register called the Individual Address Register. The IA Register is part of an address filtering mechanism that includes bits from a number of other registers. This address filter mechanism determines which frames will pass the CS8900A-CQ receive portal and be placed in the CS8900A-CQ receive buffer. The description of the physical IA Register and some of the supporting code have already been covered.

For the Easy Ethernet CS8900A to respond to an incoming address, the very first bit of the destination address should be 0. The reason for this is that if the first bit is not a 0, the address is not a physical address. In the Easy Ethernet CS8900A firmware using bits in the Receiver Control Register, we tell the CS8900A-CQ that the DA (Destination Address) on the incoming frame must match the physical address in the CS8900A-CQ IA Register.

If the first bit of the incoming DA is a 1, then the frame is a multicast frame and the address is logical, not physical. The CS8900A-CQ uses a hash technique to determine if it should accept the incoming multicast frame. If you look at the RXCTL_SETUP definition in the Easy Ethernet CS8900A source code, you will find that the multicast (RXCTL_MCAST_A) bit is not included in the OR (|) scheme and as a result is not set. Thus, our implementation of the Easy Ethernet CS8900A will ignore multicast addresses. Note also that the Easy Ethernet CS8900A's onboard CS8900A-CQ has been instructed to accept broadcast addresses.

CS8900A-CQ Receive and Transmit Frame Locations

These areas are used to transfer Ethernet frames between the CS8900A-CQ and the PIC16F877. Only one receive and transmit frame are available at any time and the space for each is dynamically allocated. The Easy Ethernet CS8900A runs in 8-bit mode and all of the data is transferred to and from the CS8900A-CQ buffer memory area via the PacketPage I/O ports.

Did It Register?

That does it for the CS8900A-CQ register set. As you can see, there are a multitude of variables and configurations that can be had by setting and clearing the right set of bits in the right set of registers.

We've covered a large amount of information and some of it may not make sense to you right now. I can tell you that in the beginning of my experiences with the CS8900A-CQ, it took a few readings of the CS8900A-CQ datasheet and application notes for some of the concepts to become clear to me. Don't worry—if we left anything out or set a bit incorrectly, the problem will surface as we apply the firmware to the hardware and walk through receiving and transmitting frames with the CS8900A-CQ.

Writing the CS8900A-CQ Firmware

The best way to learn the ways of the CS8900A-CQ is to write some code to drive the CS8900A-CQ. You've been introduced to the "do's and don'ts" of using the CS8900A-CQ in 8-bit mode and you've been schooled in Ethernet etiquette. So, dust off your hexadecimal calculator because it's time to bring the CS8900A-CQ section of the Easy Ethernet CS8900A to life.

LINE CONTROL REGISTER

7	6	5	4	3	2	1	0
SerTxON	SerRxON	010011					

F	E	D	C	B	A	9	8
	LoRxSquelch	2partDefDis	PloarityDis	ModBackoffE		AutoAUI/10BT	AUIonly

Figure 8.1: You can always cross-reference a CS8900A-CQ register by the number in the lower six bits of the register.

As we make our way through the hows and whys of the Easy Ethernet CS8900A firmware, I'll reference the contents of certain registers. For example, note that in Figure 8.1 the value of the first six bits of the Line Control Register equate to decimal 13. A look back at our discussion of CS8900A-CQ registers, CS8900A-CQ Register 13 is the Line Control Register. You and I can't overwrite these first six bits as they are used internally by the CS8900A-CQ to positively identify registers. To test this point, I actually commented out the original bit_set lines of code and replaced them with code in Code Snippet 8.1 that would erase the lower byte of the Line Control Register and replace it with a mask of 0xC0 (11000000 binary), which would normally turn on the SerTxON and SerRxON bits and write zeroes to the lower five bits.

```
//*****************************************************************
//*   TEST CODE TO PROVE THAT LOWER BITS OF LOWER BYTE ARE PERMANENT
//*****************************************************************
      // ORIGINAL CODE RUNS NORMALLY UP TO HERE
      // bit_set(data_L,LINECTL_RX_ON_BIT);
      //bit_set(data_L,LINECTL_TX_ON_BIT);
```

```
//TEST CODE
   data_L &= 0x00;    //clear the data_L byte just read in to 0x00
   data_L |= 0xC0;    //set both SerTxON and SerRxON bits
//CODE CONTINUES NORMALLY FROM HERE
```

Code Snippet 8.1: Some things just don't change.

The Easy Ethernet CS8900A was still able to turn on the receiver and transmitter after loading and running this temporary change.

OK...you should have a pretty good idea of what we want to accomplish. We're going to take all of the concepts that we have discussed and put them to work. With that, let's dive into the Easy Ethernet CS8900A code.

The First Step

The *setup_xxxx* portion of the *Absolute Start Point* code is automatically generated by the Custom Computer Services C Compiler. Those lines of source code could also have been generated manually, but why reinvent the wheel when the functionality is put there for you. You'll notice that we've turned off many of the goodies that are part of the PIC16F877. That doesn't mean you can't use them. The PIC16F877 internal peripherals are simply not used in this spin of the code.

The Easy Ethernet CS8900A purposely pins out the entire bit structure of PORTA. If the A/D resources of the PIC16F877 were needed, we could actually get rid of the first two lines of code and enable the analog circuitry of the PIC16F877 just as easily as we've disabled it.

All of the microcontroller's initial port directions are set up with the Custom Computer Services C Compiler "set_tris_x"(0xYY)" function. For the Microchip PIC, a '1' says that the bit position is an input pin and a '0' denotes an output pin. Check the set_tris_x bit patterns against the schematic to find out which pins are set for input and which pins are set for output.

The series of set/clr macros that follow the set_tris functions are directed at preparing the CS8900A-CQ for reset and initialization. If the macros look familiar, it's because we discussed them earlier. The idea is to set all of the CS8900A-CQ I/O control pins to their inactive states.

The final printf statement is optional and was used in the code development and debugging process.

```
//*******************************************************************
//*   Absolute Start Point
//*******************************************************************
void main() {
   int16 scratch16;
```

```
   setup_adc_ports(NO_ANALOGS);
   setup_adc(ADC_OFF);
   setup_psp(PSP_DISABLED);
   setup_spi(FALSE);
   setup_counters(RTCC_INTERNAL,WDT_18MS);
   setup_timer_1(T1_DISABLED);
   setup_timer_2(T2_DISABLED,0,1);

   set_tris_a(0x00);
   set_tris_b(0xE0);   //11100000
   set_tris_c(0xBD);   //10111101
   set_tris_d(0x00);
   set_tris_e(0x00);
   clr_reset;
   set_ior;
   set_iow;
   set_aen;
   clr_hex;
   //printf("Starting CS8900A-CQ Initialization.\r\n");
```

Code Snippet 8.2: The clr_hex statement is clearing a flag bit that is used in the UDP application code.

Reset the CS8900A-CQ

In Code Snippet 8.3, the reset bit within *ppageSelfCTL* is set with the code generated by the *WPP(ppageSelfCTL,SELFCTL_RESET)* function and a 10 millisecond delay is implemented to allow the CS8900A-CQ to calibrate its on-chip analog circuitry.

```
//*****************************************************************
//*   PacketPage Self Control Register Bit Definitions
//*****************************************************************
#define  SELFCTL_RESET     0x0040
//*****************************************************************
//*   Reset the CS8900
//*****************************************************************
   WPP(ppageSelfCTL,SELFCTL_RESET);
   do
   {
      delay_ms(10);
      RPP(ppageSelfStatus);
   }while(!(bit_test(data_L,SELFSTAT_INIT_DONE_BIT)));
   //printf("CS8900A-CQ is RESET.\r\n");
```

Code Snippet 8.3: The delay_ms() function within the Custom Computer Services C Compiler saves the programmer from having to write complex and tricky delay routines from scratch.

The ppageSelfCTL RESET bit (bit 6 of the Self Control Register) is an Act-Once bit. That is, it is set and cleared automatically by the action it initiates. Notice in the code that a mask of 0x40 (the SELFCTL_RESET value) was written to the lower byte of the Self Control Register, and only bit 6 (RESET) was affected.

SELF CONTROL REGISTER

7	6	5	4	3	2	1	0
	RESET	010101					

F	E	D	C	B	A	9	8
HCB1	HCB0	HC1E	HC0E		HW Standby	HWSleepE	SW Suspend

While we're talking about the Self Control Register, note that the SELFCTL_RESET mask also determines how the CS8900A-CQ LINKLED pin will operate. By clearing the HC0E bit in the Self Control Register, we turn the LINKLED function on with the upper byte of the SELFCTL_RESET mask. If either the HC0E or HC1E bits were set, bits HCB0 and HCB1 allow the microcontroller to have control of the pins used to drive the indicator LEDs.

SELF STATUS REGISTER

7	6	5	4	3	2	1	0
INITD	3.3V Active	010110					

F	E	D	C	B	A	9	8
			EEsize	ELPresent	EEPROM OK	EEPROM present	SIBUSY

After 10 milliseconds, a bit check is run against the INITD bit in the Self Status Register (SELFSTAT_INIT_DONE_BIT). When this bit clears, the global CS8900A-CQ reset is complete. Again, I've added an optional printf statement to signal a successful CS8900A-CQ reset sequence.

Load the CS8900A-CQ Basic Parameters

The first line of code in the *Load the CS8900 Basic Parameter* code block dictates that the ether will be 10BASE-T (LINECTL_10BASET) and the CS8900A-CQ will pump Manchester-encoded bits onto the ether in Full Duplex mode (TESTCTL_FDX).

LINE CONTROL REGISTER

7	6	5	4	3	2	1	0
SerTxON	SerRxON	010011					

F	E	D	C	B	A	9	8
	LoRxSquelch	2partDefDis	PloarityDis	ModBackoffE		AutoAUI/10BT	AUIonly

TEST CONTROL REGISTER

7	6	5	4	3	2	1	0
DisableLT		011001					

F	E	D	C	B	A	9	8
	FDX			Disable Backoff	AUIloop	ENDEC loop	

Here's what the mnemonics within the RXCTL_SETUP definition specify:

- The CS8900A-CQ accepts frames with correct CRC and length only (RXCTL_RX_OK_A)
- The Destination Address in the packet header must match the IA address found in ppageIA (RXCTL_IND_A)
- Broadcast frames with a Destination Address of FFFF FFFF FFFF hexadecimal are accepted (RXCTL_BCAST_A)

```
//********************************************************************
//*   PacketPage Line Control Bit Definitions
//********************************************************************
#define  LINECTL_10BASET      0x0000
//********************************************************************
//*   PacketPage Test Control Bit Definitions
//********************************************************************
#define  TESTCTL_FDX          0x4000
//********************************************************************
//*   PacketPage Receiver Control Register Bit Definitions
//********************************************************************
#define  RXCTL_SETUP (RXCTL_RX_OK_A|RXCTL_IND_A|RXCTL_BCAST_A)
//********************************************************************
//*   Load the CS8900 Basic Parameters
//*   10BaseT/Full Duplex/accept broadcast /individual addresses
//********************************************************************
      WPP(ppageLineCTL,LINECTL_10BASET);
      WPP(ppageTestCTL,TESTCTL_FDX);
      WPP(ppageRxCTL,RXCTL_SETUP);
      WPP(ppageRxCFG,RXCFG_NOBUF_CRC);
   //printf("CS8900A-CQ Basic Parameters SET.\r\n");
```

Code Snippet 8.4: Remember that to choose the 10Base-T option, both bits 8 and 9 of the Line Control Register must be cleared. The PacketPage write to the ppageRxCFG (Receiver Configuration Register) is our insurance code that makes sure the BufferCRC bit inside the Receiver Configuration Register is clear.

Let's check our work. In the code we logically OR (that's what the | between the bit fields RXCTL_RX_OK_A|RXCTL_IND_A|RXCTL_BCAST_A means) the hexadecimal values within the RXCTL_SETUP definition and apply the mask to the Receiver Control Register.

```
                                              HEX      BINARY
              RXCTL_RX_OK_A              0x0100 = 00000001 00000000
              RXCTL_IND_A                0x0400 = 00000100 00000000
              RXCTL_BCAST_A              0x0800 = 00001000 00000000
 Resulting mask applied to Receiver Control Register = 00001101 00000000
```

RECEIVER CONTROL REGISTER

7	6	5	4	3	2	1	0
PromiscuousA	IAHashA	000101					

F	E	D	C	B	A	9	8
	ExtradataA	RuntA	CRCerrorA	BroadcastA	IndividualA	MulticastA	RxOKA

Matching the logically OR'ed bits to the high byte of the Receiver Control Register sets bits 8, A and B, which instructs the CS8900A-CQ to accept frames with the correct CRC and length (bit 8), match the frame's DA (Destination Address) to the Individual Address and accept broadcast frames. That's exactly what we want to happen when a packet is received.

Load the CS8900A-CQ Individual Address Register Set

We've already covered the next code module. For the sake of continuity, I'll show it to you again.

```
//*****************************************************************
//*    HARDWARE (MAC) ADDRESS DEFINITION
//*    YOU MAY CHANGE THIS TO ANY VALID MAC ADDRESS
//*****************************************************************
int8 MYMAC[6] = { 0,0,'E','D','T','P' };
```

MYMAC	5	4	3	2	1	0
CS REG	0X15D	0X15C	0X15B	0X15A	0X159	0X158
	P	T	D	E	0	0

```
//*****************************************************************
//*    PacketPage Internal Register Definitions
//*****************************************************************
#define ppageIA    0x0158  //Individual Address
int16 scratch16;
//*****************************************************************
//*    Load the CS8900 IA
//*    INDIVIDUAL ADDRESS LAYOUT IN CS8900
//*****************************************************************

      scratch16 = make16(MYMAC[1],MYMAC[0]);
      WPP(ppageIA,scratch16);
      scratch16 = make16(MYMAC[3],MYMAC[2]);
      WPP(ppageIA+2,scratch16);
      scratch16 = make16(MYMAC[5],MYMAC[4]);
      WPP(ppageIA+4,scratch16);


//uncomment this code to see the MAC address as it has been entered
      //RPP(ppageIA);
      //printf("%x%x \r\n",data_H,data_L);
```

```
        //RPP(ppageIA+2);
        //printf("%x%x \r\n",data_H,data_L);
        //RPP(ppageIA+4);
        //printf("%x%x \r\n",data_H,data_L);
//end commented code
        //printf("CS8900A-CQ MAC Address LOADED.\r\n");
```

Code Snippet 8.5: I inserted the CS8900A-CQ Individual Address layout into the snippet to refresh your memory as to how the MAC bytes are stored inside the CS8900A-CQ.

The code in Code Snippet 8.5 uses the services of the WPP macro to load the IA (Individual Address) into the CS8900A-CQ's IA register set. In Code Snippet 8.4, we set bit A in the Receiver Control Register to instruct the CS8900A-CQ to receive packets addressed to the IA that was loaded in Code Snippet 8.5.

Enable the CS8900A-CQ Transmitter and Receiver

Everything register-wise needed to get the CS8900A-CQ online is set and ready to go. All that stands between the CS8900A-CQ and the LAN segment is the enabling of the CS8900A-CQ's transmitter and receiver in Code Snippet 8.6.

```
//******************************************************************
//*   PacketPage Internal Register Definitions
//******************************************************************
#define  ppageINT            0x0022  //Interrupt number (0,1,2, or 3)
//******************************************************************
//*   PacketPage Bus Control Bit Definitions
//******************************************************************
#define  BUSCTL_INT_ENBL_BIT  0x0007
//******************************************************************
//*   PacketPage Line Control Bit Definitions
//******************************************************************
#define  LINECTL_RX_ON_BIT    0x0006
#define  LINECTL_TX_ON_BIT    0x0007
//******************************************************************
//*   Enable CS8900 TRANSMITTER AND RECEIVER
//******************************************************************

      RPP(ppageLineCTL);
      bit_set(data_L,LINECTL_RX_ON_BIT);
      bit_set(data_L,LINECTL_TX_ON_BIT);
      PPWrite();
         //printf("CS8900A-CQ Ethernet Transceiver ENABLED.\r\n");
         printf("Easy Ethernet CS8900A Version 03.08.02\r\n");
```

Code Snippet 8.6: The RPP (Read PacketPage) macro reads a PacketPage register into the data_H and data_L variables. Our targeted bits are in data_L. Once the desired bits are twiddled, the PPWrite (PacketPage Write) macro puts the newly revised register contents (data_H and data_L) back into the PacketPage register they originated from.

We set the link type to 10Base-T in a previous code segment by clearing some bits in the Line Control Register. To activate the CS8900A-CQ transmitter and receiver, we need to set bits 6 and 7 of the Line Control Register. The easiest way to set the bits inside the register without disturbing the bits we've already twiddled is to read the Line Control Register, set the SerTxON and SerRxON bits in the contents that we read from the Line Control Register and write the new bit settings back into the Line Control Register. Of course there are macros to perform the register reads (RPP macro) and writes (PPWrite macro). I added a printf statement to assure you that the Easy Ethernet CS8900A's RS-232 port is functional, and to inform you of the version of firmware your Easy Ethernet CS8900A is running.

Let's stop for a moment and collect our thoughts. We've completed quite a bit of work towards our goal of putting the Easy Ethernet CS8900A on a LAN. Up to this point we've:

- Effected the PIC16F877-to-CS8900A-CQ interface
- Reset and Initialized the CS8900A-CQ
- Loaded transmit and receive parameters into the CS8900A-CQ
- Loaded the MAC address as *00EDTP*
- Enabled the CS8900A-CQ's transmitter and receiver

We've also covered the important CS8900A-CQ registers and their roles in passing a packet from one Ethernet node to another on a LAN. All of the Easy Ethernet CS8900A hardware is primed and ready to go. It's time to venture out into the ether…

The Main Service Loop

You were probably expecting lots of activity inside the "Main Service Loop" you see in Code Snippet 8.7. Without a doubt, there are other things we could be doing with the microcontroller inside the Main Service Loop but right now our focus is on transmitting and receiving Ethernet packets. Once we accomplish that, then we can spend our time coding the applications.

The Main Service Loop consists of two do-while loops running inside of a never-ending while loop. The first do-while loop performs the polling function.

RECEIVER EVENT REGISTER

7	6	5	4	3	2	1	0
Dribblebits	IAHash	000100					

F	E	D	C	B	A	9	8
	Extradata	Runt	CRCerror	Broadcast	Individual Adr	Hashed	RxOK

The RPP (Read PacketPage) macro continually reads the Receiver Event Register. Bit 8 of the Receiver Event Register is set when a valid frame has been loaded into the CS8900A-CQ's receive buffer. So, following the read of the Receiver Event Register, the first do-while loop checks the state of the RxOK bit in the Receiver Event Register. If the RxOK bit is not set, the polling do-while loop repeats the read/bit test procedure until the RxOK bit is set by the CS8900A-CQ.

To receive a frame, the CS8900A-CQ must accept it using the IA Register and Destination Address filter. We have already specified that only broadcast addresses and the matching IA Register address will be recognized by our CS8900A-CQ. Once the packet is accepted, the preamble and Start of Frame Delimiter are stripped off, and the bits following the SFD are loaded into the CS8900A-CQ receive buffer area. Remember that a packet is the entire message including the preamble, the Start of Frame Delimiter (SOF), the Destination Address, the Source Address, the length and packet type information, the data, any necessary padding and the FCS (Frame Check Sequence) or CRC (Cyclic Redundancy Check) value. A frame is a packet without the preamble and SOF.

Earlier we cleared the BufferCRC bit inside the Receiver Configuration Register. Clearing the BufferCRC bit means we will not include the CRC bytes in the receive buffer contents or the length calculations.

```
//******************************************************************
//*   PacketPage Internal Register Definitions
//******************************************************************
#define  ppageRxEvent          0x0124   //Receiver Event
//******************************************************************
//*   PacketPage Receiver Event Register Bit Definitions
//******************************************************************
#define  RXEVENT_RX_OK_BIT     0x0000
//******************************************************************
//*   MAIN SERVICE LOOP
//******************************************************************
   while(1)
   {

   do{
      RPP(ppageRxEvent);
      }while(!(bit_test(data_H,RXEVENT_RX_OK_BIT)));
   do{
         get_frame();
         RPP(ppageRxEvent);
      }while(bit_test(data_H,RXEVENT_RX_OK_BIT));

   }
```

Code Snippet 8.7: Other processes can be included inside the Main Service Loop, but one must crawl before he or she begins to walk.

Once a valid frame is received by the CS8900A-CQ and the RxOK bit is set, the second do-while loop in Code Snippet 8.7 takes over. The second do-while loop transfers the frame from the CS8900A-CQ receive queue to the PIC16F877 buffer, which is actually a memory array called *packet*. The frame is then processed by the PIC16F877. After the processing of the current frame is complete, the second do-while loop reads the Receiver Event Register and checks the status of the RxOK bit. If the RxOK bit is set, there is another frame waiting to be retrieved from the CS8900A-CQ receive queue. The second do-while loop continually retrieves frames from the CS8900A-CQ until the RxOK bit is cleared by the CS8900A-CQ. When the last frame is transferred and the RxOK bit clears, the first do-while loop once again begins the polling operation. The two do-while loops run forever polling for frames and transferring frames.

A Frame Under the Microscope

The only function that is called from the Main Service Loop is the get_frame function. The get_frame function does exactly what its name implies. It gets frames.

I've set up an MPLAB ICE 2000 and the Sniffer to show you what goes on inside the get_frame function. I'll kick off the capture using a PING command. I'll issue "ping 192.168.0.150" from the personal computer on the LAN segment with the Easy Ethernet CS8900A. You should recognize the IP address of 192.168.0.150 as it belongs to the Easy Ethernet CS8900A.

Here's what should happen. The personal computer issuing the ping request only knows the Easy Ethernet CS8900A's IP address, which I enter in the ping command. The personal computer, not knowing the IA or hardware address of the Easy Ethernet CS8900A, will issue an ARP request. The ARP request is a broadcast message asking for a hardware address from the owner of the IP address in the ping command. The Easy Ethernet CS8900A will see a broadcast IA and allow the packet to be retrieved for processing. Once the Easy Ethernet CS8900A sees that the ARP request belongs to the Easy Ethernet CS8900A's IP address, the Easy Ethernet CS8900A will generate an ARP reply. I'm going to stop the capture just before the Easy Ethernet CS8900A assembles and sends the ARP reply. The MPLAB ICE 2000 capture results are shown in Screen Capture 8.1.

Let's use the get_frame function code and the hex dump you see in Screen Capture 8.1 to see if what I said should happen after issuing the PING command really happens. We'll start with the conditions that satisfy the do-while loops in the Main Service Loop that ultimately call the get_frame function.

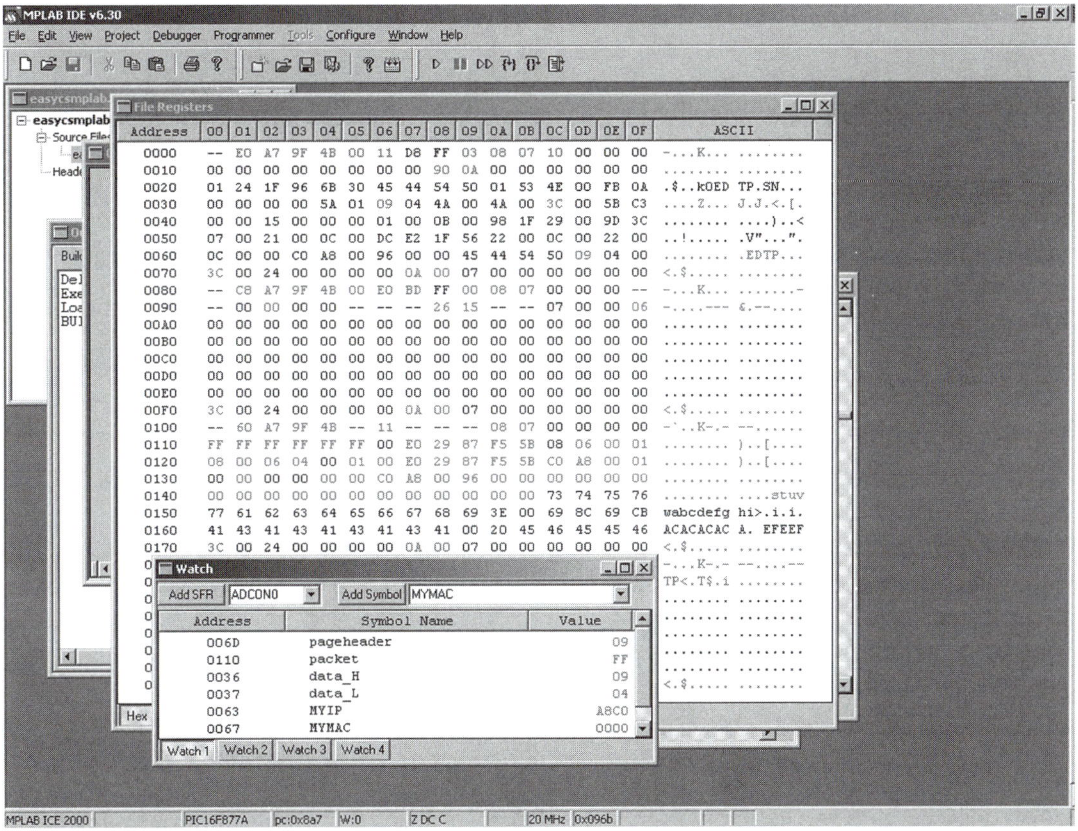

Screen Capture 8.1: This graphic gives you an idea of the level of detail provided by the MPLAB ICE 2000. I'll supplement screen shots like this with the hex dump data in print format to make it a bit easier to read and follow.

Address	Symbol Name	Value
006D	pageheader	09
0110	packet	FF
0036	data_H	09
0037	data_L	04
0063	MYIP	A8C0
0067	MYMAC	0000

Table 8.1: This is a text version of the Watch Window in Figure 8.1.

The first do-while loop performed the last read of a PacketPage register before passing control to the get_frame function. That means that the data in the data_H and data_L global variables is the content of the Receive Event Register. Using the Watch Window as our guide, we see the data_H and data_L global variables keep their values at addresses 0x0036 and 0x0037, respectively. Let's lay the contents of the data_H and data_L variables into the Receiver Event Register bit scheme.

RECEIVER EVENT REGISTER

7	6	5	4	3	2	1	0
Dribblebits	IAHash	000100					

F	E	D	C	B	A	9	8
	Extradata	Runt	CRCerror	Broadcast	Individual Adr	Hashed	RxOK

Laying the data_L value (0x04) into bits 0:7 of the Receiver Event Register tells us that no bits dribbled in after the packet reception was completed. Also, since we're not using the CS8900A-CQ hash filter to determine which packets to accept, the IAHash bit is zero. That leaves only one bit set in the Receiver Event Register identifier, which gives the lower byte of the Receiver Event Register a value of 0x04.

Turning our attention to the upper byte of the Receiver Event Register and laying in 0x09 informs us that the incoming frame is a broadcast frame and it was received into the CS8900A-CQ receiver queue without encountering any problems.

Take another look at Table 8.1, as it identifies the microcontroller memory area (the packet array starting at address 0x0110) that holds the frame data that gets transferred from the CS8900A-CQ receive buffer.

```
Address   00 01 02 03 04 05 06 07 08 09 0A 0B 0C 0D 0E 0F       ASCII

  0110    FF FF FF FF FF FF 00 E0 29 87 F5 5B 08 06 00 01 ........ )..[....
  0120    08 00 06 04 00 01 00 E0 29 87 F5 5B C0 A8 00 01 ........ )..[....
  0130    00 00 00 00 00 00 C0 A8 00 96 00 00 00 00 00 00 ........ ........
  0140    00 00 00 00 00 00 00 00 00 00 00 00 00 00 00 00 ........ ........
  0150    00 00 00 00 00 00 00 00 00 00 00 00 00 00 00 00 ........ ........
  0160    00 00 00 00 00 00 00 00 00 00 00 00 00 00 00 00 ........ ........
```

Hex Dump 8.1: Pretty soon, you'll be picking out specific protocol fields in hex dumps like this without external visual aids.

Note that the first 6 bytes of Hex Dump 8.1 beginning at address 0x0110 contain 0xFF. This series of 0xFF bytes is a broadcast address, and we've already been told that this is a broadcast packet by the Receiver Event Register.

The "packet" area in Hex Dump 8.1 is a byte array that contains all of the frame data. The neat thing about the packet array is that every byte of data is in order. Since this is frame data, there is no preamble and no SOF delimiter. So, the packet array begins with the DA, or Destination Address. In our Easy Ethernet CS8900A code, the DA begins at location 0x00 of the packet array and continues for 6 bytes ending at packet array location 0x05. The DA begins at location 0x0110 in Hex Dump 8.1, which is the beginning of the packet byte array, and contains 6 bytes of 0xFF, which we now know is a broadcast address.

OK…now the get_frame function is called, and the first thing that happens is a read of the RxStatus word that is the same information that is contained in the Receiver Event Register we read in the Main Service Loop do-while loop. The next word following the RxStatus word is the RxLength, which is the length of the incoming frame. The RxStatus and RxLength words must be read before any of the frame data can be transferred from the CS8900A-CQ to the PIC16F877 packet array. The RxStatus and RxLength values are kept in another PIC16F877 array called pageheader. Table 8.1 tells us that the pageheader array begins at PIC16F877 memory location 0x006D.

```
Address 00 01 02 03 04 05 06 07 08 09 0A 0B 0C 0D 0E 0F      ASCII

 0060   0C 00 00 C0 A8 00 96 00 00 45 44 54 50 09 04 00 ........ .EDTP...
 0070   3C 00 24 00 00 00 00 0A 00 07 00 00 00 00 00 00 <.$..... ........
```

Hex Dump 8.2: If you look closely, you can pick out the Easy Ethernet CS8900A's IP (0x0063:0x0066) and MAC (0x0067:0x006C) addresses.

In Hex Dump 8.2, you can see the high-order byte of the RxStatus word at 0x006D and the low-order byte of the RxStatus word at 0x006E. The RxLength word follows, beginning at PIC16F877 memory location 0x006F.

A look at the beginning of the get_frame function in Code Snippet 8.8 shows the code collecting the RxStatus and RxLength values from the CS8900A-CQ and storing them in the PIC16F877's pageheader array.

```
//******************************************************************
//*   Ethernet Header Layout
//******************************************************************
int8  pageheader[4];
#define  enetpacketstatusH    0x00
#define  enetpacketstatusL 0x01
#define  enetpacketLenH 0x02
#define  enetpacketLenL 0x03
//******************************************************************
//*   PacketPage I/O Port Definitions
//******************************************************************
#define  pageport_RxTxData0      0x00  //Receive/Transmit data Port 0
//******************************************************************
//*   Receive a Frame
//******************************************************************
```

```
void get_frame()
{
   int8 i;
   dataport_in;
      RpppH(pageport_RxTxData0,pageheader[enetpacketstatusH]);
      RpppL(pageport_RxTxData0,pageheader[enetpacketstatusL]);
      RpppH(pageport_RxTxData0,pageheader[enetpacketLenH]);
      RpppL(pageport_RxTxData0,pageheader[enetpacketLenL]);
rxlen = make16(pageheader[enetpacketLenH],pageheader[enetpacketLenL]);
//printf("rxlen=%lu\r\n",rxlen);
```

Code Snippet 8.8: The dataport_in macro configures the PIC16F877 databus pins as inputs.

After the PIC16F877's data port is put into input mode, the receive status is read high-order byte first using the RpppH (Read PacketPage Port High) macro. Notice that the receive status is stored for later use, but never used. Even so, it still must be read.

The next read brings in the length of the frame that will be transferred from the CS8900A-CQ to the PIC16F877. This is the total length of bytes starting with the DA and ending with the last byte before the 4-byte CRC value. The RxLength word is used to provide a count value for retrieving the rest of the frame from the CS8900A-CQ receive queue.

```
//******************************************************************
//*   PacketPage I/O Port Definitions
//******************************************************************
#define  pageport_RxTxData0      0x00  //Receive/Transmit data Port 0
int8  packet[96];    //50 bytes of UDP data available
//******************************************************************
//*   Receive a Frame
//******************************************************************
   for(i=0;i<rxlen;i+=2)

         {
            //dump any bytes that will overrun the receive buffer
            if(i < 96)
            {
               RpppL(pageport_RxTxData0,packet[i]);
               RpppH(pageport_RxTxData0,packet[i+1]);
            }
         }
```

Code Snippet 8.9: Any byte after byte number 96 will be tossed into the bit bucket and discarded. You can experiment with the maximum frame value as it depends on how much free RAM you have to work with.

We know what rxlen will resolve to as the RxLength value becomes the rxlen value and the RxLength value is stored in the pageheader array. RxLength is equal to 0x003C, which translates to a frame length of 60 bytes. So, no bits in this frame will be tossed into the bit bucket.

The CS8900A-CQ receive buffer is then read in low byte/high byte order (Code Snippet 8.9) into the PIC16F877 packet array memory area for the length of the buffered frame, which is determined by the value of rxlen. The 60 bytes of this frame are shown in Hex Dump 8.1.

Through trial and error, I determined that the PIC16F877's frame buffer area extends safely out to be 96 bytes. There is plenty of leftover RAM in the PIC16F877 in other banks that can be used, and for starters, the 96 bytes allocated to the packet array buffer area is plenty. To keep things from getting out of hand, we must only accept frames of 96 bytes or less, and in transmit mode only fill our PIC16F877 packet buffer area with 96 bytes maximum. If you're on a LAN with personal computers running Windows, the 96-byte incoming frame length limit is certain to be exceeded. The good news is that we can throw away any incoming bytes above number 96. In the get_frame function if incrementing the *i* variable in the *for(i=0;i<rxlen;i+=2)* loop, which is pointing to the next available packet array memory location, puts the "i" byte count 96, the rest of the remaining bytes are read from the CS8900A-CQ receive buffer and dumped into the bitbucket. Instead of throwing away bytes beyond number 96 in the CS8900A-CQ receive buffer, we could use the receive event data to trigger the CS8900A-CQ to move on to the next frame in its buffer after byte number 96, but since we're reading the entire frame there's no need to do that. By not prematurely trashing the frame ensures that the entire frame is read whether we will use all of the data or not.

Now that the frame has been transferred from the CS8900A-CQ buffer memory to the PIC16F877's buffer memory, the contents of the frame must be processed. We already know quite a bit about the frame we just transferred, and thanks to the MPLAB ICE 2000, we have all of the clues needed to complete the analysis of the newly acquired frame data. This is where the second half of the get_frame function comes into play.

Earlier, I predicted that an ARP request would be the first message received by the Easy Ethernet CS8900A from the personal computer sharing the LAN segment.

```
//******************************************************************
//*   Ethernet Header Layout
//******************************************************************
int8  pageheader[4];
#define  enetpacketstatusH 0x00
#define  enetpacketstatusL 0x01
#define  enetpacketLenH    0x02
#define  enetpacketLenL    0x03
int8  packet[96];        //50 bytes of UDP data available
#define  enetpacketDest0   0x00  //destination mac address
```

```
#define   enetpacketDest1    0x01
#define   enetpacketDest2    0x02
#define   enetpacketDest3    0x03
#define   enetpacketDest4    0x04
#define   enetpacketDest5    0x05
#define   enetpacketSrc0     0x06   //source mac address
#define   enetpacketSrc1     0x07
#define   enetpacketSrc2     0x08
#define   enetpacketSrc3     0x09
#define   enetpacketSrc4     0x0A
#define   enetpacketSrc5     0x0B
#define   enetpacketType0    0x0C   //type/length field
#define   enetpacketType1    0x0D
#define   enetpacketData     0x0E   //IP data area begins here
//*******************************************************************
//*   Receive a Frame
//*******************************************************************
   if(packet[enetpacketType0] == 0x08 && packet[enetpacketType1] == 0x06)
   {
      if(packet[arp_hwtype+1] == 0x01 &&
      packet[arp_prtype] == 0x08 && packet[arp_prtype+1] == 0x00 &&
      packet[arp_hwlen] == 0x06 && packet[arp_prlen] == 0x04 &&
      packet[arp_op+1] == 0x01 &&
      MYIP[0] == packet[arp_tipaddr] &&
      MYIP[1] == packet[arp_tipaddr+1] &&
      MYIP[2] == packet[arp_tipaddr+2] &&
      MYIP[3] == packet[arp_tipaddr+3] )
      arp();
   }
```

Code Snippet 8.10: I purposely overloaded the packet array dependencies to illustrate the location of the bytes that determine if the packet received by the CS8900A-CQ carried an ARP frame.

If we can satisfy the *if* statement in Code Snippet 8.10, the data contained in Hex Dump 8.1 is an ARP frame of some sort. The layout of the PIC16F877 packet array says that we want to look at memory locations 0x0C and 0x0D in the packet array. If the enetpacketType0 value is 0x08 and the value of enenpacketType1 is 0x06, this is an ARP frame. Checking Hex Dump 8.1 verifies just that. We have a broadcast frame in the PIC16F877 memory that so far has the markings of an ARP request. The code goes on to check other fields of the packet that would confirm this is really an ARP request frame. In addition, the ARP request check code in Code Snippet 8.9 checks the destination IP address within the ARP frame to make sure the ARP request frame is addressed to the Easy Ethernet CS8900A.

If you browse through the PIC16F877 packet byte array, you'll notice that the bytes within the packet array match with the fields found in an Ethernet frame. The only exceptions being the Pad and FCS fields, which are not kept in the microcontroller buffer memory.

Remember that earlier we coded the appropriate CS8900A-CQ registers so that the Pad and FCS are generated by the CS8900A-CQ at transmit time. In reality, our packet array is really an Ethernet frame array. I used the name packet because ultimately that's what the collection of frame data will end up as. As you're comparing bytes of the packet array to fields in the Ethernet packet graphic, the bytes are transmitted left to right according to our Ethernet packet graphic and bytes are transmitted beginning at packet array location 0x00 in our code.

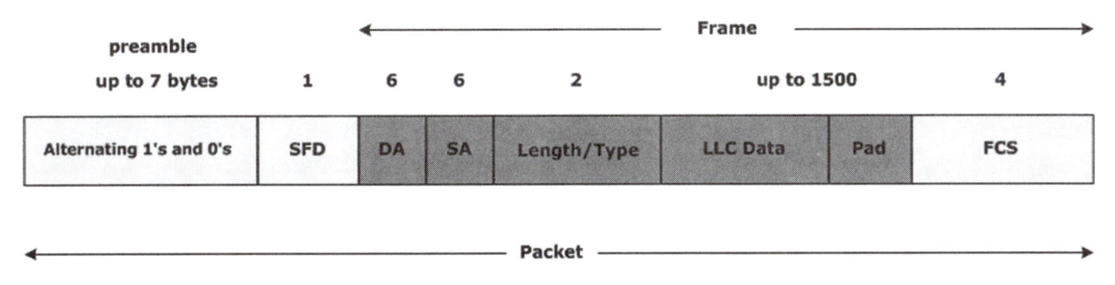

Figure 8.1: Everything in red (gray area) is contained in the incoming frame data shown in Hex Dump 8.1.

OK…where are we? So far, we've:

- Effected the PIC16F877-to-CS8900A-CQ interface
- Reset and Initialized the CS8900A-CQ
- Loaded transmit and receive parameters into the CS8900A-CQ
- Loaded the MAC address as *00EDTP*
- Enabled the CS8900A-CQ's transmitter and receiver
- Received what seems to be an ARP request

Let's find out a little more about ARP.

The Art of ARP

Let's assume another host such as a personal computer is trying to communicate with the Easy Ethernet CS8900A. Let's also assume that no other communication sessions between the host personal computer and the Easy Ethernet CS8900A have occurred. Going with our assumptions, the personal computer has to first find out how to contact the Easy Ethernet CS8900A. Since there was no prior contact with the Easy Ethernet CS8900A, the only information the host personal computer has is the Easy Ethernet CS8900A's IP address, which has been provided by the human user or via a program procedure. One more assumption must be made here. The host personal computer knows the Easy Ethernet CS8900A's IP address and that is all it knows about the Easy Ethernet CS8900A. To communicate with the Easy Ethernet CS8900A, the host must learn the Easy Ethernet CS8900A's hardware address. The Easy Ethernet CS8900A's hardware address is known to you as its MAC address or

Individual Address (00EDTP). The host personal computer learns the Easy Ethernet CS8900A's hardware address by issuing an ARP request to the Easy Ethernet CS8900A. ARP is short for Address Resolution Protocol. Let's look further at the data in Hex Dump 8.1 and try to determine if the frame in our PIC16F877's packet buffer is an ARP request frame. I've placed a copy of Hex Dump 8.1 in the text here for your convenience.

```
Address   00 01 02 03 04 05 06 07 08 09 0A 0B 0C 0D 0E 0F       ASCII

  0110    FF FF FF FF FF FF 00 E0 29 87 F5 5B 08 06 00 01 ........ )..[....
  0120    08 00 06 04 00 01 00 E0 29 87 F5 5B C0 A8 00 01 ........ )..[....
  0130    00 00 00 00 00 00 C0 A8 00 96 00 00 00 00 00 00 ........ ........
  0140    00 00 00 00 00 00 00 00 00 00 00 00 00 00 00 00 ........ ........
  0150    00 00 00 00 00 00 00 00 00 00 00 00 00 00 00 00 ........ ........
  0160    00 00 00 00 00 00 00 00 00 00 00 00 00 00 00 00 ........ ........
```

We already know that the first 6 bytes represent a broadcast address. A quick look at our Ethernet packet graphic and the Easy Ethernet CS8900A source code tells us the next 6 bytes are the sender's hardware address (source address or SA). The SA may not be as meaningful as our homebrewed MAC address, but if you break down the bytes and check them against the IEEE registration in Figure 8.2, you would find that they have a meaning. I can tell you that the SA is from an Ethernet NIC sold by SMC. That's more than you need to know about the SA's bytes at this point.

```
00-E0-29  (hex)          STANDARD MICROSYSTEMS CORP.
00E029    (base 16)      STANDARD MICROSYSTEMS CORP.
                         6 HUGHES
                         IRVINE CA 92718
                         UNITED STATES
```

Figure 8.2: You can find the owner of any OUI by visiting the following IEEE web site: http://standards.ieee.org/regauth/oui/index.shtml. The OUI text for the SMC NIC in my personal computer, that is included in this figure, was taken directly from a page on the IEEE web site.

Moving on, the next 2 bytes at addresses 0x0C and 0x0D are the type/length bytes and hold some significance in our search for the purpose of this frame. We only have to go as far as the beginning of the second half of the get_frame function in the source code to gain some valuable knowledge about the yet unknown frame.

We already know that the enetpackeType0 and enetpackeType1 bytes satisfy the criteria set forth by the *if(packet[enetpacketType0] == 0x08 && packet[enetpacketType1] == 0x06)* statement that begins the second half of our get_frame function. The combination of 0x0806 signals that the frame is an Ethernet ARP frame.

An ARP frame has a distinct layout. So, in the Easy Ethernet CS8900A source code, I've laid out the contents of an ARP frame just as it would appear in microcontroller memory (Code Snippet 8.11). The ARP bytes are located inside the Ethernet frame data area.

```
//*******************************************************************
//*   Ethernet Header Layout
//*******************************************************************
#define  enetpacketData         0x0E  //IP data area begins here
//*******************************************************************
//*   ARP Layout
//*******************************************************************
#define  arp_hwtype             0x0E
#define  arp_prtype             0x10
#define  arp_hwlen              0x12
#define  arp_prlen              0x13
#define  arp_op                 0x14
#define  arp_shaddr             0x16  //arp source mac address
#define  arp_sipaddr            0x1C  //arp source ip address
#define  arp_thaddr             0x20  //arp target mac address
#define  arp_tipaddr            0x26  //arp target ip address
```

Code Snippet 8.11: Notice that the ARP bytes lie inside the IP data area.

The first real ARP bytes begin at memory location 0x011E in the microcontroller memory and location 0x0E in the packet array. The 0x01 value at memory location 0x011F denotes the hardware type, which is 10 MB Ethernet. Since an IP address was used to find the Easy Ethernet CS8900A, the next 2 bytes beginning at address 0x0120 (0x0800) identify the protocol as IP (Internet Protocol). Also, since the value of the Length/Type field is greater than 0x0600, the field is defined to be a Type field.

We know that a MAC or hardware address is 6 bytes in length and an IP address is 4 bytes in length. This is verified by the 2 bytes beginning at memory address 0x0122. The byte at address 0x0122 represents the length of the hardware address in bytes (6 bytes), and the value of the byte at address 0x0123 is the length of the protocol address, which in the case of IP is 4 bytes. The next 2 bytes nail it down. The 0x0001 in this position tells us the frame is an ARP request.

```
Address  00 01 02 03 04 05 06 07 08 09 0A 0B 0C 0D 0E 0F     ASCII

  0110                                             00 01 ........ )..[....
  0120   08 00 06 04 00 01 00 E0 29 87 F5 5B C0 A8 00 01 ........ )..[....
```

Hex Dump 8.3: I've removed some preceding bytes for clarity. Was I right about the ARP or what?

Now that we know the frame is a request for the Easy Ethernet CS8900A's MAC address, it would be nice to know who requested our hardware address. In addition, every host on the LAN is processing this request up to this point. If this ARP request isn't pointed toward our Easy Ethernet CS8900A, it would be nice to know that before we process a return frame. Fortunately, the sending host includes both its IP and hardware address in the ARP request frame. Actually, we already know where the packet came from in the hardware address sense as the SA is the sending host's hardware or MAC address. Beginning at microcontroller memory address 0x0126, the sender's hardware address is enumerated and is immediately followed by the sender's IP address. Decoding the bytes beginning at microcontroller memory address 0x012C yields a sender IP address of 192 (0xC0).168 (0xA8).0 (0x00).1 (0x01). The Easy Ethernet CS8900A knows it is responsible for answering this ARP request because its IP address exists at memory location 0x0136. Earlier in the CS8900A-CQ initialization process, we placed our IP address in an array called MYIP. All we have to do to see if this ARP request is for us, is compare the bytes in the MYIP array with the destination IP address buried within the ARP request frame. Let's decode the Easy Ethernet CS8900A IP address just for fun:

```
192(0xC0).168(0xA8).0(0x00).150(0x96).
```

If you're wondering what the 6 bytes of 0x00 beginning at memory address 0x0130 are for in Hex Dump 8.4, that's a neat thing about the ARP frame. We'll put our hardware address there, change a byte or two here and there, and send the packet back to the sender.

```
Address 00 01 02 03 04 05 06 07 08 09 0A 0B 0C 0D 0E 0F      ASCII

 0110   FF FF FF FF FF FF 00 E0 29 87 F5 5B 08 06 00 01 ........ )..[....
 0120   08 00 06 04 00 01 00 E0 29 87 F5 5B C0 A8 00 01 ........ )..[....
 0130   00 00 00 00 00 00 C0 A8 00 96 00 00 00 00 00 00 ........ ........
```

Hex Dump 8.4: An ARP response is a swap-this-for-that-fill-in-the-blanks process that returns vital information about a remote host to the ARP requester.

Although it's fun reading hex dumps and solving problems byte by byte, there is a really nifty tool out there to help sort out what's going on inside host devices participating in an Ethernet LAN. It's called a Sniffer and the one I will be using throughout this book is manufactured by Network Associates. The Sniffer is a piece of software that runs on a personal computer. You can also purchase certified Ethernet NICs that will allow the Sniffer software to capture hardware errors as well as packet errors on the LAN.

I used the Sniffer extensively when writing the microcontroller driver firmware for the CS8900A-CQ. The Sniffer was an indispensable tool in my firmware development of the CS8900A-CQ firmware and I want to use it to give you a deeper insight of what is going on inside an Ethernet frame. Sniffer Screen Capture 8.2 is a Sniffer view of the ARP request hex dump we just analyzed.

In Sniffer Screen Capture 8.2, the DLC Header (Data Link Control Header) is a part of the service that is provided by the Data Link Layer. The Data Link Layer is responsible for providing reliable data transfer across one physical link or communications path within a network. The Data Link Layer consists of two sublayers: the Logical Link Control layer and the Media Access Control layer. The Logical Link Control layer concentrates on flow control and error control. You already know a bit about the Media Access Control, or MAC layer. The MAC layer allows us to share the LAN with a number of hosts using unique hardware addresses. As you can see from its contents, the DLC Header is part of the Ethernet protocol.

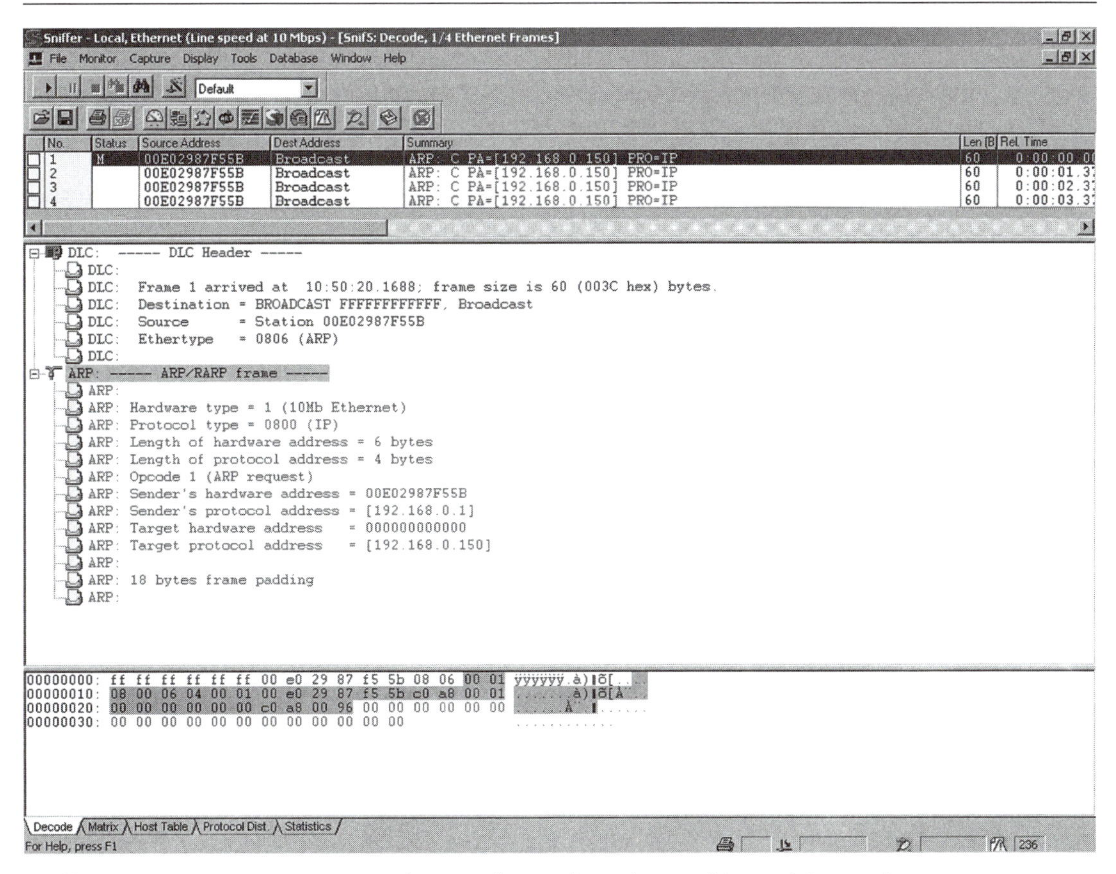

Sniffer Screen Capture 8.2: Don't know what a "broadcast address" is? You do now...

```
Address 00 01 02 03 04 05 06 07 08 09 0A 0B 0C 0D 0E 0F       ASCII

 0060   0C 00 00 C0 A8 00 96 00 00 45 44 54 50 09 04 00 ........ .EDTP...
 0070   3C 00 24 00 00 00 00 0A 00 07 00 00 00 00 00 00 <.$..... ........
```

Hex Dump 8.5: DLC is short for Data Link Control.

The frame length is displayed at locations 0x6F and 0x70 in the PIC16F877 hex frame dump as shown in Screen Capture 8.1 and Hex Dump 8.5. The frame length values were gleaned from the RxLength header bytes we read from the CS8900A-CQ. When the frame is captured using a Sniffer, no searching through a hex dump is necessary as the Sniffer capture simply tells you the frame size in the DLC Header area.

Starting at location 0x0110 in the PIC16F877 ARP request frame dump (Hex Dump 8.1), you can follow along byte for byte in the Sniffer hex frame dump area, which resides at the bottom of the Sniffer capture window. The Sniffer gives us a byte detail view of the frame that is not immediately obvious in the PIC16F877 ARP request frame dump. Let's use the Sniffer capture as a pointer to the associated data in the PIC16F877 ARP request frame dump.

In the DLC area of the Sniffer capture, a frame size of 60 tells us that there's probably been some padding of the original data to meet the minimum frame length requirements, and if you're not quick with decimal-to-hexadecimal conversion, the Sniffer gives you the frame size in both hex and decimal.

Just in case you aren't a networking guru and don't know what a broadcast address is, the Sniffer makes sure you're properly educated on broadcast addressing in the DLC Header area. A look at the Sniffer hex dump and the DLC Header descriptive area indicates that the DA (Destination Address) is filled with 0xFF's, which as we already know is a broadcast address. The Sniffer has also isolated the source address (SA) that belongs to a personal computer running an SMC NIC on the local LAN segment. We know it's an SMC NIC because we looked it up on the IEEE OUI registration web page. By the way, you won't find an IEEE entry for our homebrew OUI *00EDTP*.

The Sniffer Ethertype is an element of data that the Easy Ethernet CS8900A firmware will key on. These 2 bytes tell the Easy Ethernet CS8900A firmware that the frame transferred from the CS8900A-CQ to the PIC16F877 is an ARP frame. The rest of the Easy Ethernet CS8900A ARP firmware checks every field you see in the Sniffer ARP/RARP frame. If it all matches the criteria we programmed into our ARP algorithm (which it does), then an ARP request has been tendered by a personal computer host and the Easy Ethernet CS8900A must assemble and transmit an ARP reply to the sender. We also know that the real length of the original ARP request frame was 42 bytes, as the Sniffer capture tells us that 18 bytes of the frame are padding. Now you know why I love my Sniffer.

The top of the Sniffer capture is a short-form description of the frames that are held in the Sniffer's capture buffer. Remember that I set a break point in the Easy Ethernet CS8900A firmware that stops the Easy Ethernet CS8900A code just after the ARP request is received. So, the events you see in the Sniffer capture are ARP requests from the personal computer, which were never answered by the Easy Ethernet CS8900A. A look at the destination address next to the summary box gives us a clue as to if it is an ARP request or an ARP reply. Since the DA is all 0xFF's, which translates to a broadcast address, it's a safe bet this frame is an ARP request frame.

I want to show you a successful ARP request/ARP reply sequence. So, I removed the breakpoint I set in the Easy Ethernet CS8900A firmware using the MPLAB ICE 2000. I also issued an *arp –d ** command to clear the personal computer's ARP cache. With the Easy Ethernet CS8900A running normally and the personal computer ignorant to the existence of the Easy Ethernet CS8900A, I issued another PING command from the personal computer.

Screen Capture 8.3 looks much like the ARP request Sniffer capture with the exception of the events area.

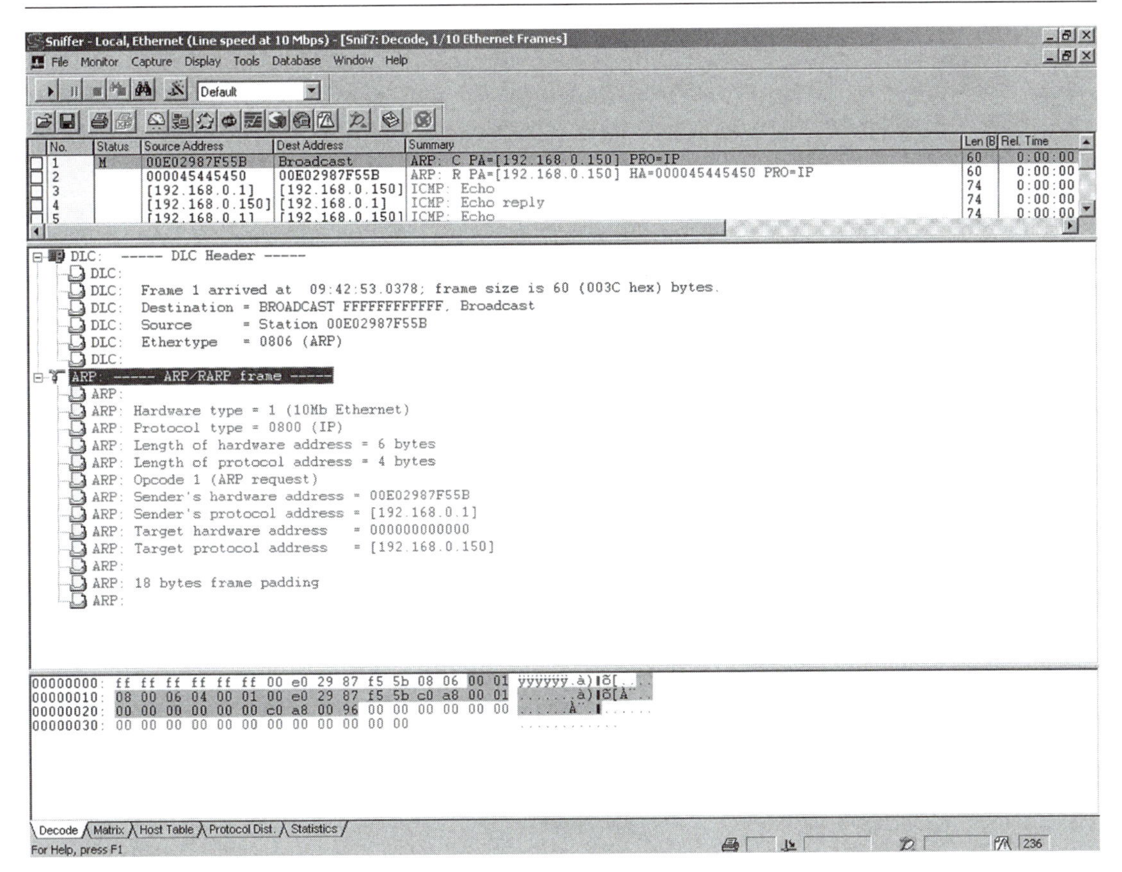

Sniffer Screen Capture 8.3: You can see what's on the way and who's doing who in the Summary Window.

In Screen Capture 8.4, note that there is an additional hardware address supplied by the Easy Ethernet CS8900A and an opcode in the ARP frame denoting the frame as an ARP reply. This whole ARP process is somewhat like an algebra problem. You use the known (IP addresses) to solve for the unknowns (hardware addresses).

An ARP frame has been successfully received by the CS8900A-CQ and transferred to the PIC16F877 on the Easy Ethernet CS8900A. For the ARP process to be successful, the Easy Ethernet CS8900A must answer the ARP request and provide the requested information to the personal computer host. Let's take a look at the firmware that supplies the Easy Ethernet CS8900A's ARP reply, beginning with Code Snippet 8.12.

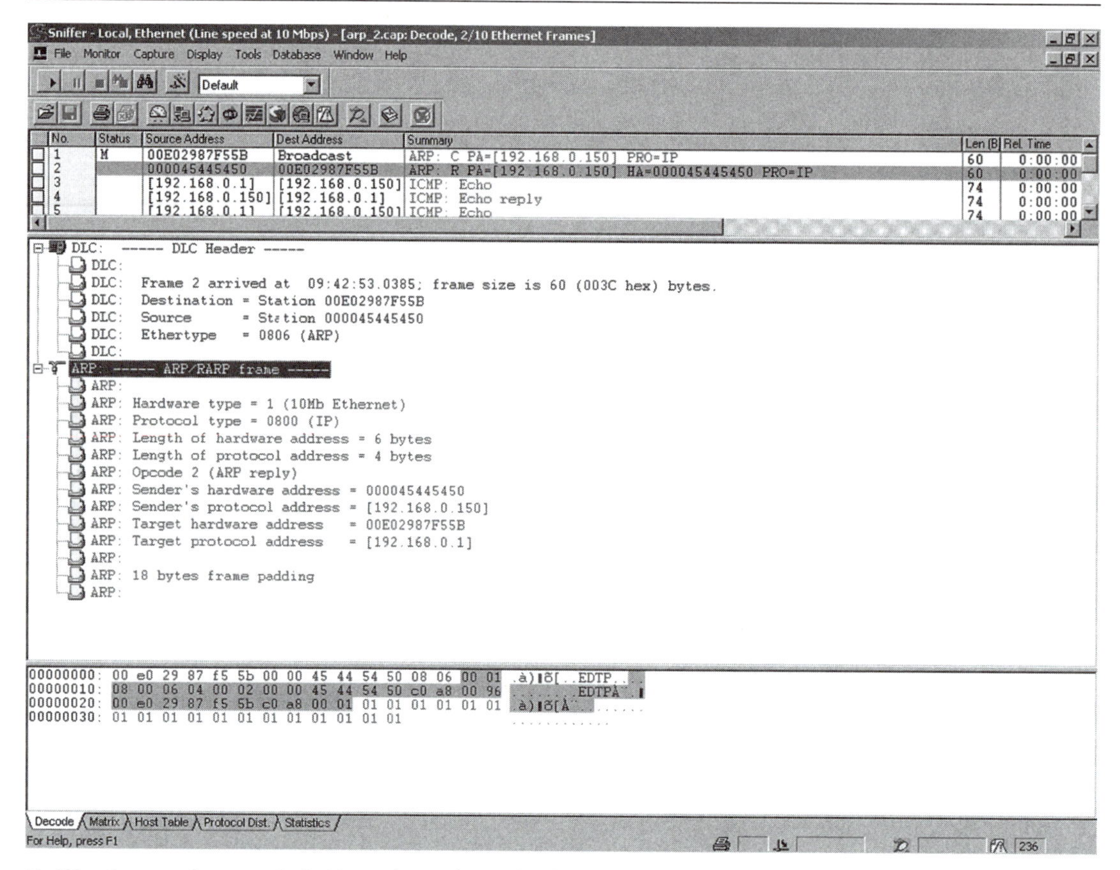

Sniffer Screen Capture 8.4: Note that what is highlighted in the detail area is also highlighted in the hex dump area.

```
//*******************************************************************
//*   PacketPage I/O Port Definitions
//*******************************************************************
#define  pageport_TxCmd    0x04  //Transmit Command
#define  pageport_TxLen    0x06  //Transmit Length
//*******************************************************************
//*   PacketPage Internal Register Definitions
//*******************************************************************
#define  ppageBusStatus    0x0138  //Bus Status
//*******************************************************************
//*   PacketPage Bus Status Bit Definitions
//*******************************************************************
#define  BUSSTA_RDY4TXNOW_BIT 0x0000
//*******************************************************************
//*   SEND ARP RESPONSE
//*******************************************************************
```

```
void arp()
{
   dataport_out;
   WpppL(pageport_TxCmd,TXCMD_AFTER_ALL);
   WpppH(pageport_TxCmd,0);
   WpppL(pageport_TxLen,0x2A);
   WpppH(pageport_TxLen,0);
   do{
      RPP(ppageBusStatus);
   }while(!(bit_test(data_H,BUSSTA_RDY4TXNOW_BIT)));
```

Code Snippet 8.12: The WpppL/WpppH (Write PacketPage Port) macros are issuing a bid for transmit buffer space on the CS8900A-CQ. Once the bid is accepted, the CS8900A-CQ allows the transfer of data into the newly allocated CS8900A-CQ transmit buffer area.

In the SEND ARP RESPONSE code (Code Snippet 8.12), the data port of the PIC16F877 is commanded to become an output port. The CS8900A-CQ is instructed to wait until all bytes have been transferred to its transmit buffer from the PIC16F877 microcontroller before beginning the transmit operation. 0x2A or 42 bytes of ARP response frame buffer area are requested from the CS8900A-CQ. After issuing the bid to the CS8900A-CQ, the PIC16F877 polls the CS8900A-CQ Bus Status Register looking for permission to start the data transfer.

BUS STATUS REGISTER

7	6	5	4	3	2	1	0
TxBidErr		011000					

F	E	D	C	B	A	9	8
							Rdy4TxNOW

Once the CS8900A-CQ allocates the space and sets the RDY4TXNOW_BIT in the CS8900A-CQ Bus Status Register, the bytes of our ARP reply flow out of the PIC16F877's packet buffer into the CS8900A-CQ transmit buffer in the order shown in the Ethernet packet graphic (Figure 8.1).

```
//*****************************************************************
//*   PacketPage I/O Port Definitions
//*****************************************************************
#define  pageport_RxTxData0      0x00  //Receive/Transmit data Port 0
//*****************************************************************
//*   SEND ARP RESPONSE
//*****************************************************************
// GENERATE THE ARP RESPONSE DA
   dataport_out;
```

```
WpppL(pageport_RxTxData0,packet[enetpacketSrc0]);
WpppH(pageport_RxTxData0,packet[enetpacketSrc1]);
WpppL(pageport_RxTxData0,packet[enetpacketSrc2]);
WpppH(pageport_RxTxData0,packet[enetpacketSrc3]);
WpppL(pageport_RxTxData0,packet[enetpacketSrc4]);
WpppH(pageport_RxTxData0,packet[enetpacketSrc5]);
```

Code Snippet 8.13: Here's an example of some swap-this-for-that code. Imagine answering a letter. You send your reply to the return address on the envelope. The return address in this case is the SA (Source Address) in the ARP request frame.

The Easy Ethernet CS8900A firmware derives the ARP reply DA from the SA of the ARP request in Code Snippet 8.13. The 6 bytes of the ARP request SA were stored in the PIC16F877's packet array using the enetpacketSrcX elements. All we have to do is to simply copy the original SA bytes into the CS8900A-CQ buffer as if they were the DA bytes.

The next field that we should transfer to the CS8900A-CQ in the Ethernet frame order is the SA. The SA is the Easy Ethernet CS8900A's hardware address and that is already stored as an array within the PIC16F877 we named MYMAC. The Easy Ethernet CS8900A's MAC address is written into the CS8900A-CQ buffer memory in order as *00EDTP* in Code Snippet 8.14.

```
//******************************************************************
//*   PacketPage I/O Port Definitions
//******************************************************************
#define  pageport_RxTxData0      0x00  //Receive/Transmit data Port 0
//******************************************************************
//*   HARDWARE (MAC) ADDRESS DEFINITION
//*   YOU MAY CHANGE THIS TO ANY VALID MAC ADDRESS
//******************************************************************
int8 MYMAC[6] = { 0,0,'E','D','T','P' };
//******************************************************************
//*   SEND ARP RESPONSE
//******************************************************************
// GENERATE THE ARP RESPONSE SA

   WpppL(pageport_RxTxData0,MYMAC[0]);
   WpppH(pageport_RxTxData0,MYMAC[1]);
   WpppL(pageport_RxTxData0,MYMAC[2]);
   WpppH(pageport_RxTxData0,MYMAC[3]);
   WpppL(pageport_RxTxData0,MYMAC[4]);
   WpppH(pageport_RxTxData0,MYMAC[5]);
```

Code Snippet 8.14: This code, when compared to answering a letter, is the letter answerer's return address. In the ARP process, this is what the ARP requesting host is asking for.

`Since this is an ARP reply, the Ethernet frame type is still ARP and is denoted by loading 0x0806 into the CS8900A-CQ buffer memory immediately following the SA. This value could have been obtained from the packet array area (enetpacketType0/1) as well, as its value doesn't change between ARP reply and ARP response. The same can be said of the hardware type field (arp_hwtype), the protocol type field (arp_prtype) and the protocol and hardware address length fields (arp_hwlen and arp_prlen). You can simply copy these fields out to the CS8900A-CQ if you wish. I chose to code the explicit values for clarity in Code Snippet 8.15.

```
//******************************************************************
//*   PacketPage I/O Port Definitions
//******************************************************************
#define  pageport_RxTxData0      0x00  //Receive/Transmit data Port 0
//******************************************************************
//*   SEND ARP RESPONSE
//******************************************************************
// GENERATE THE ARP RESPONSE

   //Ethertype = ARP
   WpppL(pageport_RxTxData0,0x08);
   WpppH(pageport_RxTxData0,0x06);

   //Hardware type = 10Mb Ethernet
   WpppL(pageport_RxTxData0,0x00);
   WpppH(pageport_RxTxData0,0x01);

   //Protocol type = IP
   WpppL(pageport_RxTxData0,0x08);
   WpppH(pageport_RxTxData0,0x00);

   //Hardware Address Length/Protocol Address Length
   WpppL(pageport_RxTxData0,0x06);
   WpppH(pageport_RxTxData0,0x04);
```

Code Snippet 8.15: Specifying IP as the protocol implies that the ARP messages travel within the confines of the IP data area.

There are two fields that change values in the ARP response that is generated by the Easy Ethernet CS8900A. The first is the frame opcode. Instead of an ARP request opcode of 0x0001, in Code Snippet 8.16, the frame opcode is replaced with an ARP response opcode of 0x0002.

```
//*********************************************************************
//*    SEND ARP RESPONSE
//*********************************************************************
// GENERATE THE ARP RESPONSE

    WpppL(pageport_RxTxData0,0x00);
    WpppH(pageport_RxTxData0,0x02);
```

Code Snippet 8.16: So far, everything we've coded follows the byte order in Sniffer Screen Capture 8.3. Remember, we're still loading the CS8900A-CQ transmit buffer and nothing has been passed to the ether as of yet.

We already know the sender's full address, as we can find the sender's hardware address in the ARP request SA and within the sender's hardware address area of the ARP request frame. The sender's protocol or IP address can be found in the sender's protocol address area of the ARP request frame.

In Hex Dump 8.6, the sender's hardware address (MAC address) is found to begin at memory offsets 0x0116 and 0x0126.

```
Address 00 01 02 03 04 05 06 07 08 09 0A 0B 0C 0D 0E 0F       ASCII

 0110   FF FF FF FF FF FF 00 E0 29 87 F5 5B 08 06 00 01 ........ )..[....
 0120   08 00 06 04 00 01 00 E0 29 87 F5 5B C0 A8 00 01 ........ )..[....
 0130   00 00 00 00 00 00 C0 A8 00 96 00 00 00 00 00 00 ........ ........
```

Hex Dump 8.6: The Easy Ethernet CS8900A's IP address is also conveniently positioned inside the ARP request frame for use by the ARP responder.

The sender's IP address in Hex Dump 8.6 can be found beginning at memory offset 0x012C.

The idea of an ARP request is to contact the host you want to talk to and request its hardware address with the expectation that you will receive the remote host's hardware address in an ARP reply. The host requesting the ARP reply will use the address information within the ARP reply frame to form a full address for the host it is trying to reach. A full address includes both an IP and MAC address.

With that idea in mind, let's fill in the blanks for the requesting host beginning with the sender's hardware address. In an ARP reply, the Easy Ethernet CS8900A becomes the sender. So, the sender's hardware address will be the Easy Ethernet CS8900A MAC address that is stored in the MAC address array MYMAC.

The requesting host already knew the Easy Ethernet CS8900A's IP address, but it is protocol to include it in the sender's protocol address field. Like the MAC address, the Easy Ethernet CS8900A's protocol or IP address is also stored in the PIC16F877's memory as an array called MYIP.

```
//***************************************************************
//*   SEND ARP RESPONSE
//***************************************************************
// GENERATE THE ARP RESPONSE

   //Sender's hardware address
   WpppL(pageport_RxTxData0,MYMAC[0]);
   WpppH(pageport_RxTxData0,MYMAC[1]);
   WpppL(pageport_RxTxData0,MYMAC[2]);
   WpppH(pageport_RxTxData0,MYMAC[3]);
   WpppL(pageport_RxTxData0,MYMAC[4]);
   WpppH(pageport_RxTxData0,MYMAC[5]);

   //Sender's protocol address
   WpppL(pageport_RxTxData0,MYIP[0]);
   WpppH(pageport_RxTxData0,MYIP[1]);
   WpppL(pageport_RxTxData0,MYIP[2]);
   WpppH(pageport_RxTxData0,MYIP[3]);
```

Code Snippet 8.17: The sender's hardware address begins at memory offset 0x0126 of Hex Dump 8.6. The sender's IP address begins immediately following the sender's MAC address at memory offset 0x012C.

Keep in mind that we are actually building the ARP reply packet inside the CS8900A-CQ transmit buffer and not within the PIC16F877's packet array. In Code Snippet 8.17, we're simply taking known values from the PIC16F877 RAM and transferring them to the CS8900A-CQ in the correct Ethernet frame order.

In the ARP reply process the Easy Ethernet CS8900A has become the sending host and the requesting host has become the target host. The last fields of the ARP reply require the target host's hardware and protocol addresses. Fortunately, those addresses are identified and stored from the original ARP request frame within the PIC16F877's packet array memory area. The original ARP request SA is stored as elements of the packet array (enetpacketSrcX). So, we can use the original SA to fill the target hardware address field within the CS8900A-CQ's transmit buffer. We also captured and stored the ARP requester's protocol address from within the ARP request frame. The packet array elements called *arp_sipaddr* in Code Snippet 8.18 hold the 4 bytes of the ARP requester's IP address.

```
//***************************************************************
//*   SEND ARP RESPONSE
//***************************************************************
// GENERATE THE ARP RESPONSE

   //Target hardware address
   WpppL(pageport_RxTxData0,packet[enetpacketSrc0]);
   WpppH(pageport_RxTxData0,packet[enetpacketSrc1]);
```

```
WpppL(pageport_RxTxData0,packet[enetpacketSrc2]);
WpppH(pageport_RxTxData0,packet[enetpacketSrc3]);
WpppL(pageport_RxTxData0,packet[enetpacketSrc4]);
WpppH(pageport_RxTxData0,packet[enetpacketSrc5]);

//Target Protocol address
WpppL(pageport_RxTxData0,packet[arp_sipaddr]);
WpppH(pageport_RxTxData0,packet[arp_sipaddr+1]);
WpppL(pageport_RxTxData0,packet[arp_sipaddr+2]);
WpppH(pageport_RxTxData0,packet[arp_sipaddr+3]);
```

Code Snippet 8.18: Again we've used captured data from the ARP request frame to fill in the swap-this-for-that blanks in the ARP reply.

If you want to take a look back at what we've discussed so far, you'll see that the Easy Ethernet CS8900A firmware has stuffed the fields needed to reply to an ARP request into the CS8900A-CQ buffer in Ethernet protocol order. This fact becomes clear if you compare what we've fed the CS8900A-CQ with the hex dump at the bottom of Sniffer Screen Capture 8.3.

Once all of the bytes the CS8900A-CQ was told would be sent are transferred from the Easy Ethernet CS8900A's PIC16F877 microcontroller and collected in the CS8900A-CQ transmit buffer, the CS8900A-CQ looks to see if the ether is clear. If there is no traffic on the ether and the CS8900A-CQ determines that its receiver is not receiving data, the CS8900A-CQ then generates a preamble that is immediately followed by an SFD and, assuming no collision occurs, our ARP reply hits the ether followed by a CS8900A-CQ generated CRC. All of the packet building and packet transmission "magic" is performed by the CS8900A-CQ. Once the CS8900A-CQ is initialized, all we have to do is supply some address information and data to the CS8900A-CQ's transmit buffer to pass information between hosts on an Ethernet LAN segment.

CHAPTER 9

PINGing the Easy Ethernet CS8900A

You're well on the way. Now that your Easy Ethernet CS8900A can identify itself to others, let's see if we can get the Easy Ethernet CS8900A to raise its hand when called upon.

I started this whole ARP thing with a PING command. If you know absolutely nothing about Internet protocols, you've probably heard someone talking about "pinging" someone else's computer. PING is actually an application of sorts that is based on the ICMP protocol. It's a quick and nasty way to establish that a remote host is online.

We'll use what we've learned about Ethernet frames and reference the Sniffer capture of the ICMP frame when we move into unknown territory. Hex Dump 9.1 should look familiar. It's a hex dump of the ICMP frame as it appears inside the PIC16F877 microcontroller. All of the Easy Ethernet CS8900A code that you've been introduced to that is used to collect frames from the CS8900A-CQ has brought us to this point (Code Snippet 9.1).

```
//*****************************************************************
//*   Receive a Frame
//*****************************************************************
void get_frame()
{
   int8 i;
   dataport_in;
      RpppH(pageport_RxTxData0,pageheader[enetpacketstatusH]);
      RpppL(pageport_RxTxData0,pageheader[enetpacketstatusL]);
      RpppH(pageport_RxTxData0,pageheader[enetpacketLenH]);
      RpppL(pageport_RxTxData0,pageheader[enetpacketLenL]);
   rxlen = make16(pageheader[enetpacketLenH],pageheader[enetpacketLenL]);
   //printf("rxlen=%lu\r\n",rxlen);
   for(i=0;i<rxlen;i+=2)

            {
               //dump any bytes that will overrun the receive buffer
               if(i < 96)
               {
                  RpppL(pageport_RxTxData0,packet[i]);
                     RpppH(pageport_RxTxData0,packet[i+1]);
               }
            }
```

```
//process an ARP packet
if(packet[enetpacketType0] == 0x08 && packet[enetpacketType1] == 0x06)
{
   if(packet[arp_hwtype+1] == 0x01 &&
   packet[arp_prtype] == 0x08 && packet[arp_prtype+1] == 0x00 &&
   packet[arp_hwlen] == 0x06 && packet[arp_prlen] == 0x04 &&
   packet[arp_op+1] == 0x01 &&
   MYIP[0] == packet[arp_tipaddr] &&
   MYIP[1] == packet[arp_tipaddr+1] &&
   MYIP[2] == packet[arp_tipaddr+2] &&
   MYIP[3] == packet[arp_tipaddr+3] )
   arp();
}
//process an IP packet
else if(packet[enetpacketType0] == 0x08 &&
      packet[enetpacketType1] == 0x00 &&
      packet[ip_destaddr] == MYIP[0] &&
      packet[ip_destaddr+1] == MYIP[1] &&
      packet[ip_destaddr+2] == MYIP[2] &&
      packet[ip_destaddr+3] == MYIP[3])
{
   if(packet[ip_proto] == PROT_ICMP)
   {
      // WE ARE HERE
      i = 0; // this line added to support an emulator break point
      icmp();
   }
   else if(packet[ip_proto] == PROT_UDP)
      udp();
   else if(packet[ip_proto] == PROT_TCP)
      tcp();
   }
```

Code Snippet 9.1: The MPLAB ICE 2000 is great when it comes to stopping "time" in the microcontroller frame of reference.

Now that our personal computer has the Easy Ethernet CS8900A's MAC address in its ARP cache, let's tear the ICMP or PING request in Hex Dump 1 apart byte by byte.

```
Address  00 01 02 03 04 05 06 07 08 09 0A 0B 0C 0D 0E 0F      ASCII

 0110    00 00 45 44 54 50 00 E0 29 87 F5 5B 08 00 45 00 ..EDTP.. )..[..E.
 0120    00 3C 02 33 00 00 80 01 B6 A6 C0 A8 00 01 C0 A8 .<.3.... ........
 0130    00 96 08 00 0A 5C 02 00 41 00 61 62 63 64 65 66 .....\.. A.abcdef
 0140    67 68 69 6A 6B 6C 6D 6E 6F 70 71 72 73 74 75 76 ghijklmn opqrstuv
 0150    77 61 62 63 64 65 66 67 68 69 00 00 00 00 00 00 wabcdefg hi......
```

Hex Dump 9.1: Lots of this should be familiar by now. Even though this frame does something other than ARP, it's still an Ethernet frame.

I've let the cat out of the bag and told you that Hex Dump 9.1 is an ICMP frame. Let's pretend you don't know that yet. So, thus far the CS8900A-CQ has received yet another unknown frame that bears the Easy Ethernet CS8900A's hardware and protocol address. We've just processed an ARP. So, it's safe to assume that the host that requested the Easy Ethernet CS8900A's hardware address wants to communicate with the Easy Ethernet CS8900A and that the unknown frame we just received is from the host that ARPed earlier.

Thus far, a remote host has sent a frame addressed to the Easy Ethernet CS8900A. The CS8900A-CQ signaled the Easy Ethernet CS8900A's microcontroller that a properly addressed frame is ready to be transferred from the CS8900A-CQ's receive queue to the Easy Ethernet CS8900A's PIC16F877 microcontroller's RAM. The Easy Ethernet CS8900A's PIC16F877 microcontroller executed the get_frame function, and we stopped the Easy Ethernet CS8900A's microcontroller at the "WE ARE HERE" point shown in the get_frame source code in Code Snippet 9.1.

If all is as it should be in an Ethernet frame, the first six bytes of the hex frame dump in Hex Dump 9.1 should be a hardware or physical destination address (DA). Decoding the first

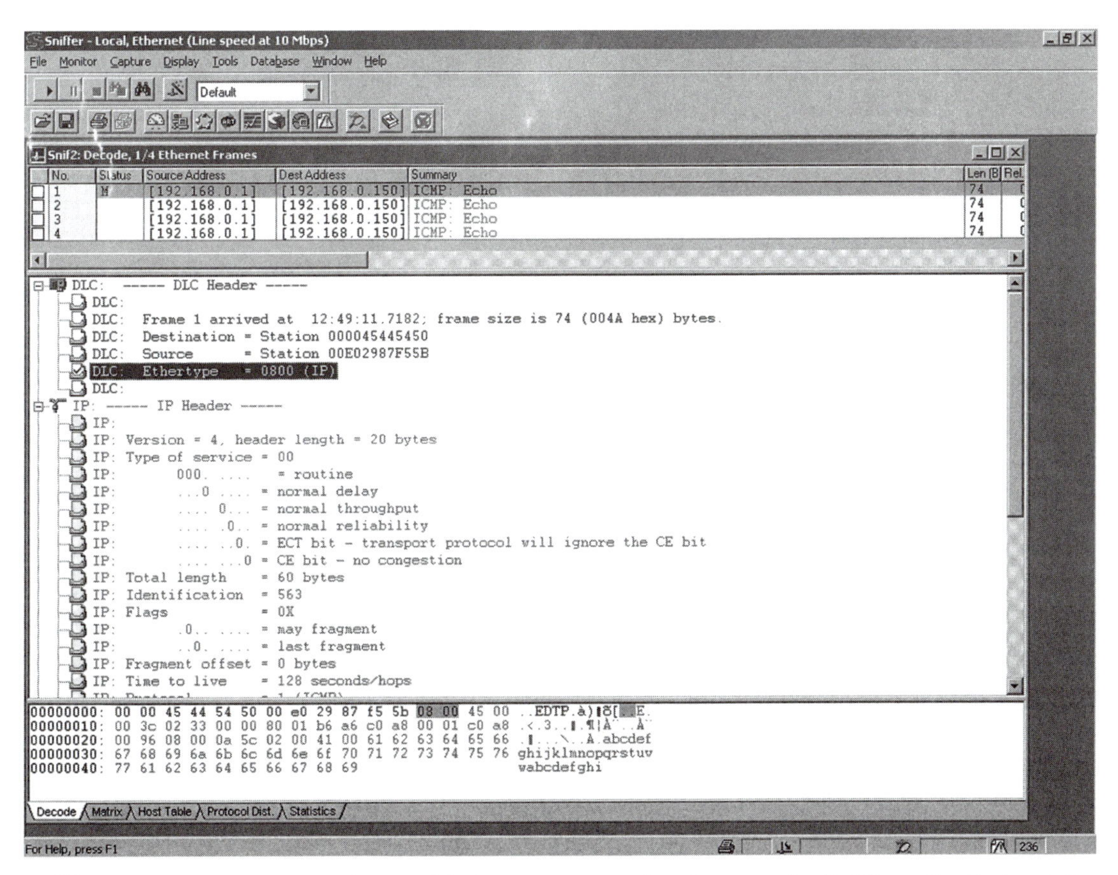

Sniffer Screen Capture 9.1: You could almost stop reading here as the Sniffer tells all about the ICMP frame.

six bytes yields *00EDTP*, which happens to be the Easy Ethernet CS8900A's MAC (hardware) address. Assuming the hosts on our network segment haven't changed, the source address (SA) looks just like the one that ARPed the Easy Ethernet CS8900A earlier. The Ethernet packet type field of 0x0800 doesn't jive with 0x0806, which we know as an ARP frame indicator. At this point, we could consult the one thousand plus pages of technical documentation that describe Internet protocols or we could simply take a look at Sniffer Screen Capture 9.1.

The 0x0800 in the Ethernet packet type field is telling us that the frame is an IP frame. And, to further complicate matters, there's something called and IP Header where our ARP data used to be. This is all by design and is called *encapsulation*. In fact, the ARP data that occupied this space in the ARP process was encapsulated too.

The Ethernet packet is simply a box that all of the Internet protocols can travel in. The IP data area begins at the same point that denotes the beginning of the Ethernet frame data area. Since and ARP packet will never coexist with an IP packet within an Ethernet frame, they can share the data space in a mutually exclusive kind of way. A look at the layout of the PIC16F877's packet array ARP data and IP header starting locations in Code Snippet 9.2 makes this point.

```
//******************************************************************
//*   Ethernet Header Layout
//******************************************************************
#define  enetpacketType0   0x0C  //type/length field
#define  enetpacketType1   0x0D
#define  enetpacketData 0x0E  //IP data area begins here
//******************************************************************
//*   ARP Layout
//******************************************************************
#define  arp_hwtype     0x0E
//******************************************************************
//*   IP Header Layout
//******************************************************************
#define  ip_vers_len    0x0E  //IP version and header length
```

Code Snippet 9.2: You'll soon see that all of the Internet protocols can be shipped in the Ethernet box.

We're obviously not going any further until we crack the contents of the IP header. So, let's do what we do best; tear it down byte by byte.

Figure 9.1 is a graphical depiction of the contents of the IP header. The top row represents bit positions within each 32-bit field of the IP header. Each row in the figure represents 32 bits or 4 bytes.

0..3	4..7	8 .. 15	16 ..	19 .. 23	24 .. 31
Version	IHL	Type of Service	Total Length		
Identification			Flags	Fragment Offset	
Time To Live		Protocol	Header Checksum		
Source IP Address					
Destination IP Address					
Options					Padding
Data					

Figure 9.1: The IP header looks really busy until you break it down into fields and their purposes.

IP packets are called *datagrams*. The definition of a datagram implies that a datagram is an independent entity that can carry a message on a network but cannot guarantee the safe arrival of that message.

The first 4 bits of the IP header contain the datagram's IP version, which is currently 4. The next 4 bits of the IP header represent the length of the IP header. The IP header length is calculated by multiplying the IHL (IP header length) by 4. The IHL value is actually the number of 32-bit words in the IP header.

The Type of Service field is one of those fields that makes the Sniffer shine. Instead of digging through tons of documentation to find out about the bits within the Type of Service field, it only takes a look at the Sniffer capture to get the total breakdown. The Easy Ethernet CS8900A firmware doesn't care about any of the bits in the Type of Service field. That doesn't mean you can twiddle them if you want. As you can see from the Sniffer capture descriptions, the Type of Service field specifies how upper-layer protocols want the datagram handled. Note there aren't any special handling instructions for our datagram.

The Total Length field of the IP header is a field that the Easy Ethernet CS8900A firmware will use. This field's value represents the length of the entire IP packet including the data and the header. The total length value could come in handy when loading an IP datagram into the CS8900A-CQ transmit buffer.

The Identification field is a number that represents the current datagram. It's used to reorder fragments. I noticed that when I issued a PING with the Easy Ethernet CS8900A disconnected from the network the Identification number incremented by one for each ICMP echo request issued by the host personal computer.

The Flags field is used to determine if the packet can be fragmented. The low order bit controls fragmentation, while the middle bit signals if this is the last fragment in a series of fragmented packets. The high order bit of the Flags field is not used. This is another field that makes you happy to have a Sniffer. According to our Sniffer screen capture, fragmentation is not allowed.

Even though the process of sending datagrams is reliable to a great extent, there still looms the possibility of having a "bad" datagram bouncing around uncontrolled on a network. The Time to Live value is really a counter that gradually decrements until its value reaches zero. When Time to Live equals zero, the life of the packet ends and it is discarded.

The Protocol field is another field that is used by the Easy Ethernet CS8900A firmware. The value in this field represents an upper-layer protocol that will be the recipient of the IP packet after it is processed. From the Easy Ethernet CS8900A's source code, you can see that the Easy Ethernet CS8900A will service the ICMP, TCP and UDP protocols. A value of 0x01 in this field indicates that the ICMP protocol is in control.

The IP header checksum is very important, as it is a way of checking the integrity of the IP header. I'll hold the theory behind calculating the IP header checksum until we get to the checksum firmware description.

The source and destination addresses are the protocol addresses of the players in the ICMP echo communications session. Note that the IP header is the only place you will see an IP address in an IP datagram. No options have been specified for our IP datagram. So, the destination IP address field gives way directly to the IP data area.

Encapsulation is an important concept to understand when dealing with Ethernet and IP. The IP header is encapsulated within the data area of an Ethernet packet.

Here's how the IP header is arranged in the Easy Ethernet CS8900A's PIC16F877's internal memory. Note the IP data area in Code Snippet 9.3. Carving out this data area implies that it has been put there to hold something else. That's what encapsulation is all about.

```
//*******************************************************************
//*   IP Header Layout
//*******************************************************************
#define  ip_vers_len    0x0E  //IP version and header length
#define  ip_tos         0x0F  //IP type of service
#define  ip_pktlen      0x10  //packet length
#define  ip_id          0x12  //datagram id
#define  ip_frag_offset 0x14  //fragment offset
#define  ip_ttl         0x16  //time to live
#define  ip_proto       0x17  //protocol (ICMP=1, TCP=6, UDP=11)
#define  ip_hdr_cksum   0x18  //header checksum
#define  ip_srcaddr     0x1A  //IP address of source
#define  ip_destaddr    0x1E  //IP addess of destination
#define  ip_data        0x22  //IP data area
```

Code Snippet 9.3: The IP header consists of 20 bytes. The IP header is like the markings on a package or letter as it tells the "postman" how to handle the package, where to send and who sent it. The "goods" ride inside the IP data area.

ICMP messages serve many purposes. For instance, we're using ICMP to echo a packet from one host to another and back. An ICMP message may also be sent when a datagram cannot reach its destination. ICMP messages are primarily used to provide feedback about problems that exist with datagrams in the communication environment. The only thing an ICMP message can't do is tell on itself when it's bad.

A "ping" is really an application that issues an ICMP echo request packet. Basically, a ping sends some data to a remote host and expects the remote host to echo it back.

The ICMP header and its data are encapsulated within the IP data area, which is encapsulated within the Ethernet packet data area (Figure 9.2).

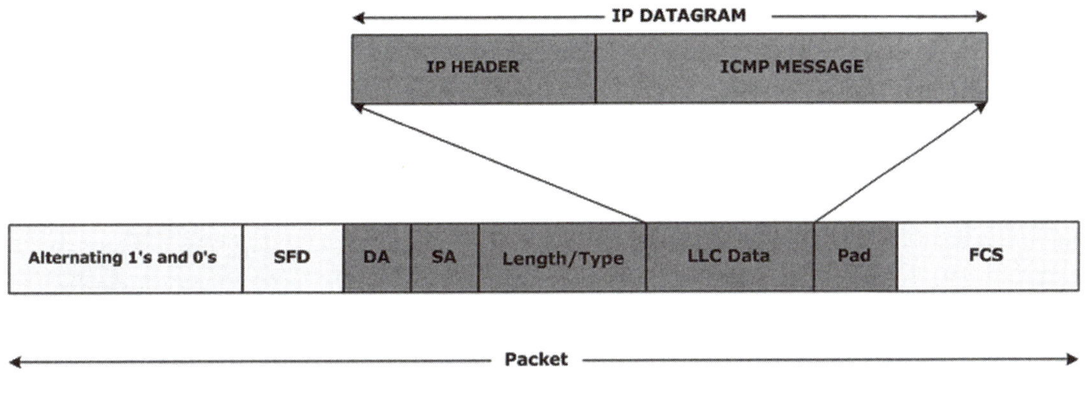

Figure 9.2: Here's a graphical view of the ICMP message riding in the IP box, which is riding in the Ethernet box. That's encapsulation.

In Code Snippet 9.4, I've included a snippet of the source code that describes how and where the ICMP header is laid out in the Easy Ethernet CS8900A's microcontroller's memory.

```
//******************************************************************
//*   IP Header Layout
//******************************************************************
#define  ip_srcaddr     0x1A  //IP address of source
#define  ip_destaddr    0x1E  //IP addess of destination
#define  ip_data        0x22  //IP data area
//******************************************************************
//*   ICMP Header
//******************************************************************
#define  ICMP_type      ip_data
#define  ICMP_code      ICMP_type+1
#define  ICMP_cksum     ICMP_code+1
#define  ICMP_id        ICMP_cksum+2
#define  ICMP_seqnum    ICMP_id+2
#define  ICMP_data      ICMP_seqnum+2
```

Code Snippet 9.4: The entire ICMP header and the message it carries in the ICMP data area ride encapsulated inside the IP data area.

0..7	8..15	16..31
Type	Code	Checksum
Data (depends on type and code)		

Figure 9.3: ICMP messages come in many colors. What's carried in the data area depends on the type of ICMP message.

Rather than describe the fields of the ICMP header in Figure 9.3, let's look at the code that drives the response to the ping request. In Sniffer Screen Capture 9.2, I've also included the rest of the Sniffer capture that details the meaning of the fields inside the ICMP header, which is riding inside the IP data area.

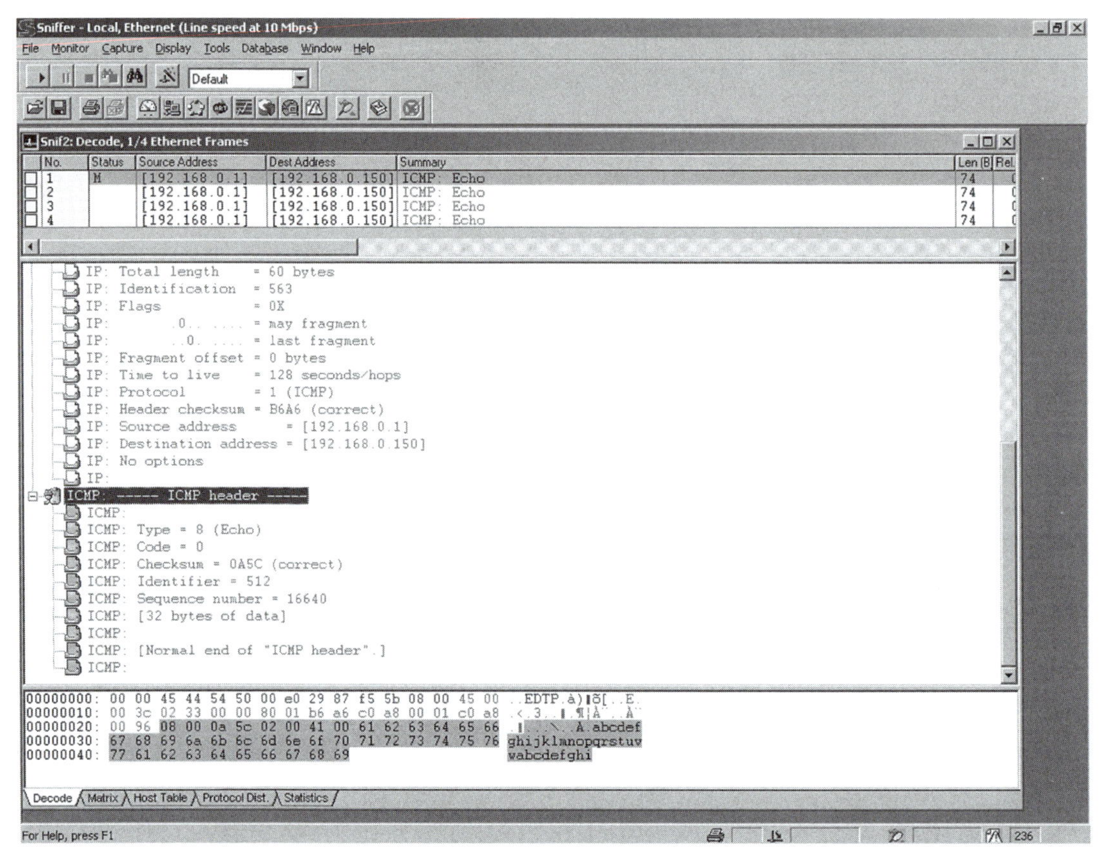

Sniffer Screen Capture 9.2: When it comes to tearing apart an Ethernet frame, it just doesn't get any better than this.

After the get_frame function pulls in the ICMP frame, it's up to the Easy Ethernet CS8900A firmware and the PIC16F877 to put together a frame to echo in response to the ICMP echo request. The echo response begins in earnest inside the ICMP function in Code Snippet 9.5.

```
//*******************************************************************
//*    PING
//*******************************************************************
void icmp()
{
   //set echo reply
   packet[ICMP_type]=0x00;
   packet[ICMP_code]=0x00;

   //clear the ICMP checksum
   packet[ICMP_cksum ]=0x00;
   packet[ICMP_cksum+1]=0x00;
```

Code Snippet 9.5: Like the ARP response, the ICMP echo reply is a swap-this-for-that procedure but requires a bit more work from the firmware and the PIC16F877 microcontroller.

The ICMP_type field contains a 0x08, which represents an ICMP echo request. So, the very first thing we have to do is change that field to a 0x00, which says this is an echo response. The ICMP_code field will remain at 0x00 for both the echo request and echo reply. The identifier and sequence number in the ICMP header may be used to identify and match echo requests and echo replies. A 0x00 in the code field allows a sequence number or identifier to be 0x00.

The next step in preparing to answer the ICMP echo request is to clear the checksum word. The ICMP checksum word should always be cleared before beginning the calculation of a new ICMP checksum value.

```
//*******************************************************************
//*    PING //
******************************************************************
   //setup the IP header
   setipaddrs();
```

Code Snippet 9.6: This function call is much like the ARP function call except the ICMP echo reply is actually built inside the PIC16F877's RAM that holds the packet array.

You didn't think I would take you through all of that IP header stuff without having a good reason, did you? After clearing the ICMP checksum fields and before calculating any checksums, we must assemble our IP datagram. All of the fields for the Ethernet packet, the IP header and the ICMP header are logically arranged in the Easy Ethernet CS8900A's PIC16F877 internal RAM. All we have to do is place our data and checksums into the correct elements of the packet array and bid for some transmit buffer space on the Easy Ethernet CS8900A's resident CS8900A-CQ. The setipaddrs function called in Code Snippet 9.6 handles putting the right IP stuff in the right packet array slots. Let's analyze the setipaddrs function module by module.

```
//**********************************************************************
//*   IP ADDRESS DEFINITION
//*   YOU MAY CHANGE THIS TO ANY VALID IP ADDRESS
//**********************************************************************
int8 MYIP[4] = { 192,168,0,150 };
//**********************************************************************
//*   IP Header Layout
//**********************************************************************
#define  ip_srcaddr          0x1A  //IP address of source
#define  ip_destaddr         0x1E  //IP addess of destination
//**********************************************************************
//*   Ethernet Header Layout
//**********************************************************************
int8  packet[96];                  //50 bytes of UDP data available
//**********************************************************************
//*   Do IP and MAC Housekeeping SNIPPET
//**********************************************************************
void setipaddrs()
{
   //move IP source address to destination address
   packet[ip_destaddr]=packet[ip_srcaddr];
   packet[ip_destaddr+1]=packet[ip_srcaddr+1];
   packet[ip_destaddr+2]=packet[ip_srcaddr+2];
   packet[ip_destaddr+3]=packet[ip_srcaddr+3];

   //make Easy Ethernet CS8900A module IP address the source address
   packet[ip_srcaddr]=MYIP[0];
   packet[ip_srcaddr+1]=MYIP[1];
   packet[ip_srcaddr+2]=MYIP[2];
   packet[ip_srcaddr+3]=MYIP[3];
```

Code Snippet 9.7: Much like the ARP reply, we assemble the ICMP echo response using data that has been acquired and data in the PIC16F877 vault.

We are assembling an echo. So, most of what will be sent back to the requester is just what the requester has sent to us. The first module in Code Snippet 9.7 simply takes the source IP address that was included in the ICMP echo request (packet[ip_srcaddr]) and loads it into the IP header destination address fields (packet[ip_destaddr]).

Since we already know what our local IP address is (it's preloaded into a section of the PIC16F877 RAM), we can safely overwrite the IP header destination address we received in the echo request IP header. The IP header source IP address will be drawn from the MYIP array that is internal to the PIC16F877. With that, by completing the execution of the second code module in Code Snippet 9.7, we've logically addressed our echo reply datagram.

```
//*****************************************************************
//*   HARDWARE (MAC) ADDRESS DEFINITION
//*   YOU MAY CHANGE THIS TO ANY VALID MAC ADDRESS
//*****************************************************************
int8 MYMAC[6] = { 0,0,'E','D','T','P' };
//*****************************************************************
//*   Ethernet Header Layout
//*****************************************************************
int8  packet[96];        //50 bytes of UDP data available
#define  enetpacketDest0   0x00  //destination mac address
#define  enetpacketDest1   0x01
#define  enetpacketDest2   0x02
#define  enetpacketDest3   0x03
#define  enetpacketDest4   0x04
#define  enetpacketDest5   0x05
#define  enetpacketSrc0    0x06  //source mac address
#define  enetpacketSrc1    0x07
#define  enetpacketSrc2    0x08
#define  enetpacketSrc3    0x09
#define  enetpacketSrc4    0x0A
#define  enetpacketSrc5    0x0B
//*****************************************************************
//*   Do IP and MAC Housekeeping SNIPPET
//*****************************************************************
   //move hardware source address to destination address
   packet[enetpacketDest0]=packet[enetpacketSrc0];
   packet[enetpacketDest1]=packet[enetpacketSrc1];
   packet[enetpacketDest2]=packet[enetpacketSrc2];
   packet[enetpacketDest3]=packet[enetpacketSrc3];
   packet[enetpacketDest4]=packet[enetpacketSrc4];
   packet[enetpacketDest5]=packet[enetpacketSrc5];

   //make Easy Ethernet CS8900A MAC address the source address
   packet[enetpacketSrc0]=MYMAC[0];
   packet[enetpacketSrc1]=MYMAC[1];
   packet[enetpacketSrc2]=MYMAC[2];
   packet[enetpacketSrc3]=MYMAC[3];
   packet[enetpacketSrc4]=MYMAC[4];
   packet[enetpacketSrc5]=MYMAC[5];
```

Code Snippet 9.8: If you're having trouble getting your code to run, remember that the "00" in 00EDTP *MAC address are zeroes.*

To perform the hardware addressing, we turn to the physical (hardware) addresses that were transmitted in the echo request. In Code Snippet 9.8, again, we sacrifice the already stored destination MAC address (packet[enetpacketDest0:5]) and load the contents of the source MAC address (packet[enetpacketSrc0:5]) into its packet array memory slots. And, since we have the Easy Ethernet CS8900A's MAC coded into the Easy Ethernet CS8900A's firmware, we simply pull the values for the source MAC address from the MYMAC array.

Now that all of the physical and logical addressing is completed, it's time to compute the IP header checksum. The IP header checksum is computed on the IP header fields only. The IP checksum is defined as the 16-bit one's complement of the one's complement sum of all 16-bit words in the header. Got that? It sounds more confusing than it really is. Let's break the language down into something that we can all understand and write some code to.

```
//*******************************************************************
//*   Ethernet Header Layout
//*******************************************************************
int8  packet[96];       //50 bytes of UDP data available
//*******************************************************************
//*   IP Header Layout
//*******************************************************************
#define  ip_vers_len       0x0E  //IP version and header length
#define  ip_hdr_cksum      0x18  //header checksum
int8 *addr;
int8 data_H,data_L;
int16 chksum16,hdrlen;
int32 hdr_chksum;
//*******************************************************************
//*   Do IP and MAC Housekeeping SNIPPET
//*******************************************************************
   //calculate the IP header checksum
   packet[ip_hdr_cksum]=0x00;
   packet[ip_hdr_cksum+1]=0x00;

   hdr_chksum =0;
   hdrlen = (packet[ip_vers_len] & 0x0F) * 4;
   addr = &packet[ip_vers_len];
   cksum();
   chksum16= ~(hdr_chksum + ((hdr_chksum & 0xFFFF0000) >> 16));
   packet[ip_hdr_cksum] = make8(chksum16,1);
   packet[ip_hdr_cksum+1] = make8(chksum16,0);
```

Code Snippet 9.9: If you think this code is "complicated," check out the assembler version of this function. If that doesn't make you want to learn to code microcontrollers in C, nothing will.

The first order of IP header checksum business in Code Snippet 9.9 is to clear the IP header checksum array elements. Once the checksum array slots are cleared, we can use the lower nibble of the first byte of the IP header to compute the length of the IP header. According to our hex frame dump in Hex Dump 9.1 and the hex dump at the bottom of Sniffer Screen Captures 9.1 and 9.2, the first byte of the IP header is 0x45. To compute our value for the hdrlen variable in Code Snippet 9.9, we simply multiply the lower nibble of the first byte of the IP header (0x05) by 4 (32 bits = 4 bytes). Thus, hdrlen is loaded with the value of 20 decimal. This number is reinforced by Sniffer Screen Capture 9.1, which also calculated the IP header length at 20 bytes.

Now, let's repeat the definition of the IP header checksum and think about what it is telling us. You may want to read this out loud:

> *The IP checksum is defined as the 16-bit one's complement of the one's complement sum of all 16-bit words in the header.*

The first step towards a successful checksum calculation is to satisfy the *all 16-bit words in the header* substatement. To satisfy the word *all*, we load the pointer *addr* with the address of the beginning byte of the IP header.

The next important word in the IP checksum definition is *sum*, which when added to *of all 16-bit words in the header* means to add every pair of bytes within the IP header. That makes our checksum function in Code Snippet 9.10 easy to write.

```
int8 *addr;
int8 data_H,data_L;
int16 chksum16,hdrlen;
int32 hdr_chksum;
//*******************************************************************
//*   CHECKSUM CALCULATION ROUTINE
//*******************************************************************
void cksum()
{
   while(hdrlen > 1)
   {
      data_H=*addr++;
      data_L=*addr++;
      chksum16=make16(data_H,data_L);
      hdr_chksum = hdr_chksum + chksum16;
      hdrlen -=2;
   }
   if(hdrlen > 0)
   {
      data_H=*addr;
      data_L=0x00;
```

```
        chksum16=make16(data_H,data_L);
        hdr_chksum = hdr_chksum + chksum16;
    }
}
```

Code Snippet 9.10: The make16 *built-in function makes the checksum code a bit less hairy to write.*

As long as the hdrlen variable is an even number, we can simply move our addr pointer through the IP header one byte at a time and combine every other byte (data_L) with the byte before it (data_H) to make a 16-bit number (chksum16). Each individual 16-bit number is then added together to form a final total (hdr_chksum). If by chance the IP header length is an odd number, the lower byte of the last 16-bit number (chksum16) will be filled with the value of 0x00. I didn't make that up. The addition of the padding 0x00 value is actually part of the rules for the IP header checksum algorithm.

```
//******************************************************************
//*   Ethernet Header Layout
//******************************************************************
int8  packet[96];        //50 bytes of UDP data available
#define      ip_hdr_cksum       0x18  //header checksum
int16 chksum16,hdrlen;
int32 hdr_chksum;
//******************************************************************
//*   Do IP and MAC Housekeeping SNIPPET
//******************************************************************
   chksum16= ~(hdr_chksum + ((hdr_chksum & 0xFFFF0000) >> 16));
   packet[ip_hdr_cksum] = make8(chksum16,1);
   packet[ip_hdr_cksum+1] = make8(chksum16,0);
```

Code Snippet 9.11: Lots of mumbo jumbo in the three lines of code that make up this snippet, but this is where the rubber meets the road as the IP header checksum is installed in the appropriate packet array slots.

At this point, the 32-bit variable hdr_chksum contains the sum of all of the 16-bit words in the IP header. Let's talk out loud one more time:

> *The IP checksum is defined as the 16-bit one's complement sum of all 16-bit words in the header.*

Reading out loud, we can conclude that we haven't satisfied the *one's complement* rule that is applied against the sum we accumulated in the hdr_chksum variable. A one's complement sum is calculated by summing all of the numbers and adding the sum of the carry bits to the result. We know that our final checksum figure must be a 16-bit word. Therefore, the carry sum will be the value of the upper 16-bits of the 32-bit hdr_chksum variable that holds the sum of the IP header words. So, we simply add each bit that was carried over into the upper 16-bits of hdr_chksum back into hdr_chksum. This satisfies the substatement *one's complement sum of all 16-bit words in the header*, as the one's complement sum of all 16-bit words in the header is now the contents of the variable hdr_chksum. The "~" in CS8900A-CQ 11 performs a one's complement of the one's complement sum of all 16-bit words in the header (hdr_chksum). By using chksum16 to hold the one's complement sum of all 16-bit words in the header, we only take the lower 16-bits of hdr_chksum, which satisfies the IP checksum definition. The value of chksum16 is placed into the ip_hdr_cksum memory slots inside the packet array memory area. Doing the checksum trick in PIC assembler is a bit more tricky than manipulating the numbers with C. If you're into pain, I've included the assembler checksum algorithm on the CD-ROM that accompanies this book.

Calculating the IP header checksum is the last task the setipaddrs function has to perform. The completion of the setipaddrs function returns us to the ICMP echo reply function where the next step is to compute and place the ICMP header checksum as I've done with code in Code Snippet 9.12.

```
//*******************************************************************
//*   PING SNIPPET
//*******************************************************************

    //calculate the ICMP checksum
    hdr_chksum =0;
    hdrlen = (make16(packet[ip_pktlen],packet[ip_pktlen+1])) -
                              (packet[ip_vers_len] & 0x0F) * 4);
    addr = &packet[ICMP_type];
    cksum();
    chksum16= ~(hdr_chksum + ((hdr_chksum & 0xFFFF0000) >> 16));
    packet[ICMP_cksum] = make8(chksum16,1);
    packet[ICMP_cksum+1] = make8(chksum16,0);
```

Code Snippet 9.12: As one famous British rock group would say, "Second verse same as the first."

The ICMP checksum is defined as:

The ICMP checksum is the 16-bit one's complement of the one's complement sum of the ICMP message starting with the ICMP Type. If the total length is odd, the received data is padded with one octet of zeros for computing the checksum.

The process of computing the ICMP checksum is no different than what we just did for the IP header checksum. There's nothing in the ICMP header to tell us how long the ICMP header is. So, we calculate the ICMP header length by simply subtracting the IP header length from the total length of the frame. The total length of the IP datagram is 60 bytes. The total length value is found in the total length field of the IP header. We previously calculated the IP header length as 20 bytes. So, 60 bytes – 20 bytes leaves 40 bytes as the ICMP message. This checks out against the numbers in Sniffer Screen Capture 9.2. Sniffer Screen

Capture 9.2 tells us directly that there are 32 bytes of data in the ICMP message. Counting the ICMP header bytes totals to 8 bytes in the ICMP header. So, 32 + 8 is 40 total bytes in the ICMP message.

Now, all that's left to do is point to the ICMP_type memory slot and walk through 40 bytes and adding them as if they were 20 16-bit values. The 16-bit ICMP checksum is then placed in the ICMP checksum array slots in the packet array memory area using the same 16-bit one's complement algorithm we used for the IP header checksum. The final function performed by the ICMP reply code, *echo_packet*, puts us back on familiar ground in Code Snippet 9.13.

```
//*******************************************************************
//*   PING SNIPPET
//*******************************************************************
   echo_packet();

//*******************************************************************
//*   ECHO THE PACKET
//*******************************************************************
void echo_packet()
{
   dataport_out;
   WpppL(pageport_TxCmd,TXCMD_AFTER_ALL);
   WpppH(pageport_TxCmd,0x00);
   WpppL(pageport_TxLen,pageheader[enetpacketLenL]);
   WpppH(pageport_TxLen,pageheader[enetpacketLenH]);
   do{
         RPP(ppageBusStatus);
      }while(!(bit_test(data_H,BUSSTA_RDY4TXNOW_BIT)));

   dataport_out;
   txlen = make16(pageheader[enetpacketLenH],pageheader[enetpacketLenL]);
   for(i=0;i<txlen;i+=2)
   {
      WpppL(pageport_RxTxData0,packet[i]);
      WpppH(pageport_RxTxData0,packet[i+1]);
   }
}
```

Code Snippet 9.13: Wondering what British group I quoted in Code Snippet 12? How about Herman's Hermits singing "I'm Henry the VIII, I Am."

At this point, we've touched every necessary field to send an echo reply packet back to the requester. All we have to do now is bid for some CS8900A-CQ transmit buffer space and roll out the bytes from the Easy Ethernet CS8900A's PIC16F877 microcontroller to the CS8900A-CQ transmit buffer in the order they are packed into the PIC16F877's packet array. There's nothing in the *echo_packet* function in Code Snippet 9.13 that you don't already know about unless you didn't get the British band brainteaser I offered up in Code Snippet 9.12.

Hex Dump 9.2 is a PIC16F877 RAM dump of the ICMP echo request frame that originated from the personal computer on the LAN segment with the Easy Ethernet CS8900A. Hex Dump 9.3 is the PIC16F877 memory dump captured just before the ICMP echo reply data was to be transferred from the PIC16F877 microcontroller to the CS8900A-CQ transmit buffer.

```
Address   00 01 02 03 04 05 06 07 08 09 0A 0B 0C 0D 0E 0F      ASCII

 0110     00 00 45 44 54 50 00 E0 29 87 F5 5B 08 00 45 00 ..EDTP.. )..[..E.
 0120     00 3C 02 33 00 00 80 01 B6 A6 C0 A8 00 01 C0 A8 .<.3.... ........
 0130     00 96 08 00 0A 5C 02 00 41 00 61 62 63 64 65 66 .....\.. A.abcdef
 0140     67 68 69 6A 6B 6C 6D 6E 6F 70 71 72 73 74 75 76 ghijklmn opqrstuv
 0150     77 61 62 63 64 65 66 67 68 69 00 00 00 00 00 00 wabcdefg hi......
```

Hex Dump 9.2: Before the swap-this-for-that...

```
Address   00 01 02 03 04 05 06 07 08 09 0A 0B 0C 0D 0E 0F      ASCII

 0110     00 E0 29 87 F5 5B 00 00 45 44 54 50 08 00 45 00 ..)..[.. EDTP..E.
 0120     00 3C 02 33 00 00 80 01 B6 A6 C0 A8 00 96 C0 A8 .<._.... .z......
 0130     00 01 00 00 06 5C 02 00 41 00 61 62 63 64 65 66 .....\.. M.abcdef
 0140     67 68 69 6A 6B 6C 6D 6E 6F 70 71 72 73 74 75 76 ghijklmn opqrstuv
 0150     77 61 62 63 64 65 66 67 68 69 00 00 00 00 00 00 wabcdefg hi......
```

Hex Dump 9.3: After the PIC16F877 microcontroller has manipulated the addresses and computed new checksums.

Now that you're a bit more familiar with what should be where, I'll bet you can pick out the echo request and echo reply address inversions in the MAC and IP address fields in Hex Dumps 9.2 and 9.3. Can you also see that the ICMP checksum is different in the before and after hex dumps? The ICMP checksum changed because we changed the ICMP_type byte from 0x08 in the echo request frame to 0x00 in the echo reply frame. The results of our work thus far can be seen in Sniffer Screen Capture 9.3.

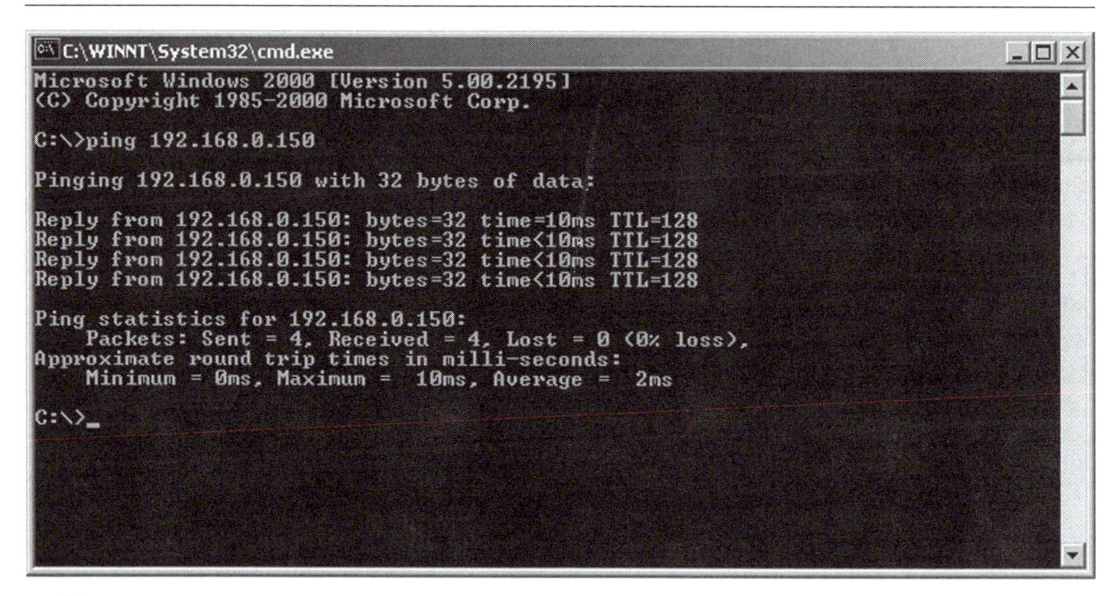

Sniffer Screen Capture 9.3: It's ALIVE!

CHAPTER 10

UDP and the Easy Ethernet CS8900A

UDP is the closest thing to RS-232 communications you'll get in an internet protocol. I love UDP because it just doesn't care. You can send just about anything with it at any time you wish. Of all of the Internet protocols, I think UDP is the easiest to implement. UDP is lots of fun as well. Now that our Easy Ethernet CS8900A can speak, let's do some UDP coding.

If you thought seeing that PING request being answered by your Easy Ethernet CS8900A was exciting, you're gonna love working with UDP. UDP is short for User Datagram Protocol. You could also unofficially call it Unreliable Delivery Protocol. Like IP, UDP has absolutely no means of ensuring that a data packet will arrive in one piece or arrive at all. However, you'll find that it is reliable enough for most tasks it's used for.

UDP is a very simple protocol. Basically, UDP takes a message from an application and tags on a checksum and source and destination port numbers before flinging the UDP segment to IP for encapsulation. IP does its best to deliver the UDP segment since there is nothing to guarantee that the UDP segment will arrive intact.

Logically, a UDP host transmits a UDP datagram through a source port to a UDP recipient's destination port. The destination port number and destination IP address are used to route the UDP segment to the correct application once the segment arrives at its destination. By using port numbers, various applications can be using the services of UDP simultaneously. This is called *multiplexing*. The combination of the IP address and the port number is called a *socket*.

A UDP transmission can occur at any time without the need to establish a communications session with the remote host. Since there is no handshaking or predetermined contact between UDP hosts, UDP is defined as a connectionless protocol. This is similar to RS-232 communications.

Despite the shortcomings that UDP appears to emanate, UDP does have advantages over its cousin TCP. For instance, UDP does not have to establish a formal connection and as a result, is a faster way to send a message. As you will see later, TCP uses a three-way handshake to establish a communications session before transmitting any data, and TCP does an awful lot of housekeeping compared to none for UDP.

UDP is able to send messages as fast as the microcontroller and application it is involved with can run. The only thing that slows UDP down is the limitations of the hardware it is running on and the bandwidth of the LAN it is riding on. Unlike UDP, TCP has built in rev limiters that throttle the data rate to relieve congestion on the LAN segment. UDP segments with any kind of problems are simply discarded.

The bottom line is that both UDP and TCP have their place depending on what the application demands. The application of UDP that we're about to explore successfully transferred data on the internet between an Easy Ethernet CS8900A located in Florida and a personal computer in Australia. That's pretty danged good for an unreliable protocol!

```
//******************************************************************
//*   Ethernet Header Layout
//******************************************************************
#define  enetpacketType0    0x0C  //type/length field
#define  enetpacketType1    0x0D
#define  enetpacketData     0x0E  //IP data area begins here
//******************************************************************
//*   UDP Header
//******************************************************************
#define  UDP_srcport     ip_data
#define  UDP_destport    UDP_srcport+2
#define  UDP_len         UDP_destport+2
#define  UDP_cksum       UDP_len+2
#define  UDP_data        UDP_cksum+2
```

Code Snippet 10.1: UDP is yet another protocol that lies within the bowels of the IP data area.

The layout of the UDP header in Code Snippet 10.1 shows that the UDP segment rides in the IP data area. Thus, the UDP datagram is encapsulated within the IP datagram (Figure 10.1). The UDP source and destination ports are 16 bits in length, which allows port numbers to range from 0 to 65535. The UDP header is only 8 bytes long, compared to the 20 bytes that make up the IP header. The UDP datagram length or size is simply the total number of bytes in the UDP header plus the number of bytes in the payload (data area).

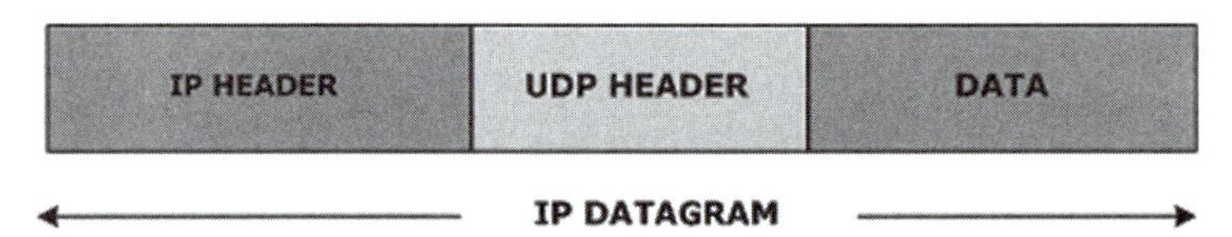

Figure 10.1: There are 28 bytes of header information in the UDP datagram. The UDP datagram is riding inside the IP datagram, which will all ride in the Ethernet frame.

UDP checksumming is optional. However, the Easy Ethernet CS8900A firmware checksums every UDP datagram. The UDP checksum is put there for use by the application as UDP itself doesn't care about it at all.

Every bit of code we produce as we go along is another networking brick in the wall. Up to this point, our Easy Ethernet CS8900A is answering ARP requests and ICMP echo requests or pings. The ICMP echo reply said its ABC's but we really haven't passed any application data through the Easy Ethernet CS8900A yet. The next module of source code will implement UDP functionality with a bit of help from a personal computer-based UDP program.

A UDP Internet Test Panel

Using Microsoft® Visual Basic® as a programming language is one of the easiest means of writing useful applications for today's personal computers. Visual Basic is full of functionality and includes network applications modules for UDP and TCP/IP. The inclusion of the networking modules makes Visual Basic a perfect UDP datagram generator.

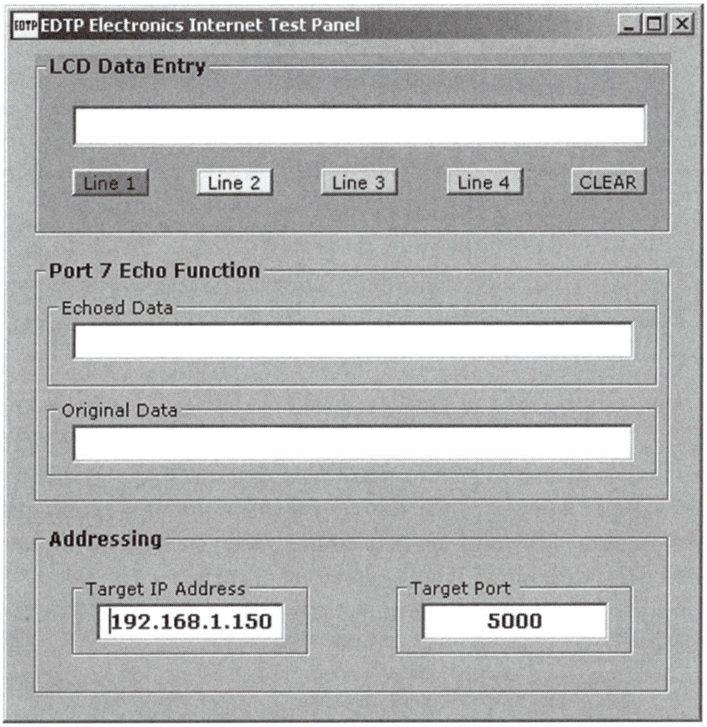

Screen Capture 10.1: You can use this program to test the operation of any of the Easy Ethernet devices depicted in this book.

The Internet Test Panel shown in Screen Capture 10.1 is a Visual Basic application that uses UDP socket programming to send a UDP datagram to a well-known port or a socket of your choice. The Internet Test Panel also includes programming to send a UDP message to an LCD-equipped UDP host. Here's how it all works.

Each window in the Internet Test Panel is associated with a name and a variable that represents the text within the window. For example, the LCD Data Entry window in Screen Capture 10.1 is named *txtlcdin* and the text within *txtlcdin* is recognized by the program as *txtlcdin.Text*.

```
Private Sub Form_Load()
On Error Resume Next
txtip.Text = "192.168.1.150"
txtport.Text = "5000"
With udp_PC
.RemoteHost = txtip.Text
.RemotePort = Val(txtport.Text)
.Bind 5002
End With
frm_B.Show
End Sub
```

Code Snippet 10.2: This code snippet is shown in Visual Basic format. I've included source that you can run inside the Visual Basic IDE that you can easily modify to build your own version of the Internet Test Panel.

The *udp_PC* mnemonic in Code Snippet 10.2 represents the Visual Basic Winsock module that provides the UDP protocol and UDP socket services. In the Internet Test Panel application, our Winsock module, *udp_PC*, is bound to local port 5002. The binding precludes any other application or Winsock module from using port 5002.

Prefixing the *RemoteHost* and *RemotePort* variables with the Winsock designator associates the IP address and port address represented by *RemoteHost* and *RemotePort* with the Winsock in the prefix and its bound port number. The *RemoteHost* variable is actually a string that represents the IP address of the remote host. A number is required for the *RemotePort* variable, and the *Val* function is used to convert the text string into a numeric value that the *RemotePort* variable will accept. For example, *udp_PC.RemoteHost* associates the IP address represented by *RemoteHost* to the Winsock named *udp_PC*,

```
Private Sub txtip_Change()
On Error Resume Next
udp_PC.RemoteHost = txtip.Text
End Sub

Private Sub txtport_Change()
On Error Resume Next
udp_PC.RemotePort = Val(txtport.Text)
End Sub
```

Code Snippet 10.3: Both of the Visual Basic subroutines in this snippet are triggered when the text inside their respective windows changes.

The Visual Basic Internet Test Panel program is event-driven. Code Snippet 10.3 shows us that text entered in the Internet Test Panel's Target IP Address window is used to set the IP address of the remote host the Internet Test Panel will communicate with using UDP datagrams. As you can see in Screen Capture 10.1, the default destination IP address is 192.168.1.150.

A destination port number is also required and according to the code in Code Snippet 10.3, that value is entered using the Target Port window of the Internet Test Panel application. The default Internet Test Panel destination port number is 5000. The Internet Test Panel's source port number is fixed at 5002. The Easy Ethernet CS8900A UDP firmware is coded to respond to ports 5000 and 7 with an IP address of 192.168.0.150.

Simply typing the desired numbers into the appropriate Internet Test Panel windows will change the default values for the Target IP Address and the Target Port. The Visual Basic functions txtip_Change and txtport_Change in Code Snippet 10.3 will sense the changes in the text within their respective text boxes (txtip and txtport) and load the Internet Test Panel application's *RemoteHost* variable with the new remote IP address and the *RemotePort* variable with the new destination port address.

```
Private Sub txtsend_Change()
On Error Resume Next
udp_PC.RemoteHost = txtip.Text
udp_PC.RemotePort = 7
udp_PC.SendData txtsend.Text
End Sub
```

Code Snippet 10.4: Sending a UDP datagram to well-known port 7 and receiving an echo of the data isn't magical. The echoing host must be running code that will complete the echo operation.

A well-known port is one that has a standard function associated with it. On today's Internet, port 7 is the well-known echo port. If a UDP message has a destination port of 7, which the code in Code Snippet 10.4 is addressed to, the data transmitted to the remote host should be echoed from the remote host back to the sender.

The txsend.Text Visual Basic variable in Code Snippet 10.4 is actually text data entered in the Original Data window (txtsend) of the Internet Test Panel. The entering of text into the Original Data window triggers the txtsend_Change function, which sends the contents of the Original Data window to port 7 of the of the remote host with the IP address that is listed in the Target IP Address window of the Internet Test Panel.

When using the echo function of the Internet Test Panel, the UDP message size increases for every byte entered in the Original Data window. For instance, if the letter "A" is entered, it is immediately sent and echoed. Entering the letter "B" would send "AB," entering "C" would send "ABC," and so forth. If all works as planned, the remote host echoes whatever is in the Original Data window back to the Internet Test Panel application.

```
Private Sub udp_PC_DataArrival(ByVal bytesTotal As Long)
On Error Resume Next
udp_PC.RemoteHost = txtip.Text
Dim strData As String
udp_PC.GetData strData
txtreceive.Text = strData
End Sub
```

Code Snippet 10.5: Setting up multiple UDP sockets in the Visual Basic program would allow multiple Easy Ethernet CS8900A's to pass information to the Visual Basic application.

The incoming bytes of echoed data trigger another event that calls the udp_PC_DataArrival function. The echoed data (*txtreceive.Text*) is retrieved using the Visual Basic GetData method and displayed in the Echoed Data window (txtreceive). After executing *udp_PC.GetData strData*, the Visual Basic program knows to route the incoming echoed data from the Easy Ethernet CS8900A to the port address and IP address tied to Winsock *udp_PC*.

```
Private Sub txtlcdin_Change()
On Error Resume Next
udp_PC.RemoteHost = txtip.Text
udp_PC.RemotePort = Val(txtport.Text)
udp_PC.SendData Right(txtlcdin.Text, 1)
```

Code Snippet 10.6: Here's an example of the Visual Basic program talking to a different socket on the Easy Ethernet CS8900A.

The LCD Data Entry window (txtlcdin) has a dual-purpose role. Primarily, the data entered into the LCD Data Entry window (*txtlcdin.Text*) in Code Snippet 10.6 is designed to be interpreted by a remote host that is driving a standard 4-line LCD module. The workings of the Visual Basic event mechanism for the LCD Data Entry window are no different than when sending data from the Original Data window. Only the socket addressing is changed. The data entered into the LCD Data Entry window is aimed at port 5000 by default, and the data doesn't accumulate like it does in the Port 7 Echo Function windows. The data shown in the LCD Data Entry window will seem to accumulate visually, but only one character is sent per event and that's always the last character in the window. The LCD Data Entry window's addressing is controlled by the values of the Target IP Address and Target Port windows.

```
Private Sub btnclear_Click()
On Error Resume Next
udp_PC.RemotePort = txtport.Text
udp_PC.SendData &H0
txtlcdin = ""
End Sub
```

```
Private Sub btnline1_Click()
On Error Resume Next
udp_PC.RemotePort = txtport.Text
udp_PC.SendData &H1
txtlcdin = ""
End Sub

Private Sub btnline2_Click()
On Error Resume Next
udp_PC.RemotePort = txtport.Text
udp_PC.SendData &H2
txtlcdin = ""
End Sub

Private Sub btnline3_Click()
On Error Resume Next
udp_PC.RemotePort = txtport.Text
udp_PC.SendData &H3
txtlcdin = ""
End Sub

Private Sub btnline4_Click()
On Error Resume Next
udp_PC.RemotePort = txtport.Text
udp_PC.SendData &H4
txtlcdin = ""
End Sub
```

Code Snippet 10.7: Obviously, the buttons in this snippet were put there for a reason. I'll show you how to use them for what they were originally designed for in the Easy Ethernet AVR section.

The Easy Ethernet CS8900A doesn't have a native LCD module but we can still use the LCD Data Entry window to our advantage with the Easy Ethernet CS8900A. Text characters sent from the LCD Data Entry window can be captured by the Easy Ethernet CS8900A and used to control selected pins on the PIC16F877's I/O ports. The same can be done with the buttons that are normally used to switch from line to line on a multilined LCD module. The Internet Test Panel buttons: Line 1, Line 2, Line 3, Line 4 and CLEAR in Code Snippet 10.7 send raw hex values of 0x0100, 0x0200, 0x0300, 0x0400 and 0x0000, respectively. The Easy Ethernet CS8900A firmware is instructed to only pick up the high byte of each word sent by the buttons.

Just because we have a new protocol to play with doesn't change anything that we have discussed thus far. The PIC16F877 microcontroller residing on the Easy Ethernet CS8900A initializes the CS8900A-CQ and polls the RxOK bit in the CS8900A's Receiver Event Register looking for the opportunity to grab a frame from the CS8900A-CQ's receive buffer.

The Internet Test Panel application is running on a host personal computer that is participating on the same LAN segment as the Easy Ethernet CS8900A (192.168.0.XXX). To allow the Internet Test Panel to operate on the same LAN segment and point to the Easy Ethernet CS8900A, the Target IP Address window of the Internet Test Panel has been changed to reflect the IP address of the Easy Ethernet CS8900A, 192.168.0.150.

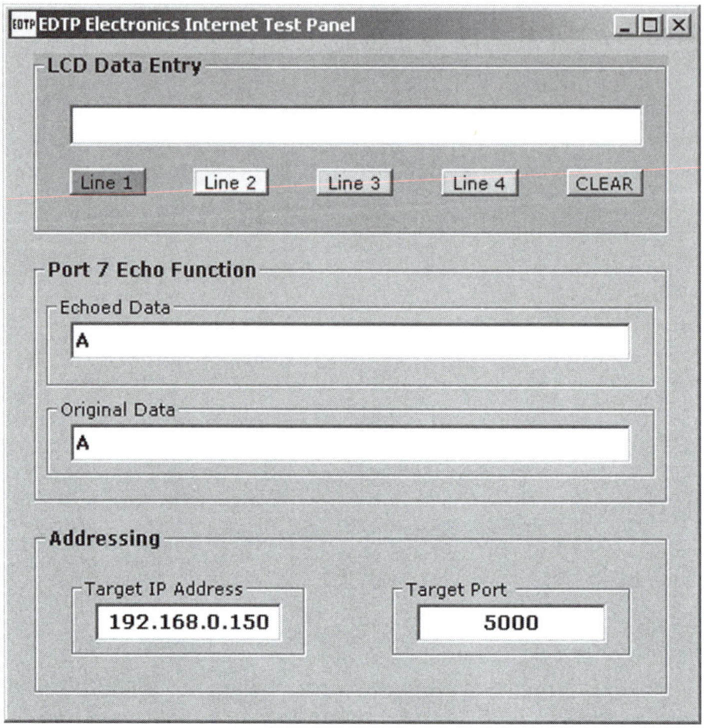

Screen Capture 10.2: This may seem to be a trivial example, but it's exciting to see that character echo back from a device you've just completed building and programming.

In Screen Capture 10.2, the letter "A" is typed into the Internet Test Panel's Original Data window. If the Easy Ethernet CS8900A's IP and MAC address are not in the Internet Test Panel's personal computer's ARP cache, an ARP request is generated. The Easy Ethernet CS8900A sees the ARP request and responds supplying its MAC address to the personal computer where it is stored in the ARP cache.

A UDP segment containing the source port value (5002 decimal), the destination port value (7 decimal), the length of the UDP segment (9 decimal), a checksum value and the data are assembled and rolled into the IP data area. The protocol field within the IP header is set to represent UDP (17 decimal), addresses are assembled, checksums calculated and the IP datagram is sent on its way. I've sniffed the personal computer side of the UDP echo message and presented it for you in Screen Capture 10.3.

```
Sniffer - Local, Ethernet (Line speed at 10 Mbps)
File  Monitor  Capture  Display  Tools  Database  Window  Help
Default

Snif10: Decode, 3/4 Ethernet Frames
No. Status Source Address     Dest Address        Summary                                                    Len (B) Rel.
1   M      00E02987F55B       Broadcast           ARP: C PA=[192.168.0.150] PRO=IP                           60
2          000045445450       00E02987F55B        ARP: R PA=[192.168.0.150] HA=000045445450 PRO=IP           60
3          [192.168.0.1]      [192.168.0.150]     UDP: D=7 S=5002   LEN=9                                    60
4          [192.168.0.150]    [192.168.0.1]       UDP: D=5002 S=7   LEN=9                                    60

IP:          .... .0.. = normal reliability
IP:          .... ..0. = ECT bit - transport protocol will ignore the CE bit
IP:          .... ...0 = CE bit - no congestion
IP: Total length     = 29 bytes
IP: Identification   = 782
IP: Flags            = 0X
IP:          .0.. .... = may fragment
IP:          ..0. .... = last fragment
IP: Fragment offset  = 0 bytes
IP: Time to live     = 128 seconds/hops
IP: Protocol         = 17 (UDP)
IP: Header checksum  = B5DA (correct)
IP: Source address      = [192.168.0.1]
IP: Destination address = [192.168.0.150]
IP: No options
IP:
UDP: ----- UDP Header -----
UDP:
UDP: Source port       = 5002
UDP: Destination port  = 7
UDP: Length            = 9
UDP: Checksum          = 2963 (correct)
UDP: [1 byte(s) of data]
UDP:

00000000: 00 00 45 44 54 50 00 e0 29 87 f5 5b 08 00 45 00  ..EDTP.à)‚õ[..E.
00000010: 00 1d 03 0e 00 00 80 11 b5 da c0 a8 00 01 c0 a8  ........µÚÀ¨..À¨
00000020: 00 96 13 8a 00 07 00 09 29 63 41 00 00 00 00 00  ........)cA.....
00000030: 00 00 00 00 00 00 00 00 00 00 00 00              ............

Decode  Matrix  Host Table  Protocol Dist.  Statistics
For Help, press F1                                                     236
```

Screen Capture 10.3: As usual, the Sniffer knows all, sees all and tells all.

The Easy Ethernet CS8900A's CS8900A-CQ receives the frame and sets the RxOK bit to let the PIC16F877 know that an Ethernet frame is waiting in the CS8900A-CQ receive buffer. Meanwhile, the Easy Ethernet CS8900A's PIC16F877 microcontroller is polling the CS8900A-CQ's Receiver Event Register and finds the RxOK bit is set. The Easy Ethernet CS8900A's get_frame function is called.

The get_frame function analyzes the new frame and concludes that it is an IP frame. Further examination reveals that the frame's protocol is UDP. It just so happens that the Easy Ethernet CS8900A contains some UDP application firmware and that's where the Easy Ethernet CS8900A's UDP datagram processing begins.

```
Address   00 01 02 03 04 05 06 07 08 09 0A 0B 0C 0D 0E 0F      ASCII

  0110    00 00 45 44 54 50 00 E0 29 87 F5 5B 08 00 45 00 .. EDTP..)..[..E.
  0120    00 1D 03 0E 00 00 80 11 B5 DA C0 A8 00 01 C0 A8 ........ ........
  0130    00 96 13 8A 00 07 00 09 29 63 41 00 00 00 00 00 ........ )cA.....
  0140    00 00 00 00 00 00 00 00 00 00 00 00 00 00 00 00 ........ ........
```

Hex Dump 10.1: If you know what you're looking for (and you do), and you don't have a Sniffer handy, the memory dumps supplied by the MPLAB ICE 2000 can be used in place of a Sniffer.

Hex Dump 10.1 is a hex dump of the PIC16F877's packet array memory and shows our data ("A") at memory location 0x013A. The UDP header begins with the source port word at memory location 0x0132. The hexadecimal number, 0x138A, at memory location 0x0132 equates to 5002 decimal, which is the source port number of our Internet Test Panel application that is running on the personal computer.

Immediately following the source port word is the destination port word, 0x0007, which is the well-known echo port that the Easy Ethernet CS8900A UDP application is coded to look for. We already know that the UDP header is 8 bytes long and adding the single data byte gives us a total UDP header length of 0x0009, which resides just ahead of the UDP checksum value.

Everything else in the hex frame dump of the UDP datagram in Hex Dump 10.1 is just as we would expect it to be. The DLC header area at the beginning of Hex Dump 10.1 contains the physical (MAC) addresses and Ethernet packet type while the protocol addressing, protocol identification and datagram handling duties are carried out by the IP header fields that follow behind the DLC bytes.

```
//******************************************************************
//*   IP Header Layout
//******************************************************************
#define  ip_data           0x22  //IP data area
//******************************************************************
//*   UDP Header
//******************************************************************
#define  UDP_srcport       ip_data
#define  UDP_destport      UDP_srcport+2

int8 data_L;
//******************************************************************
//*   UDP Function SNIPPET
//*   This function receives data from a Visual Basic UDP program and
//*   echoes the data back to the VB program and sets or resets bits
//*   on the PIC16F877's PORT A under control of the VB program.
//******************************************************************
void udp()
```

```
{
   //port 7 is the well-known echo port
   if(packet[UDP_destport] == 0x00 && packet[UDP_destport+1] ==0x07)
   {
      //build the IP header
      setipaddrs();

      //swap the UDP source and destination ports
      data_L = packet[UDP_srcport];
      packet[UDP_srcport] = packet[UDP_destport];
      packet[UDP_destport] = data_L;

      data_L = packet[UDP_srcport+1];
      packet[UDP_srcport+1] = packet[UDP_destport+1];
      packet[UDP_destport+1] = data_L;
```

Code Snippet 10.8: First the incoming frame is identified as an IP frame. Then, inspecting the contents of the incoming frame leads to it being tagged as carrying a UDP datagram destined for the Easy Ethernet CS8900A's IP address. The UDP datagram is then routed according to its port number to the Easy Ethernet CS8900A's UDP application that echoes the data back to the sending host.

The Easy Ethernet CS8900A's IP processing has already identified half of the socket, the IP address, which matches the Easy Ethernet CS8900A's IP address. The very first thing our UDP application code in Code Snippet 10.8 that is running on the Easy Ethernet CS8900A does is check the destination port number, which completes the socket address. If the port number is found to be 7, the firmware immediately flows into the process of preparing the frame to be echoed.

Using the addressing information garnered from the incoming frame and the addressing information stored inside the PIC16F877 microcontroller, the setipaddrs function points the frame in the microcontroller's packet array memory at the original sender of the frame. In addition to the IP and MAC source and destination address switcheroos, the UDP source and destination ports must be reversed as well.

```
//******************************************************************
//*    UDP Header
//******************************************************************
#define  UDP_srcport        ip_data
#define  UDP_destport       UDP_srcport+2
#define  UDP_len            UDP_destport+2
#define  UDP_cksum          UDP_len+2
//******************************************************************
//*    Ethernet Header Layout
//******************************************************************
int8  packet[96];          //50 bytes of UDP data available
```

```
//*******************************************************************
//*    IP Header Layout
//*******************************************************************
#define  ip_proto        0x17  //protocol (ICMP=1, TCP=6, UDP=11)
#define  ip_srcaddr      0x1A  //IP address of source
#define  ip_destaddr     0x1E  //IP address of destination

int32 hdr_chksum;
int16 chksum16,hdrlen;
//*******************************************************************
//*    UDP Function SNIPPET
//*    This function receives data from a Visual Basic UDP program and
//*    echoes the data back to the VB program and sets or resets bits
//*    on the PIC16F877's PORT A under control of the VB program.
//*******************************************************************
   //calculate the UDP checksum
   packet[UDP_cksum] = 0x00;
   packet[UDP_cksum+1] = 0x00;

   hdr_chksum =0;
   hdrlen = 0x08;
   addr = &packet[ip_srcaddr];
   cksum();
   hdr_chksum = hdr_chksum + packet[ip_proto];
   hdrlen = 0x02;
   addr = &packet[UDP_len];
   cksum();
   hdrlen = make16(packet[UDP_len],packet[UDP_len+1]);
   addr = &packet[UDP_srcport];
   cksum();
   chksum16= ~(hdr_chksum + ((hdr_chksum & 0xFFFF0000) >> 16));
   packet[UDP_cksum] = make8(chksum16,1);
   packet[UDP_cksum+1] = make8(chksum16,0);
```

Code Snippet 10.9: You should be getting good at checksum calculations by now. However, this one slips you a mickey.

When all of the fields within the IP and UDP headers are loaded with the correct data, the checksum calculations in Code Snippet 10.9 can begin. The UDP checksum calculation process is a bit different than what we've seen before. You would think the UDP checksum would cover only the bytes within the UDP header and data area. Not so. The UDP checksum includes some choice bytes from the IP header as well. In fact, the UDP checksum is calculated using:

- The IP source address word
- The IP destination address word

- The IP protocol byte
- The UDP length word
- The UDP header
- The UDP data

And yes, the UDP length word is used twice in the calculation of the UDP checksum. Otherwise, the UDP checksum definition is the same as the IP checksum definition, which is (repeat after me):

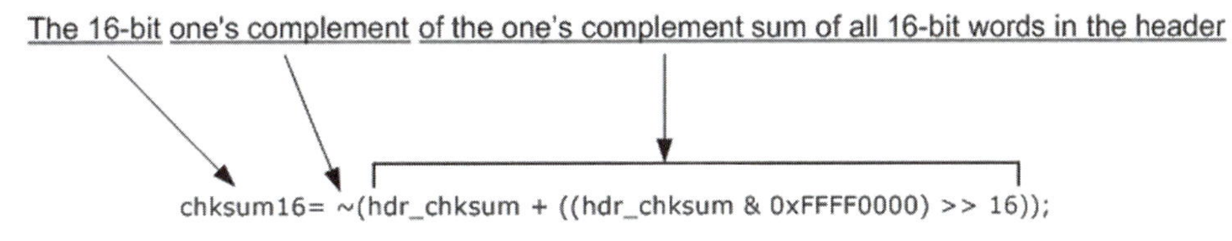

Figure 10.2: A little checksum sign language...

When all of the dust settles, the PIC16F877 has assembled a suitable UDP echo reply frame in its packet array memory area just like the one shown in Hex Dump 10.2.

```
Address  00 01 02 03 04 05 06 07 08 09 0A 0B 0C 0D 0E 0F     ASCII

0110     00 E0 29 87 F5 5B 00 00 45 44 54 50 08 00 45 00  ..)..[.. EDTP..E.
0120     00 1D 03 0E 00 00 80 11 B5 DA C0 A8 00 96 C0 A8  ........ ........
0130     00 01 00 07 13 8A 00 09 29 63 41 00 00 00 00 00  ........ )cA.....
0140     00 00 00 00 00 00 00 00 00 00 00 00 00 00 00 00  ........ ........
```

Hex Dump 10.2: The amount of data we can stuff into the UDP data area is limited by the size of the data area of an Ethernet frame. Since the PIC16F877 has less than 400 bytes of RAM, we can stuff in UDP data as long as there's RAM in the PIC16F877 to hold it.

With everything in place, the PIC16F877 executes the code contained within the *echo_packet* function (Code Snippet 10.10) and bids for some transmit buffer space on the CS8900A-CQ.

```
//******************************************************************
//*   ECHO THE PACKET
//******************************************************************
void echo_packet()
{
   dataport_out;
   WpppL(pageport_TxCmd,TXCMD_AFTER_ALL);
   WpppH(pageport_TxCmd,0x00);
```

```
    WpppL(pageport_TxLen,pageheader[enetpacketLenL]);
    WpppH(pageport_TxLen,pageheader[enetpacketLenH]);
    do{
       RPP(ppageBusStatus);
       }while(!(bit_test(data_H,BUSSTA_RDY4TXNOW_BIT)));

    dataport_out;
    txlen = make16(pageheader[enetpacketLenH],pageheader[enetpacketLenL]);
    for(i=0;i<txlen;i+=2)
    {
       WpppL(pageport_RxTxData0,packet[i]);
       WpppH(pageport_RxTxData0,packet[i+1]);
    }
}
```

Code Snippet 10.10: Since this is an echo operation, the RxLength value can be used as the length of the outgoing frame.

Now, let's put a '0' (ASCII zero or 0x30) in the LCD Data Entry window (Screen Capture 10.4).

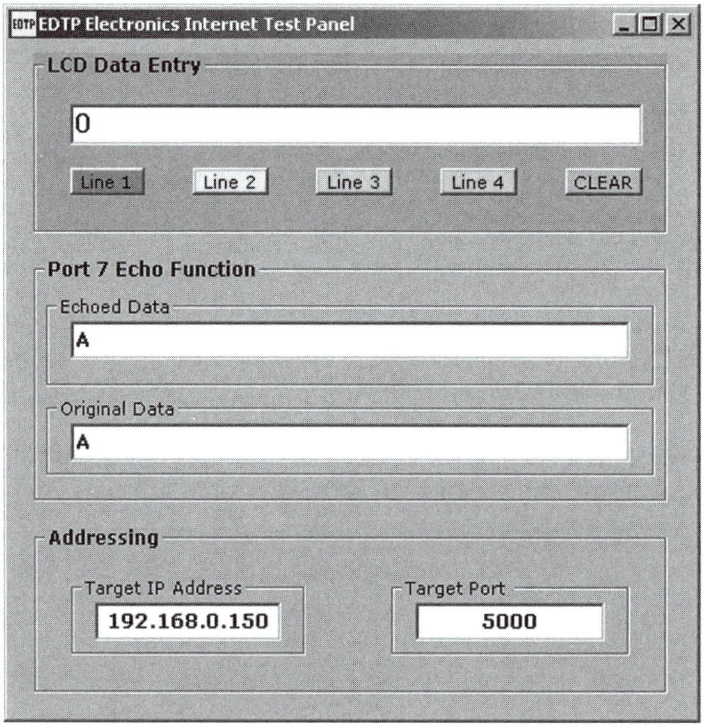

Screen Capture 10.4: The LCD Data Entry field is capable of transmitting data to a remote LCD via the Internet using UDP. Don't worry, I've included the complete source code for the LCD application on the CD-ROM.

```
//*********************************************************************
//*    UDP Function SNIPPET
//*    This function receives data from a Visual Basic UDP program and
//*    echoes the data back to the VB program and sets or resets bits
//*    on the PIC16F877's PORT A under control of the VB program.
//*********************************************************************
   //buttons on the VB GUI are pointed towards port address 5000 decimal
   else if(packet[UDP_destport] == 0x13 && packet[UDP_destport+1] == 0x88);
         {
      if(packet[UDP_data] == '0')
         //received a '0' from the VB program
         bit_clear(PORTA,5);
      else if(packet[UDP_data] == '1')
         //received a '1' from the VB program
         bit_set(PORTA,5);
      else if(packet[UDP_data] == 0x00)
         //received a 0x00 from the VB program
         output_a(0x00);
      else if(packet[UDP_data] == 0x01)
         //received a 0x01 from the VB program
         bit_set(PORTA,1);
      else if(packet[UDP_data] == 0x02)
         //received a 0x02 from the VB program
         bit_set(PORTA,2);
      else if(packet[UDP_data] == 0x03)
         //received a 0x03 from the VB program
         bit_set(PORTA,3);
      else if(packet[UDP_data] == 0x04)
         //received a 0x04 from the VB program
         bit_set(PORTA,4);
```

Code Snippet 10.11: I assigned a logical progression to the use of the Internet Test Panel buttons. The code can be changed to meet your requirements.

The PIC16F877's PORTA was conditioned to be an output port in the initialization sequence and if the Easy Ethernet CS8900A's UDP application determines that the destination port address is not 7 decimal (0x0007) but is actually 5000 decimal (0x1388), the Easy Ethernet CS8900A's UDP application looks at the data in the UDP data area. Remember the term "multiplexing"? Well, we just did it (multiplexed) when we determined that the button application in Code Snippet 10.11 was the target for the data in the UDP datagram and not the UDP echo application.

```
Address   00 01 02 03 04 05 06 07 08 09 0A 0B 0C 0D 0E 0F      ASCII

 0110    00 00 45 44 54 50 00 E0 29 87 F5 5B 08 00 45 00 ..EDTP.. )..[..E.
 0120    00 1D 03 18 00 00 80 11 B5 D0 C0 A8 00 01 C0 A8 ........ ........
 0130    00 96 13 8A 13 88 00 09 26 E2 30 00 00 00 00 00 ........ &.0.....
 0140    00 00 00 00 00 00 00 00 00 00 00 00 00 00 00 00 ........ ........
```

Hex Dump 10.3: There are a couple of logical ways to say zero. You can say ASCII 0 with a 0x30 or you can say binary 0 with 0x00. In this dump, we are speaking ASCII at memory offset 0x013A.

You're getting pretty good at reading hex frame dumps by now and you easily find the destination port value of 0x1388 (5000 decimal) and the UDP segment data, 0x30 or '0' (ASCII zero) in Hex Dump 10.3. Looking at the Easy Ethernet CS8900A UDP application source code for port 5000 and IP address 192.168.0.150 tells us that a '0' in the UDP data field controls the logic level of PORTA pin 5. This part of the Easy Ethernet CS8900A's UDP application firmware is relatively easy to follow. Clicking the Internet Test Panel's CLEAR button will take all of the PIC16F877's PORTA I/O pins to a logical low level. Clicking on an Internet Test Panel LineX button will take the PORTA I/O pin associated with the button to a logic high level.

UDP is dead easy to get up and running on a small microcontroller-based system like the Easy Ethernet CS8900A. Little is needed in addition to the necessary IP header information to pass a UDP message from host to host. I used the easy way out and used the Winsock services of Visual Basic to produce UDP datagrams. It wouldn't take too much more work to cut the personal computer out of the equation completely and have a multitude of Easy Ethernet CS8900A's swapping UDP datagrams on an Ethernet LAN segment.

Before we leave UDP I'd like to leave you with a picture of where we have been as it pertains to protocols, encapsulation and UDP. You and I have already coded every bit of what you see graphically in Figure 10.3.

The primary reason for all of this is to allow an application to send some meaningful data from Application A to Application B. In this section, we chose to use UDP, IP and Ethernet as the information conduit for the applications' data. So, we wrapped the data using UDP wrapping paper, which addressed the data to Application B and included a return address to Application A so we could receive a reply.

The freshly wrapped UDP package containing our Application A data was then put into a box that provided yet another address (IP address) that would allow our UDP wrapped package to be logically transported on the network. In addition to allowing our UDP package to ride on the network, the IP box includes a packing slip telling the receiver what's inside (a UDP packet containing the Application A data). At this point, our Application A data, which is wrapped in its UDP wrapper and boxed inside of IP, can logically get from Application A (us) to Application B (them). However, our IP package needs a physical means of going from Point A (us) to Point B (them). So, we put our IP package inside of another box (Ethernet frame), which is addressed to Point B and includes a Point A return address.

Once our trucking company talks to the folks at Point B to verify their physical address, the Ethernet frame is loaded on a truck and proceeds to travel from Point A (us) to Point B (them).

Our trucking company is a good one and the package arrives at Point B without a scratch. Our trucker hands the precious package to the loading dock receiver at Point B, who tears off the physical address (Ethernet header and trailer) and takes the IP box out of the Ethernet box. The information contained on the IP packing slip (IP header) tells the receiver to route the box to the UDP department.

Once the UDP department gets the box, it discards the IP box (IP header) and sees that the package is wrapped in a UDP wrapper (UDP header). The UDP wrapper (UDP header) tells the mail clerk to deliver the contents of the package to Application B. That's what we coded and that's how it all works. This process is not unique to UDP as you will see later.

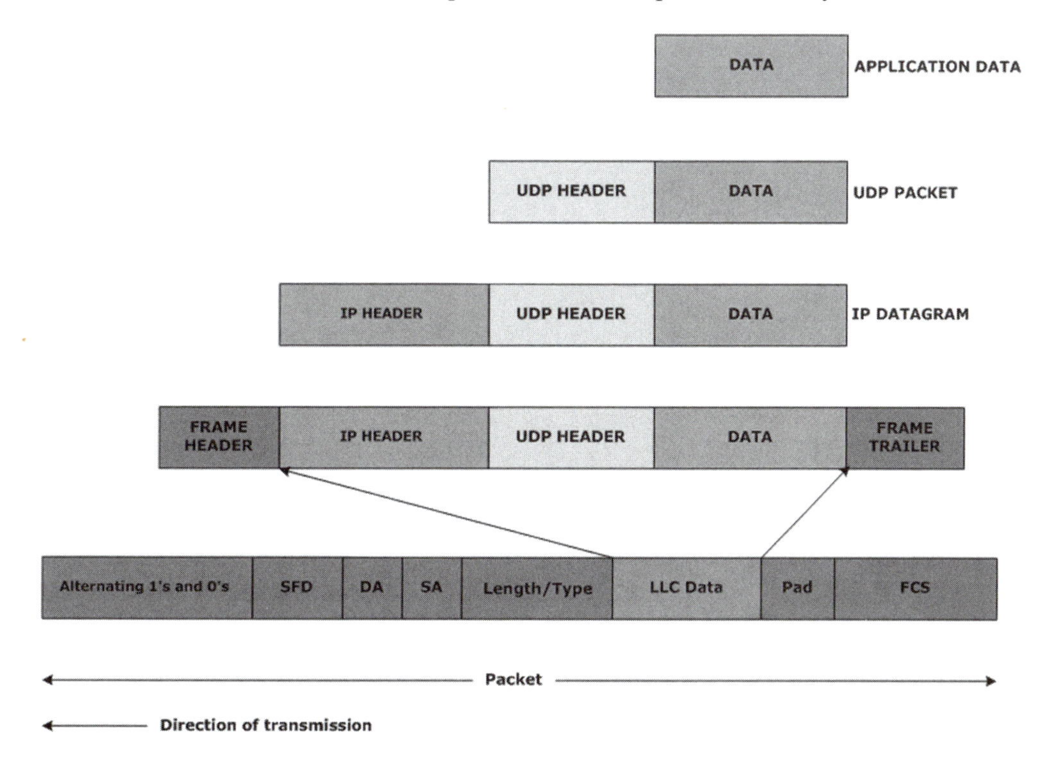

Figure 10.3: A picture is worth a thousand packets.

A situation may arise where a Visual Basic application or any Windows-based application could not be used, or was not available to communicate with a microcontroller-based LAN device like the Easy Ethernet CS8900A. If that were to happen, the alternative communications method would most likely fall to a legacy Internet application such as Telnet. UDP doesn't do Telnet, but there is one more trick in the Easy Ethernet CS8900A's bag that will, TCP.

CHAPTER 11

TCP and the Easy Ethernet CS8900A

This is the most complex of any of the Internet protocols we'll examine. There have been many books written about TCP and I'm not going to try to add my name to the list of TCP authors. Instead, I'm going to show you how to write the bare minimum of code to deploy your own microcontroller TCP/IP firmware.

The TCP/IP implementation that is effected by the Easy Ethernet CS8900A's firmware is designed to conserve both microcontroller program memory and microcontroller data memory space, and as a result is a very minimal TCP/IP model. The TCP/IP firmware is fashioned as a Telnet server that echoes characters it receives and services a PIC16F877 port I/O application.

Just like all of the other protocols up to this point, the TCP/IP module of the Easy Ethernet CS8900A's firmware is called from the get_frame function. The Ethernet type field identifies the frame inside its data area as an IP frame and once that fact is established, the protocol field inside the IP header is examined. If the value in the protocol field of the IP header is equal to 6, the Easy Ethernet CS8900A code is routed to the TCP function's code.

You've already been exposed to IP as it relates to UDP. TCP (Transmission Control Protocol) works with IP in much the same manner. The concept of encapsulation is still strong as like UDP, TCP lies within the IP data area (Figure 11.1).

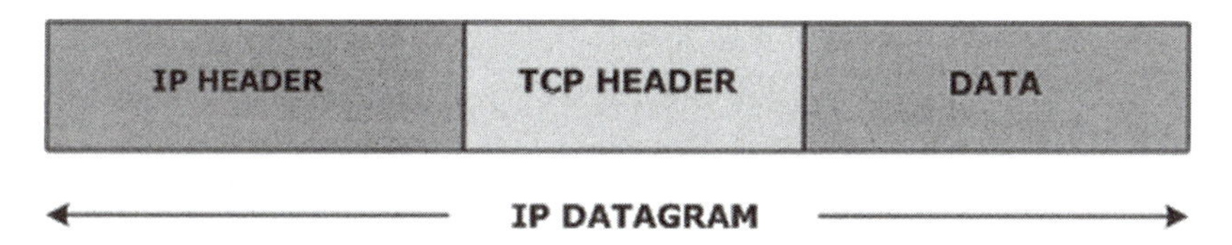

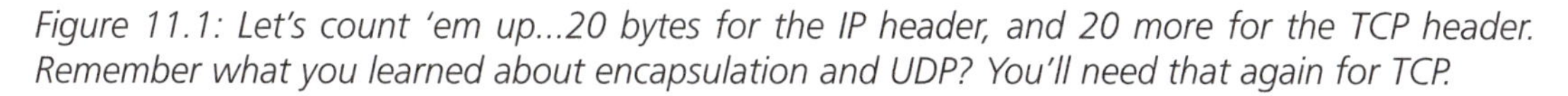

Figure 11.1: Let's count 'em up...20 bytes for the IP header, and 20 more for the TCP header. Remember what you learned about encapsulation and UDP? You'll need that again for TCP.

TCP/IP was designed from the ground up to be platform independent. That's why we can run a flavor of TCP/IP on a relatively tiny microcontroller-based system or on a mainframe complex the size of a football field. TCP/IP is a collection of protocols that are standardized across the world of networking. TCP/IP can't be used for every situation. However, while TCP/IP is not the total solution to all networking problems, TCP/IP stands as a pretty good model of what all of internetworking should be. Normally, to implement TCP/IP, one must employ the use of what is termed a "TCP/IP stack." Complete TCP/IP stacks can be very large and usually aren't easily ported fully to smaller platforms like our Easy Ethernet CS8900A.

Telnet is a protocol that is common to most users of the Internet. The purpose of the Telnet Protocol is to provide a general-purpose means of interfacing terminal devices and terminal-oriented applications to each other.

Just as you've seen with UDP network devices, every machine or device on a TCP/IP-based network, no matter how big or how small, is called a *host*. That includes clients as well as servers. It used to be that a server was always the big machine in the cloud, and the client was a workstation on someone's desk. Today, servers sit on desks and clients can be worn on your wrist or vice versa. Regardless of the host's size, the idea is that all hosts can communicate with each other. This implies that all hosts on all networks can communicate host to host across differing networks. This may sound impractical, but that's how the Internet works. Today, hosts all over the world communicate by passing messages, accessing data and transferring files between themselves sometimes using dissimilar networks.

In the TCP/IP world, messages are generally short packets of data. Just like UDP, each TCP data packet is addressed to reach a particular host on a particular network. There is no difference in the addressing scheme used for TCP and UDP. The protocol or IP address for both protocols is the familiar 4-bytes-divided-by-dots address scheme (192.168.0.150). Physical addresses or MAC addresses apply to both UDP and TCP and are used in the identical manner by both protocols.

You're probably used to hearing the term "TCP/IP" and thinking about it as one protocol. However, you know from our UDP experiences that IP is indeed its own protocol. The combination of Transmission Control Protocol and Internet Protocol was not done by accident, but by design. Just as it is in the UDP world, IP is the unreliable component of the TCP/IP pair. You will remember that IP is termed unreliable because there is no way that IP itself can guarantee that a data packet will actually be delivered to its destination. All of the hosts in the Internet or on a local LAN give their best effort to deliver an IP data packet. Despite all of the good intentions, the problem is that nobody cares how it looks when it gets where it's going. If you've ever sat and watched baggage handlers load baggage onto an aircraft, you already have a pretty good concept of how IP works. For example, relating to the IP datagram, an IP data packet is just like a piece of freight in the hands of the person loading baggage onto an airplane. They get the bag (our IP datagram) from one of those mobile baggage carts they drive around (the host) and sling it towards the moving-belt ramp (LAN, Internet). Sometimes the handlers throw a bag and it misses the conveyor. One of the other baggage handlers (other hosts) may pick up the dropped bag (corrupted IP datagram) and either sling it back on the ramp (passed it along to the next host for inspection) or put it directly on the plane (delivered the damaged IP datagram to the receiving host). The plane (receiving host), although being loaded with a bag full of broken goods, never reported back to the cart (sending host) that the bag (IP datagram) that was finally loaded had its contents damaged in transit. My little story implies that each bag or each IP datagram is independent of any other IP datagram and nobody really cares what's inside the bag while it is in transit.

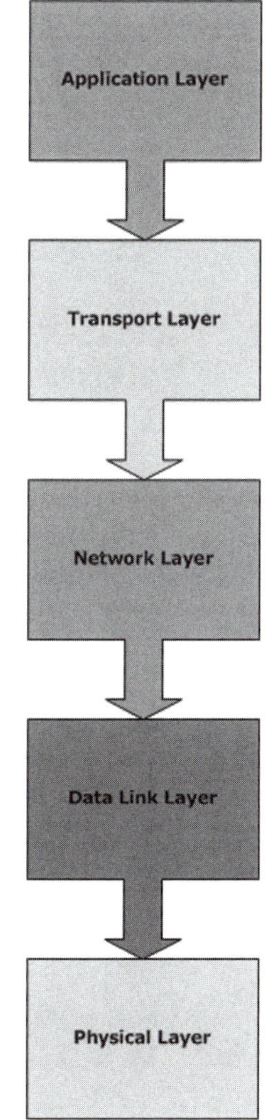

Figure 11.2: This is the "stack," which is really a bunch of layers between the wire and the application. Believe it or not, you already know more about this stack thing than you think.

The TCP/IP protocol stack is composed of five layers as shown in Figure 11.1. The Physical Layer is the simplest layer, and in my mind the hardest working layer with the Application Layer chiming in as potentially the most complex of the five layers. As you can see in the "stack" (Figure 11.2), each layer of the stack has a distinct job to do.

The confusion that could exist between the layers is eliminated by our old friend encapsulation. Each layer passes only properly formatted output to the next layer for processing. Encapsulation also allows each layer to treat the data in the way it prefers without affecting the way the data is treated in other layers. For example, the Transport Layer likes to pretend that data is entering in a constant stream while the Internet or IP layer sees data as separate connectionless datagrams. To write a successful embedded TCP/IP application it is necessary to understand the functions of each protocol layer. Let's take a look at each of them.

The Physical Layer

The Physical Layer is another way of saying hardware layer. This is the wire, cable and electronics that connect the devices and networks to each other. Physical also implies "touchable" or real. For the Easy Ethernet CS8900A, an Ethernet cable, the Ethernet isolation magnetics and some interface circuitry inside the CS8900A-CQ form the Physical Layer. The rest of the physical network could consist of an Ethernet hub or Ethernet router and any other cabling or electronic devices that tie the physical network together. The bottom line is that the Physical Layer always sees the entire packet whether it is receiving it or transmitting it and never adds or subtracts to a packet's contents.

The Data Link Layer

The Data Link Layer is only responsible for transferring a datagram from one host over a single physical link to another host. Most, if not all, of the Data Link Layer functionality resides in the implementation of the CS8900A-CQ MAC engine. The CS8900A-CQ MAC engine accepts data and wraps it into an Ethernet-compatible package that can be received and "unwrapped" by the host the data was addressed to. Remember, it's the CS8900A-CQ's MAC engine that generates the preamble and CRC when an Ethernet frame is transmitted. The CS8900A-CQ's MAC engine also makes sure the ether is clear before attempting to send a message and if a collision occurs, the CS8900A-CQ's MAC engine waits and retries the send operation. Also, recall that the CS8900A-CQ MAC engine checks the incoming frame's hardware address to determine if the incoming data belongs to it or another host on the network.

The MAC is the lower sublayer of the Data Link Layer that interfaces with the physical part of the network to help deliver a frame. The LLC (Logical Link Control) is the upper sublayer of the Data Link Layer and is not used by the Easy Ethernet CS8900A.

Just in case you're wondering where ARP belongs—ARP lives in the Data Link Layer.

The Network Layer

The Network Layer encapsulates messages passed from the Transport Layer and produces datagrams. This is where IP lives. The Network Layer encapsulates a UDP packet or TCP segment inside an IP datagram. An IP header is added, which calls out the handling of the IP datagram, and the datagram is then passed along to the Data Link Layer for transmission.

As we get further into coding some PIC16F877 TCP/IP routines, you will come to the realization that the IP datagram is the fundamental information that flows over the Internet.

ARP thrives in the Data Link Layer. ICMP hangs around in the Network Layer.

The Transport Layer

Between the Application Layer and the Network Layer lies the Transport Layer. The job of the Transport Layer is to pass data between the Application Layer and the Transport Layer using TCP or UDP protocols.

UDP lives here but TCP is what makes the Transport Layer famous. TCP, unlike UDP, uses a virtual connection to make sure that the data arrives at its destination intact and in order. TCP accomplishes this "connection" via handshaking and special codes in each data segment.

TCP and UDP receive data from the application and form segments or packets, respectively. A destination address is added before passing the packet or segment to the Network Layer.

The Application Layer

The final and topmost layer is the Application Layer. This is where the programmer reigns. There are more protocols used in this layer than I care to mention, some familiar and some homegrown. In simplest terms, data flows from the Application Layer of the originating host down through the TCP/IP stack and out the Physical Layer across to the Physical Layer of the destination host. Once the data enters the destination host's Physical Layer, the process is reversed and the data flows up through the TCP/IP stack to the Application Layer where it is processed (Figure 11.3).

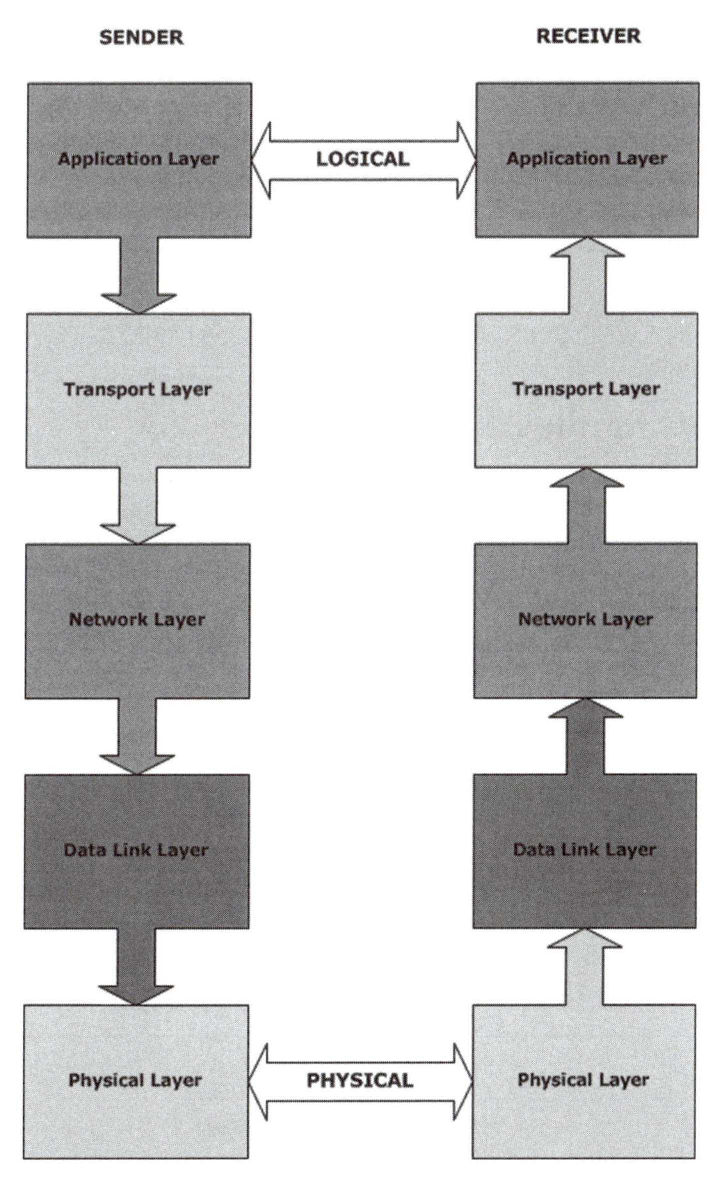

Figure 11.3: According to a very popular space exploration television series, the Vulcans were the first extraterrestrials to acknowledge their presence on Earth. This is so logical, I wonder if they were the ones that really invented this encapsulation stuff?

OK, there's only one more subject standing between us and coding our Easy Ethernet CS8900A TCP/IP application—ports. As most of you know, a host could be running multiple applications at once. In fact, we sort of did that in our UDP code. How does the TCP/IP stack know where to route the messages? Just like UDP, TCP/IP handles these situations by assigning each network connection its own protocol port. A protocol port is actually an internal TCP or UDP address. This address is passed down the stack in the header of each packet of data. IP sends logical host addresses (192.168.0.150), TCP and UDP send protocol port addresses (7, 23, 5000, and so forth).

You already know about ports and multiplexing from your experiences with UDP. So, let's go TCP/IP coding.

Coding TCP/IP for the Easy Ethernet CS8900A

Our UDP ports consisted of one well-known port (7) and one homebrewed port (5000). The Easy Ethernet CS8900A TCP/IP code will use port 0x1F98 or 8088 decimal. I've assigned the TCP/IP port address value to MY_PORT_ADDRESS in Code Snippet 11.1.

```
//******************************************************************
//* PORT ADDRESS DEFINITION
//*   YOU MAY CHANGE THIS TO ANY VALID PORT ADDRESS
//******************************************************************
#define  MY_PORT_ADDRESS      0x1F98  // 8088 DECIMAL
```

Code Snippet 11.1: I was thinking about old personal computers when I came up with the Easy Ethernet CS8900A's TCP/IP port number.

Thus far, you've seen IP headers, UDP headers and ICMP headers. It stands to reason that TCP would also have a header. A graphic depiction of the TCP header is shown in Figure 11.4.

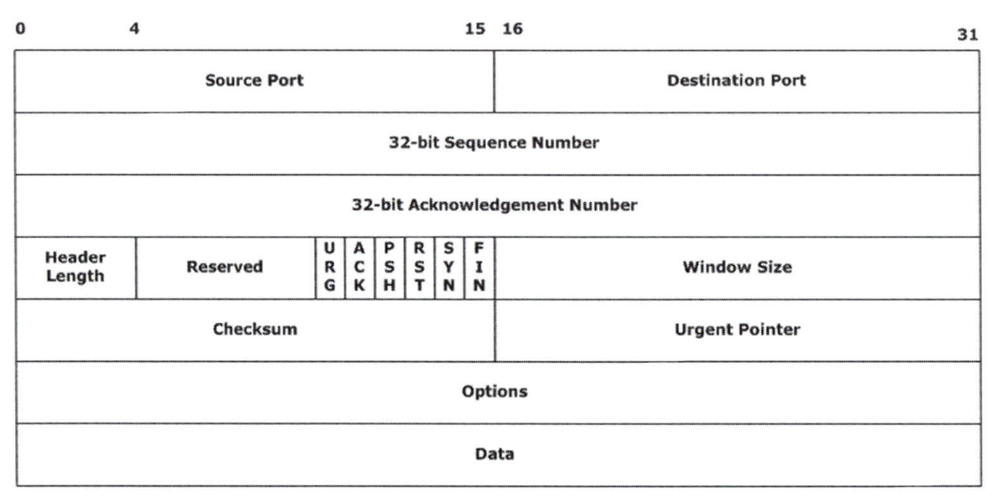

Figure 11.4: We won't be using the Options area of the TCP header in the Easy Ethernet CS8900A firmware.

And, just like UDP and the others, I've laid out the TCP header for use in the Easy Ethernet CS8900A's firmware in Code Snippet 11.2.

```
//*************************************************************
//*   Ethernet Header Layout
//*************************************************************
int8  packet[96];    //50 bytes of UDP data available
//*************************************************************
//*   TCP Header Layout
//*************************************************************
#define  TCP_srcport    0x22  //TCP source port
#define  TCP_destport   0x24  //TCP destination port
#define  TCP_seqnum     0x26  //sequence number
#define  TCP_acknum     0x2A  //acknowledgement number
#define  TCP_hdrflags   0x2E  //4-bit header len and flags
#define  TCP_window     0x30  //window size
#define  TCP_cksum      0x32  //TCP checksum
#define  TCP_urgentptr  0x34  //urgent pointer
#define  TCP_data       0x36  //option/data
```

Code Snippet 11.2: The TCP header is a bit busier than the other headers we've examined.

Looking at Code Snippet 11.3, you can see that the TCP section of the Easy Ethernet CS8900A's code is called from the get_frame function in the same manner as any of the other IP protocol's code we've discussed so far.

```
//*************************************************************
//*   IP Protocol Types
//*************************************************************
#define  PROT_ICMP   0x01
#define  PROT_TCP 0x06
#define  PROT_UDP 0x11
//*************************************************************
//*   Receive a Frame
//*************************************************************
   //process an IP packet
   else if(packet[enetpacketType0] == 0x08 &&
         packet[enetpacketType1] == 0x00 &&
         packet[ip_destaddr] == MYIP[0] &&
         packet[ip_destaddr+1] == MYIP[1] &&
         packet[ip_destaddr+2] == MYIP[2] &&
         packet[ip_destaddr+3] == MYIP[3])
```

```
{
   if(packet[ip_proto] == PROT_ICMP)
      icmp();
   else if(packet[ip_proto] == PROT_UDP)
      udp();
   else if(packet[ip_proto] == PROT_TCP)
      tcp();
}
```

Code Snippet 11.3: An incoming IP frame with the Easy Ethernet CS8900A's IP address and a pointer to the TCP protocol kicks-off the TCP function.

To have gotten to the Easy Ethernet CS8900A's TCP code, the incoming IP address matched the Easy Ethernet CS8900A's IP address and the hardware or MAC address was verified to belong to the Easy Ethernet CS8900A by the CS8900A-CQ's MAC engine. Remember that just like UDP, a TCP port address is used to select a particular application. In the case of the Easy Ethernet CS8900A, there is only one application. Nevertheless, a TCP port number is still required.

```
//******************************************************************
//*   Ethernet Header Layout
//******************************************************************
int8  packet[96];        //50 bytes of UDP data available
//******************************************************************
//*   TCP Header Layout
//******************************************************************
#define  TCP_srcport       0x22  //TCP source port
#define  TCP_destport      0x24  //TCP destination port
//******************************************************************
//*   TCP Function
//*   This function uses TCP protocol to act as a Telnet server on
//*   port 8088 decimal. The application function is called with
//*   every incoming character.
//******************************************************************
void tcp()
{
   int8 i,j;

   //assemble the destination port address from the incoming packet
   portaddr = make16(packet[TCP_destport],packet[TCP_destport+1]);
```

Code Snippet 11.4: The destination port is located in the second word of the TCP header. To invoke the Easy Ethernet CS8900A's TCP application, the destination port number should be 8088 decimal.

Code Snippet 11.4 uses the Custom Computer Services C Compiler's resident *make16* function to assemble the TCP destination port address. Just in case the port address is the one the Easy Ethernet CS8900A is looking for, the amount of data in the incoming TCP frame is determined by the code in Code Snippet 11.5.

```
//*******************************************************************
//*   Ethernet Header Layout
//*******************************************************************
int8  packet[96];        //50 bytes of UDP data available
//*******************************************************************
//*   IP Header Layout
//*******************************************************************
#define  ip_vers_len          0x0E  //IP version and header length
#define  ip_pktlen            0x10  //packet length
//*******************************************************************
//*   TCP Header Layout
//*******************************************************************
#define  TCP_hdrflags         0x2E  //4-bit header len and flags
//*******************************************************************
//*   TCP Function
//*   This function uses TCP protocol to act as a Telnet server on
//*   port 8088 decimal. The application function is called with
//*   every incoming character.
//*******************************************************************
   //calculate the length of the data coming in with the packet
   //tcpdatalen_in = incoming packet length - incoming ip header
length - incoming tcp header length
   tcpdatalen_in = (make16(packet[ip_pktlen],packet[ip_pktlen+1])) -
   ((packet[ip_vers_len] & 0x0F) * 4) - (((packet[TCP_hdrflags] & 0xF0) >> 4) * 4);
```

Code Snippet 11.5: Using information collected from here and there, we can easily find out how many bytes of data is being carried inside the TCP data area.

To get what we need to perform our length calculations, we must turn to fields in both the IP and TCP headers. Now you can see why you've always heard TCP and IP coupled as TCP/IP. In Code Snippet 11.5 the TCP data length is calculated by subtracting the IP header length and TCP header length from the total IP packet length. We already have the IP datagram length in the ip_pktlen field of the IP header. The IP header length is calculated using the ip_vers_len field of the IP header multiplied times four. In a similar fashion, the high nibble of the TCP_hdrflags field in the TCP header is multiplied by four to obtain the TCP header length. Let's see if we can pick out our header fields in Hex Dump 11.1.

```
Address   00 01 02 03 04 05 06 07 08 09 0A 0B 0C 0D 0E 0F       ASCII

 0110    00 00 45 44 54 50 00 E0 29 87 F5 5B 08 00 45 00 ..EDTP.. )..[..E.
 0120    00 30 02 E2 40 00 80 06 75 FE C0 A8 00 01 C0 A8 .0..@... u.......
 0130    00 96 04 0D 1F 98 01 C0 22 7B 00 00 00 00 70 02 ........ "{....p.
 0140    40 00 79 57 00 00 02 04 05 B4 01 01 04 02       @.yW.... ......
```

Hex Dump 11.1: This PIC16F877 hex TCP dump was taken just before the TCP function was to be called.

The destination MAC address beginning at offset 0x0110 in Hex Dump 11.1 belongs to the Easy Ethernet CS8900A (00EDTP). The source MAC address belongs to the personal computer on the LAN segment I'm using to generate the packets. 0x0800 at memory location 0x011C is the Ethernet type field and says this is an IP frame. Consulting the Easy Ethernet CS8900A source code and the layout of the packet array, which is byte-for-byte correlated with an Ethernet packet, the IP header begins at offset 0x011E. The IP datagram length is 0x0030, which is noted at offset 0x0120 of Hex Dump 11.1. That has located all of the IP header information we need so far.

The TCP header lies at the beginning of the IP data area. The easy way to find the IP data area is to locate the source and destination IP addresses in the IP header. Our LAN segment is addressed as 192.168.0.XXX. So, we can scan Hex Dump 11.1 looking for a 0xC0, which is 192 decimal. The first 0xC0 is found at offset 0x012A. Taking in the next three bytes yields "C0 A8 00 01." That's hexadecimal notation for "192 168 0 1." Add the dots between the bytes and you end up with the sending personal computer's IP address. According to the IP header layout in the Easy Ethernet CS8900A source code, this is the source address (ip_srcaddr). The next field should be the destination IP address. Scanning again for the next 0xC0, we find the pattern "C0 A8 00 96," which is "192 168 0 150." That is the Easy Ethernet CS8900A's IP address, which we assigned early on in the Easy Ethernet CS8900A source code. OK…the byte immediately following the 0x96 at offset 0x132 is the beginning of the IP data area and the beginning of the TCP header. Repeat after me: THIS IS A PRIME EXAMPLE OF ENCAPSULATION.

According to the TCP header shown in Figure 11.4, the very first word (16 bits) of the TCP header is the source port address. In Hex Dump 11.1, we see "04 0D" as the beginning word of the TCP header that begins at offset 0x0132. 0x040D equates to 1037 decimal, which is the TCP port being used by the Telnet application on the personal computer I used to generate the packet. Since I told the personal computer program to contact a device at 192.168.0.150 and use port number 8088, the next word of the TCP header should be the destination port address, which should be the TCP port address we assigned the Easy Ethernet CS8900A in the source code, which is 8088. And, indeed, 0x1F98 is the next word and that translates to 8088 decimal.

To get to the TCP header length field, we must travel across two 32-bit numbers that we will explore further in a moment. The first 32-bit word is the TCP sequence number, and the second 32-bit word is the TCP acknowledgement number. The last value we are looking for, the TCP header length field, is located at offset 0x013E of Hex Dump 11.1 (0x70). We use the upper nibble multiplied by four in our data length calculation.

Now that you've done your homework and deciphered the codes in Hex Dump 11.1, you can check your dump reading skills against Sniffer Screen Capture 11.1's interpretation of Hex Dump 11.1.

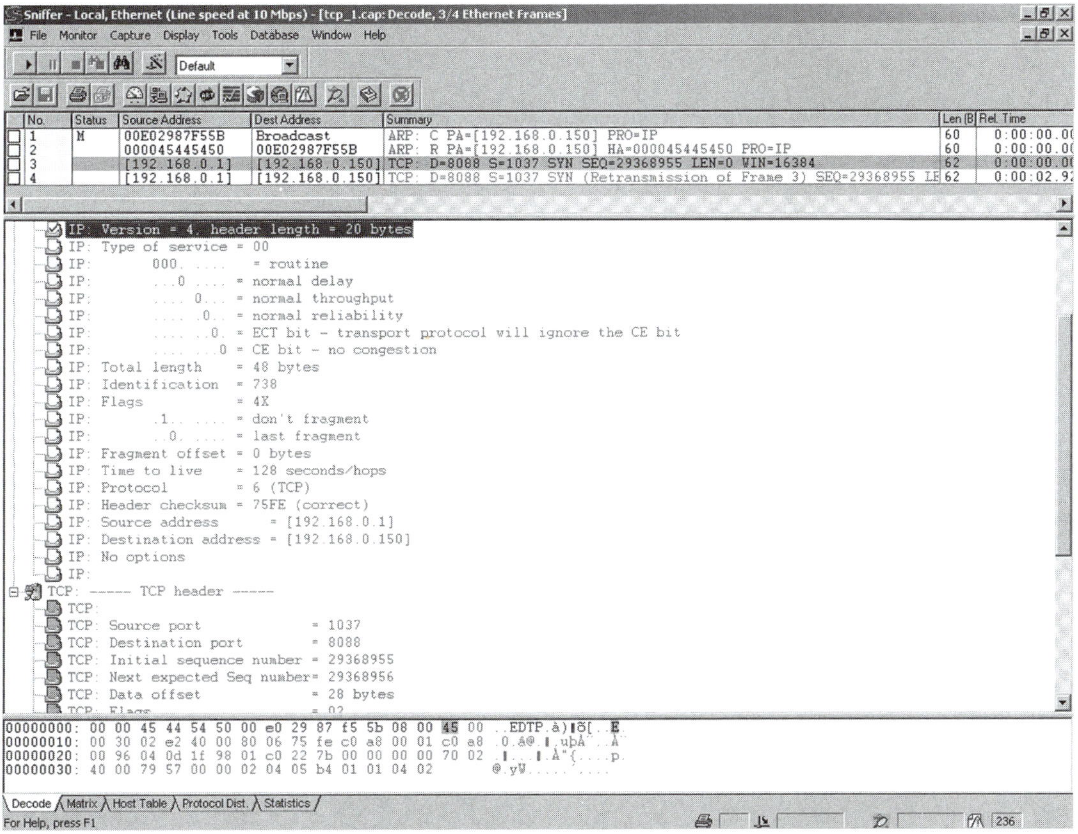

```
Sniffer - Local, Ethernet (Line speed at 10 Mbps) - [tcp_1.cap: Decode, 3/4 Ethernet Frames]
File  Monitor  Capture  Display  Tools  Database  Window  Help
Default

No. Status Source Address    Dest Address       Summary                                                          Len (B) Rel. Time
1   M      00E02987F55B      Broadcast          ARP: C PA=[192.168.0.150] PRO=IP                                 60      0:00:00.0
2          000045445450      00E02987F55B       ARP: R PA=[192.168.0.150] HA=000045445450 PRO=IP                 60      0:00:00.0
3          [192.168.0.1]     [192.168.0.150]    TCP: D=8088 S=1037 SYN SEQ=29368955 LEN=0 WIN=16384              62      0:00:00.0
4          [192.168.0.1]     [192.168.0.150]    TCP: D=8088 S=1037 SYN (Retransmission of Frame 3) SEQ=29368955 LE 62  0:00:02.9

IP: Version = 4, header length = 20 bytes
IP: Type of service = 00
IP:       000. ....   = routine
IP:       ...0 .... = normal delay
IP:       .... 0... = normal throughput
IP:       .... .0.. = normal reliability
IP:       .... ..0. = ECT bit - transport protocol will ignore the CE bit
IP:       .... ...0 = CE bit - no congestion
IP: Total length    = 48 bytes
IP: Identification  = 738
IP: Flags           = 4X
IP:       .1.. .... = don't fragment
IP:       ..0. .... = last fragment
IP: Fragment offset = 0 bytes
IP: Time to live    = 128 seconds/hops
IP: Protocol        = 6 (TCP)
IP: Header checksum = 75FE (correct)
IP: Source address      = [192.168.0.1]
IP: Destination address = [192.168.0.150]
IP: No options
IP:
TCP: ----- TCP header -----
TCP:
TCP: Source port               = 1037
TCP: Destination port          = 8088
TCP: Initial sequence number   = 29368955
TCP: Next expected Seq number= 29368956
TCP: Data offset               = 28 bytes
TCP: Flags                     = 02

00000000: 00 00 45 44 54 50 00 e0 29 87 f5 5b 08 00 45 00  ..EDTP.à)▮õ[..E.
00000010: 00 30 02 e2 40 00 80 06 75 fe c0 a8 00 01 c0 a8  .0.â@.▮.uþÀ¨..À¨
00000020: 00 96 04 0d 1f 98 01 c0 22 7b 00 00 00 00 70 02  .▮...▮.À"{....p.
00000030: 40 00 79 57 00 00 02 04 05 b4 01 01 04 02        @.yW..........

Decode  Matrix  Host Table  Protocol Dist.  Statistics
For Help, press F1                                                          236
```

Sniffer Screen Capture 11.1: If you can decipher the IP header, you can also easily figure out what's going on in the TCP header.

I stopped the Easy Ethernet CS8900A, as the personal computer was attempting to establish a TCP connection. I'll remove the PIC16F877 breakpoint and allow the code to run its course. Lots of stuff will be going on. So, I'll capture the entire session establishment with the Sniffer.

Beginning on familiar ground, in Sniffer Screen Capture 11.2, the very first thing the personal computer does after I issue the Telnet open command is to perform an ARP, which is promptly handled by the Easy Ethernet CS8900A in Sniffer Screen Capture 11.3.

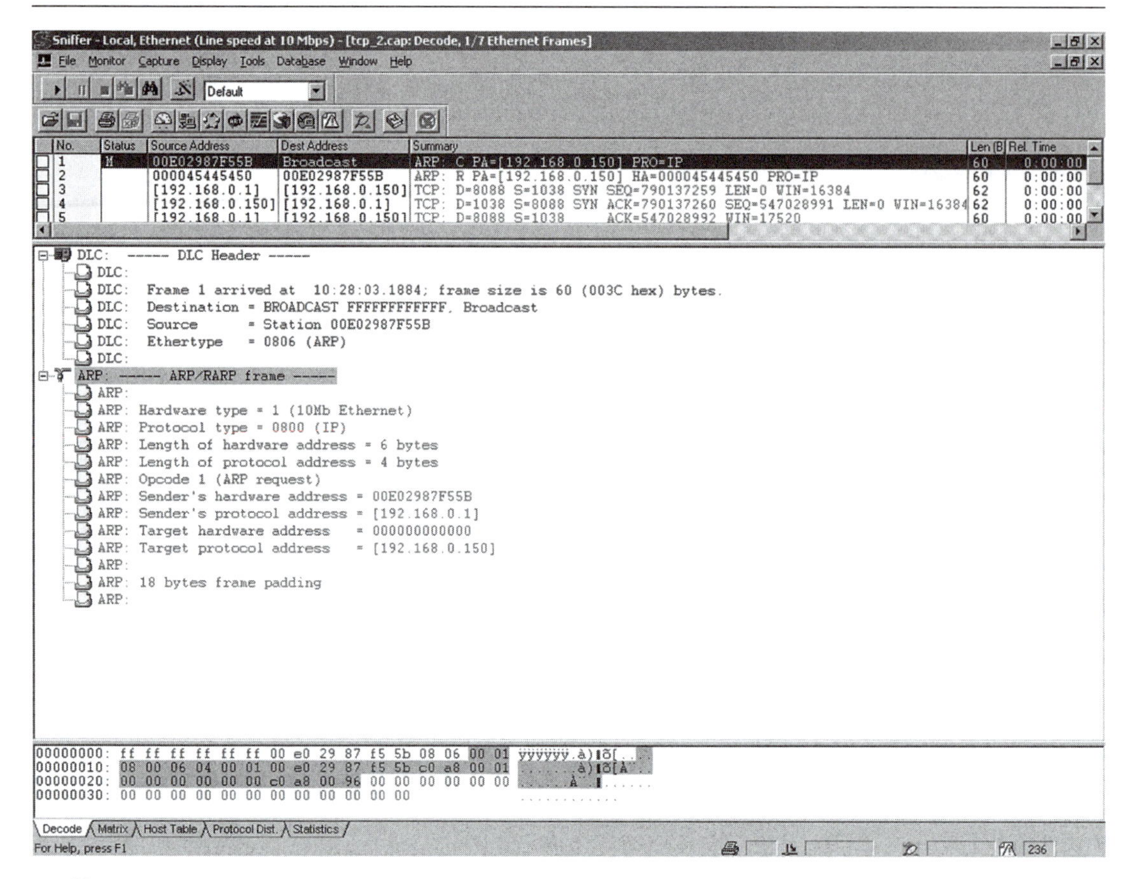

Sniffer Screen Capture 11.2: Nothing new for you here. In fact, you can probably tell this is an ARP frame by looking at the hex dump.

You should be able to work your way through an IP header. So, I'll concentrate on the TCP header and associated code for the contents of Sniffer Screen Capture 11.4. Bear in mind that TCP likes to sequence things. So, don't try to correlate any numbers in the Sniffer Screen Captures from this point on with Hex Dump 11.1. The theory remains the same but the numbers will sequence every time a new pass and new Sniffer screen capture is made.

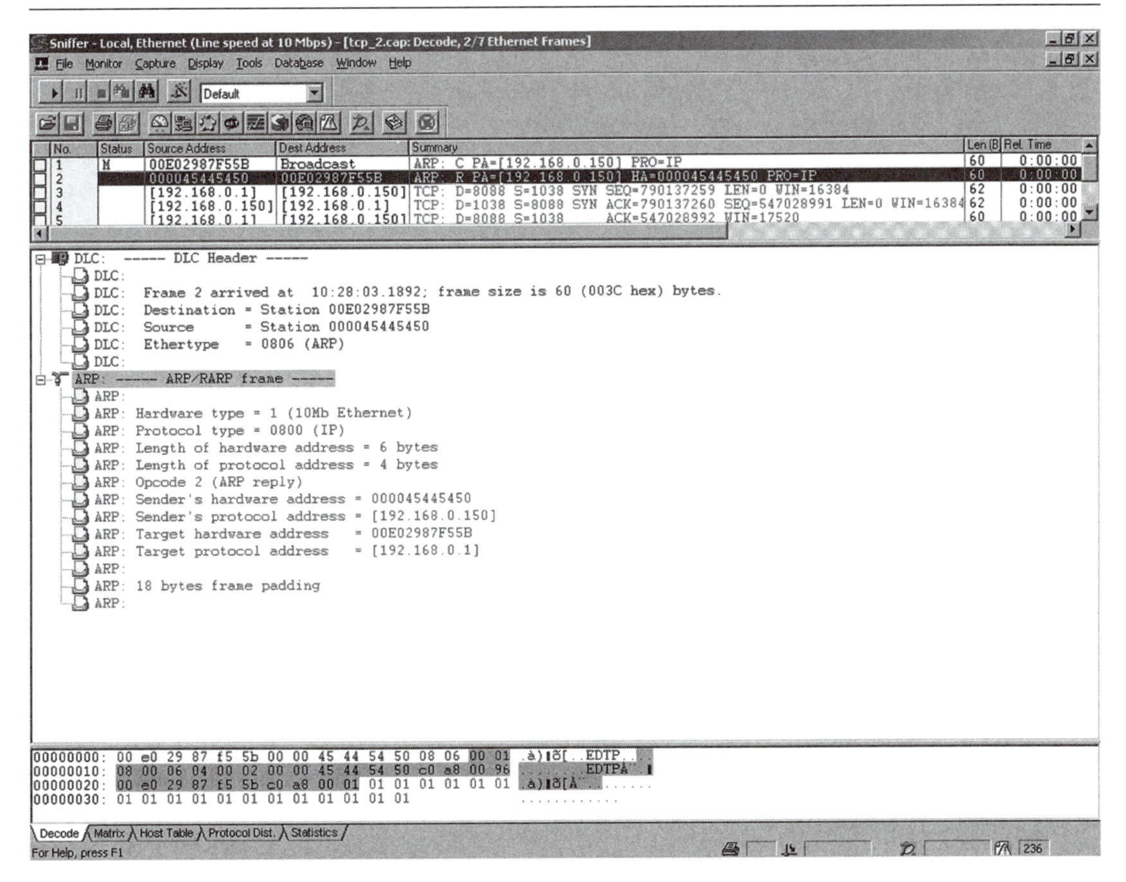

Sniffer Screen Capture 11.3: In the Summary window, notice the MAC addresses are shown for the ARP frames and IP addresses are displayed for the TCP frames.

Take some time to look over Sniffer Screen Capture 11.4 carefully. There are lots of little things that can be used on the screen that make the interpreted information more meaningful. For instance, the Summary window gives the direction of the data flow, the MAC and IP addresses of the communicating parties and a brief summary of what the packet did. Plus, using the Sniffer screen captures beats the heck out of reading dumps.

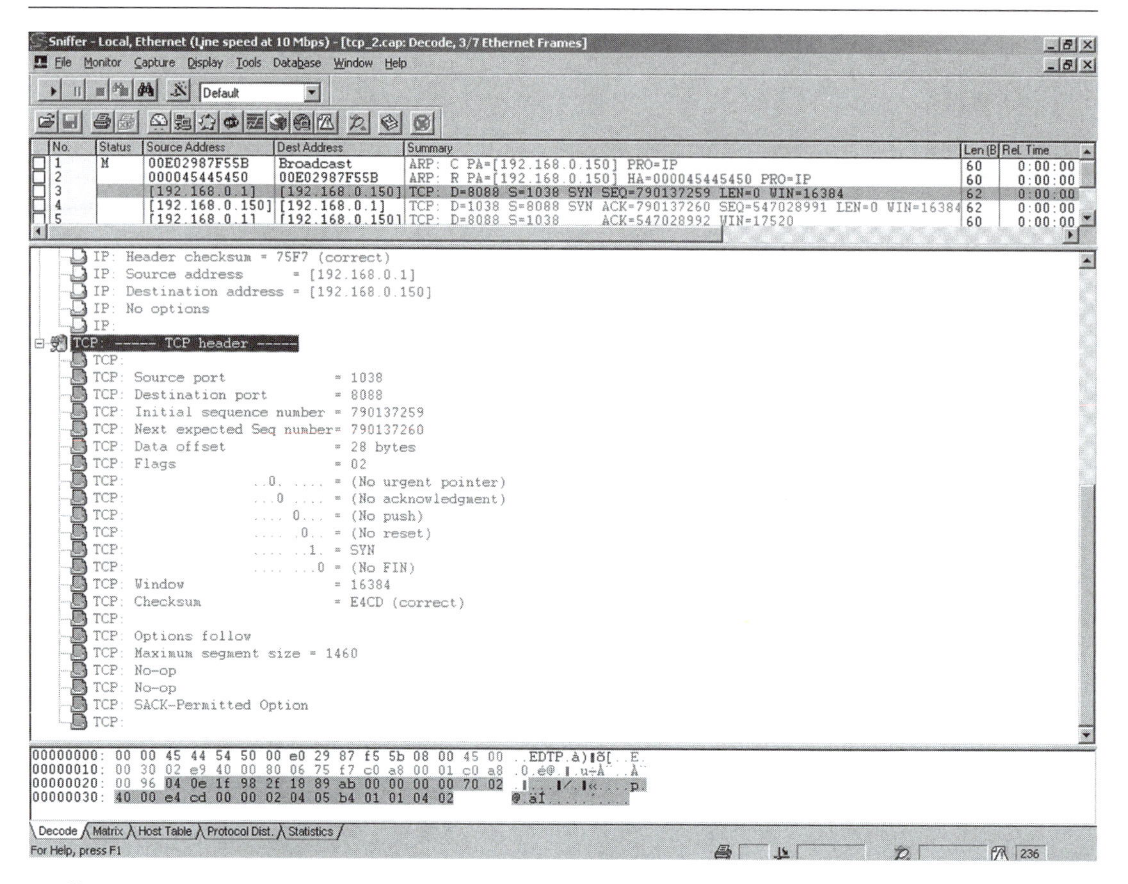

Sniffer Screen Capture 11.4: This is the first phase of the 3-way handshake process.

The Summary window for Sniffer Screen Capture 11.4, tells us at a glance, that the packet information being displayed below it is a TCP frame that originated at the personal computer (192.168.0.1) and was aimed at the Easy Ethernet CS8900A (192.168.0.150). Without looking at the detailed breakdown, the Summary area of the Summary window tells us the destination port (8088), the source port (1038) and TCP flag status for starters. Sniffer Screen Capture 11.4 is the first leg of the 3-way handshake procedure that establishes a TCP session between the personal computer and the Easy Ethernet CS8900A. I've put the associated Easy Ethernet CS8900A source code in Code Snippet 11.6.

```
//******************************************************************
//*   TELNET SERVER BANNER STATEMENT CONSTANT
//******************************************************************
int8 const telnet_banner[] = "\r\nEDTP Telnet Server>";
//******************************************************************
//* PORT ADDRESS DEFINITION
//*   YOU MAY CHANGE THIS TO ANY VALID PORT ADDRESS
//******************************************************************
#define  MY_PORT_ADDRESS      0x1F98  // 8088 DECIMAL
//******************************************************************
//*   Ethernet Header Layout
//******************************************************************
int8  packet[96];    //50 bytes of UDP data available
//******************************************************************
//*   TCP Header Layout
//******************************************************************
#define  TCP_srcport    0x22  //TCP source port
#define  TCP_destport   0x24  //TCP destination port
#define  TCP_seqnum     0x26  //sequence number
#define  TCP_acknum     0x2A  //acknowledgement number
#define  TCP_hdrflags   0x2E  //4-bit header len and flags
#define  TCP_window     0x30  //window size
#define  TCP_cksum      0x32  //TCP checksum
#define  TCP_urgentptr  0x34  //urgent pointer
#define  TCP_data       0x36  //option/data
//******************************************************************
//*   TCP Flags
//*   IN flags represent incoming bits
//*   OUT flags represent outgoing bits
//******************************************************************
#define  FIN_IN   bit_test(packet[TCP_hdrflags+1],0)
#define  SYN_IN   bit_test(packet[TCP_hdrflags+1],1)
#define  RST_IN   bit_test(packet[TCP_hdrflags+1],2)
#define  PSH_IN   bit_test(packet[TCP_hdrflags+1],3)
#define  ACK_IN   bit_test(packet[TCP_hdrflags+1],4)
#define  URG_IN   bit_test(packet[TCP_hdrflags+1],5)
#define  FIN_OUT  bit_set(packet[TCP_hdrflags+1],0);
#define  SYN_OUT  bit_set(packet[TCP_hdrflags+1],1);
#define  RST_OUT  bit_set(packet[TCP_hdrflags+1],2);
#define  PSH_OUT  bit_set(packet[TCP_hdrflags+1],3);
#define  ACK_OUT  bit_set(packet[TCP_hdrflags+1],4);
#define  URG_OUT  bit_set(packet[TCP_hdrflags+1],5);

int8 aux_data[16];         //tcp application received data area
int8 data_H,data_L;
int16 tcpdatalen_out,ISN,ip_packet_len;
```

```
int16 portaddr,chksum16,hdrlen,tcplen,tcpdatalen_in;
int32 hdr_chksum,my_seqnum,client_seqnum;
int1 synflag;
//*****************************************************************
//*   TCP Function
//*   This function uses TCP protocol to act as a Telnet server on
//*   port 8088 decimal. The application function is called with
//*   every incoming character.
//*****************************************************************
   //this code segment processes the incoming SYN from the Telnet client
   //and sends back the initial sequence number (ISN) and acknowledges
   //the incoming SYN packet
   if(SYN_IN && portaddr == MY_PORT_ADDRESS)
   {
      tcpdatalen_in = 0x01;
      synflag = 1;

      setipaddrs();

      data_L = packet[TCP_srcport];
      packet[TCP_srcport] = packet[TCP_destport];
      packet[TCP_destport] = data_L;

      data_L = packet[TCP_srcport+1];
      packet[TCP_srcport+1] = packet[TCP_destport+1];
      packet[TCP_destport+1] = data_L;

      assemble_ack();

      if(++ISN == 0x0000 || ++ISN == 0xFFFF)
         ISN = 0x1234;
      my_seqnum = make32(ISN,0xFFFF);

      set_packet32(TCP_seqnum,my_seqnum);

      packet[TCP_hdrflags+1] = 0x00;
      SYN_OUT;
      ACK_OUT;

      packet[TCP_cksum] = 0x00;
      packet[TCP_cksum+1] = 0x00;

      hdr_chksum =0;
      hdrlen = 0x08;
      addr = &packet[ip_srcaddr];
```

```
        cksum();
        hdr_chksum = hdr_chksum + packet[ip_proto];
        tcplen = make16(packet[ip_pktlen],packet[ip_pktlen+1]) -
        ((packet[ip_vers_len] & 0x0F) * 4);
        hdr_chksum = hdr_chksum + tcplen;
        hdrlen = tcplen;
        addr = &packet[TCP_srcport];
        cksum();
        chksum16= ~(hdr_chksum + ((hdr_chksum & 0xFFFF0000) >> 16));
        packet[TCP_cksum] = make8(chksum16,1);
        packet[TCP_cksum+1] = make8(chksum16,0);
```

Code Snippet 11.6: This isn't as bad as it looks.

Lots of stuff is taking place in a hurry in Code Snippet 11.6. So, let's digest it line by line, beginning with Code Snippet 11.7.

```
//******************************************************************
//*     TCP Function
//*     This function uses TCP protocol to act as a Telnet server on
//*     port 8088 decimal. The application function is called with
//*     every incoming character.
//******************************************************************
    if(SYN_IN && portaddr == MY_PORT_ADDRESS)
```

Code Snippet 11.7: A SYN flag bit and the Easy Ethernet CS8900A's TCP port address must be sensed by the Easy Ethernet CS8900A before the handshake process can begin.

The first message sent in the 3-way handshake process is a data-less TCP segment with the SYN bit set in the TCP header flags field. If there is no session established, the Easy Ethernet CS8900A's first inclination is to check for its own port address (MY_PORT_ADDRESS) and run a test on the SYN flag bit (SYN_IN) in the TCP header as shown in Code Snippet 11.8.

```
//******************************************************************
//*     TCP Function
//*     This function uses TCP protocol to act as a Telnet server on
//*     port 8088 decimal. The application function is called with
//*     every incoming character.
//******************************************************************
    tcpdatalen_in = 0x01;
    synflag = 1;

    setipaddrs();
```

Code Snippet 11.8: The SYN flag bit is considered data and is accounted for in tcpdatalen_in.

If it is determined that the remote host is trying to establish a TCP/IP session, the Easy Ethernet CS8900A prepares for part 2 of the 3-way handshake by telling itself that the incoming sequence number from the client should be incremented in the outgoing message in the server's outgoing acknowledgement number (*tcpdatalen_in = 0x01*). Let's get a better grasp on this before we move on.

The Easy Ethernet CS8900A in this scenario is the server, and the personal computer is the client. The SYN flag in the incoming frame of the client's first TCP handshake segment must be treated as a sequence number in the server's response, which is the second part of the 3-way handshake. This is best understood by examining the value of the randomly generated *Initial sequence number* field in Sniffer Screen Capture 11.4 (790137259). Note that the client personal computer (192.168.0.1 port 1038) is expecting the value of the *Initial sequence number* field to be incremented by one, as the client's *Next expected Seq number* field is set at 790137260. The Sniffer detail also points out that the SYN flag bit is set. So, the incoming data length is set for one, even though there is no data in the TCP segment. A software flag (synflag) is set to signal to the Easy Ethernet CS8900A's *TCP* function that the first part of the 3-way handshake has already taken place.

Let's pick up with the *setipaddrs* function in Code Snippet 11.8, which gets the frame ready to return from whence it came by turning around the source and destination IP and MAC addresses and calculating the IP header checksum. This is the same *setipaddrs* function we used in the UDP discussion.

```
//*******************************************************************
//*   TCP Function
//*   This function uses TCP protocol to act as a Telnet server on
//*   port 8088 decimal. The application function is called with
//*   every incoming character.
//*******************************************************************
data_L = packet[TCP_srcport];
packet[TCP_srcport] = packet[TCP_destport];
packet[TCP_destport] = data_L;

data_L = packet[TCP_srcport+1];
packet[TCP_srcport+1] = packet[TCP_destport+1];
packet[TCP_destport+1] = data_L;
```

Code Snippet 11.9: I could have picked any free variable other than data_L *to make this swapperoo.*

The same turning around of source and destination addresses has to be done for the source and destination TCP port numbers. In Code Snippet 11.9, I used *data_L* as a scratch register to hold the original TCP source port value as the *packet[TCP_srcport]* memory location gets changed in the process.

```
//*******************************************************************
//*   Assemble the Acknowledgment
//*   This function assembles the acknowledgment to send to
//*   to the client by adding the received data count to the
//*   client's incoming sequence number.
//*******************************************************************
void assemble_ack()
{
client_seqnum=make32(packet[TCP_seqnum],packet[TCP_seqnum+1],packet
[TCP_seqnum+2],packet[TCP_seqnum+3]);
   client_seqnum = client_seqnum + tcpdatalen_in;
   set_packet32(TCP_acknum,client_seqnum);
}
```

Code Snippet 11.10: A perfect example of how the Custom Computer Services C Compiler built-in make *functions make it easy to manipulate 32-bit TCP header field variables.*

The *assemble_ack* function depicted in Code Snippet 11.10 takes the client's (personal computer's) incoming sequence number and increments it by one. Since the server (Easy Ethernet CS8900A) must acknowledge with an incremented client sequence number, the client's incremented sequence number is placed in the server's outgoing acknowledgement field.

```
//*******************************************************************
//*   TCP Function
//*   This function uses TCP protocol to act as a Telnet server on
//*   port 8088 decimal. The application function is called with
//*   every incoming character.
//*******************************************************************
   if(++ISN == 0x0000 || ++ISN == 0xFFFF)
      ISN = 0x1234;
   my_seqnum = make32(ISN,0xFFFF);

   set_packet32(TCP_seqnum,my_seqnum);
```

Code Snippet 11.11: The ISN is an arbitrary number with no special meaning at this point.

Before sending an acknowledgement, the server (Easy Ethernet CS8900A) must establish an initial sequence number of its own. As you can see in the source code offered by Code Snippet 11.11, the ISN (Initial Sequence Number) is a 32-bit pseudo-random number that is ultimately placed into the server's (Easy Ethernet CS8900A) outgoing TCP sequence number header field.

```
packet[TCP_hdrflags+1] = 0x00;
SYN_OUT;
ACK_OUT;
```

Code Snippet 11.12: The TCP code isn't complicated. All we have to do is follow the TCP/IP protocol standards and fill in the blanks in our TCP header.

In this part of the 3-way handshake process, the server (Easy Ethernet CS8900A) must respond with the SYN and ACK flag bits set in the TCP header. The TCP header flags area is cleared by the first line of code in Code Snippet 11.12. If you look at the Easy Ethernet CS8900A TCP source code, you'll see that SYN_OUT and ACK_OUT are macros that set the flag bits in the server's TCP header area.

```
packet[TCP_cksum] = 0x00;
packet[TCP_cksum+1] = 0x00;

hdr_chksum =0;
hdrlen = 0x08;
addr = &packet[ip_srcaddr];
cksum();
hdr_chksum = hdr_chksum + packet[ip_proto];
tcplen = make16(packet[ip_pktlen],packet[ip_pktlen+1]) -
((packet[ip_vers_len] & 0x0F) * 4);
hdr_chksum = hdr_chksum + tcplen;
hdrlen = tcplen;
addr = &packet[TCP_srcport];
cksum();
chksum16= ~(hdr_chksum + ((hdr_chksum & 0xFFFF0000) >> 16));
packet[TCP_cksum] = make8(chksum16,1);
packet[TCP_cksum+1] = make8(chksum16,0);
```

Code Snippet 11.13: Even the dreaded checksum code is easy when you break it down into manageable bytes.

It seems that nothing we've done so far in the Ethernet and Internet worlds can live without a checksum. The TCP checksum in Code Snippet 11.13 is calculated exactly like the UDP checksum with the calculation and transmission of the TCP checksum not being an option.

We have built an image of the TCP segment we want to send to the client in the PIC16F877's packet array memory. All we have to do now is call an old friend (the *echo_packet* function) to send the Ethernet frame, which contains our brand new TCP segment, which completes part 2 of the 3-way handshake.

```
//*******************************************************************
//*   ECHO THE PACKET
//*******************************************************************
void echo_packet()
{
   dataport_out;
   WpppL(pageport_TxCmd,TXCMD_AFTER_ALL);
   WpppH(pageport_TxCmd,0x00);
   WpppL(pageport_TxLen,pageheader[enetpacketLenL]);
   WpppH(pageport_TxLen,pageheader[enetpacketLenH]);
   do{
         RPP(ppageBusStatus);
      }while(!(bit_test(data_H,BUSSTA_RDY4TXNOW_BIT)));
   dataport_out;
   txlen = make16(pageheader[enetpacketLenH],pageheader[enetpacketLenL]);
   for(i=0;i<txlen;i+=2)
   {
      WpppL(pageport_RxTxData0,packet[i]);
      WpppH(pageport_RxTxData0,packet[i+1]);
   }
}
```

Code Snippet 11.14: Most of the time what comes in must go out.

You already know about what is going on inside the *echo_packet* code contained in Code Snippet 11.14. So, take a look at what's going on in Sniffer Screen Capture 11.5, which is the Sniffer screen capture of the TCP segment we just sent back to the client.

It looks like we did everything right. The Easy Ethernet CS8900A established its ISN as 547028991 and expects 547028992 to be acknowledged from the client. Note that both the SYN and ACK flag bits are set in the Easy Ethernet CS8900A's reply TCP segment, and the Easy Ethernet CS8900A acknowledged with 790137260 just as the client expected. We know this is all correct because we just coded it.

Now that all of the sequence numbers have been exchanged, the only thing standing between the Easy Ethernet CS8900A and the client personal computer's exchange of real data is the final acknowledgement from the client. Let's check out the action in Sniffer Screen Capture 11.6 and Code Snippet 11.15.

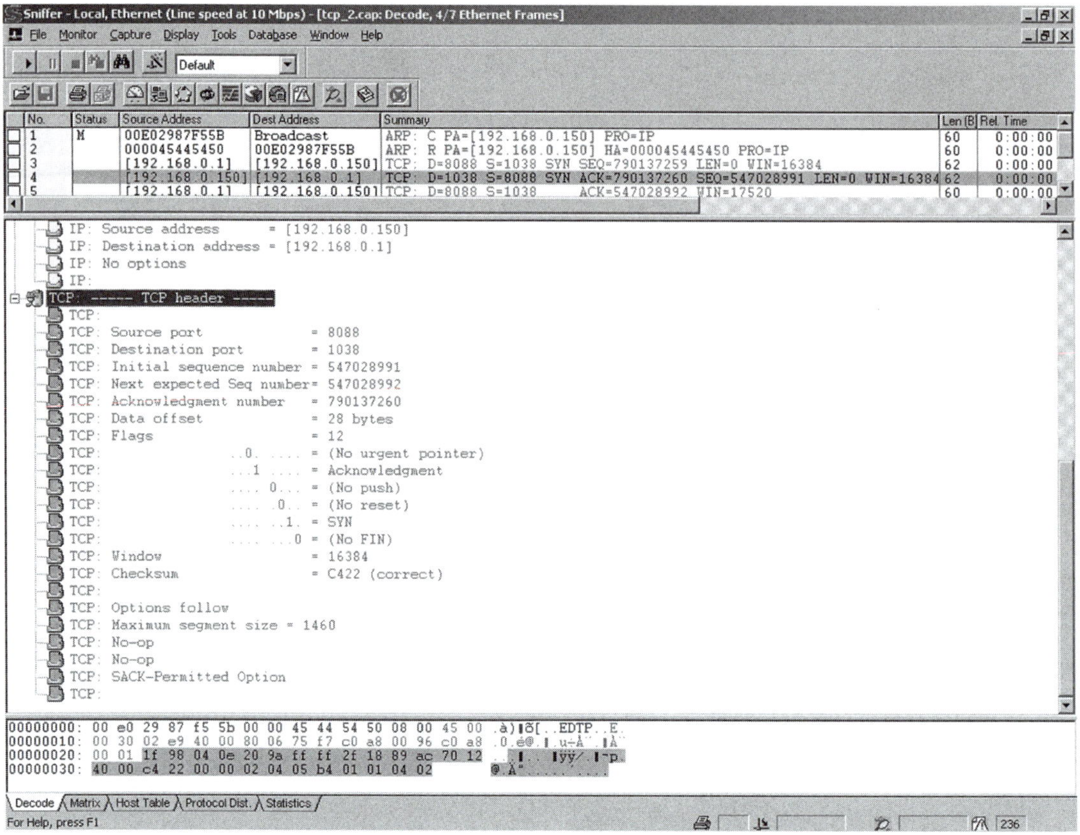

```
Sniffer - Local, Ethernet (Line speed at 10 Mbps) - [tcp_2.cap: Decode, 4/7 Ethernet Frames]
File  Monitor  Capture  Display  Tools  Database  Window  Help
Default

No. Status Source Address    Dest Address      Summary                                                                Len (B) Rel. Time
1   M      00E02987F55B      Broadcast         ARP: C PA=[192.168.0.150] PRO=IP                                       60   0:00:00
2          000045445450      00E02987F55B      ARP: R PA=[192.168.0.150] HA=000045445450 PRO=IP                       60   0:00:00
3          [192.168.0.1]     [192.168.0.150]   TCP: D=8088 S=1038 SYN SEQ=790137259 LEN=0 WIN=16384                   62   0:00:00
4          [192.168.0.150]   [192.168.0.1]     TCP: D=1038 S=8088 SYN ACK=790137260 SEQ=547028991 LEN=0 WIN=16384     62   0:00:00
5          [192.168.0.1]     [192.168.0.150]   TCP: D=8088 S=1038     ACK=547028992 WIN=17520                         60   0:00:00

IP: Source address      = [192.168.0.150]
IP: Destination address = [192.168.0.1]
IP: No options
IP:
TCP: ----- TCP header -----
TCP:
TCP: Source port               = 8088
TCP: Destination port          = 1038
TCP: Initial sequence number   = 547028991
TCP: Next expected Seq number= 547028992
TCP: Acknowledgment number     = 790137260
TCP: Data offset               = 28 bytes
TCP: Flags                     = 12
TCP:              ..0. .... = (No urgent pointer)
TCP:              ...1 .... = Acknowledgment
TCP:              .... 0... = (No push)
TCP:              .... .0.. = (No reset)
TCP:              .... ..1. = SYN
TCP:              .... ...0 = (No FIN)
TCP: Window                    = 16384
TCP: Checksum                  = C422 (correct)
TCP:
TCP: Options follow
TCP: Maximum segment size = 1460
TCP: No-op
TCP: No-op
TCP: SACK-Permitted Option
TCP:

00000000: 00 e0 29 87 f5 5b 00 00 45 44 54 50 08 00 45 00  .à)‡õ[..EDTP..E.
00000010: 00 30 02 e9 40 00 80 06 75 f7 c0 a8 00 96 c0 a8  .0.é@.€.u÷À¨..À¨
00000020: 00 01 1f 98 04 0e 20 9a ff ff 2f 18 89 ac 70 12  ..... .ÿÿ/..¬p.
00000030: 40 00 c4 22 00 00 02 04 05 b4 01 01 04 02        @.Ä"..........

Decode | Matrix | Host Table | Protocol Dist. | Statistics
For Help, press F1                                                  236
```

Sniffer Screen Capture 11.5: The story is told in both the Summary window and the detail area. Don't you love the Sniffer?

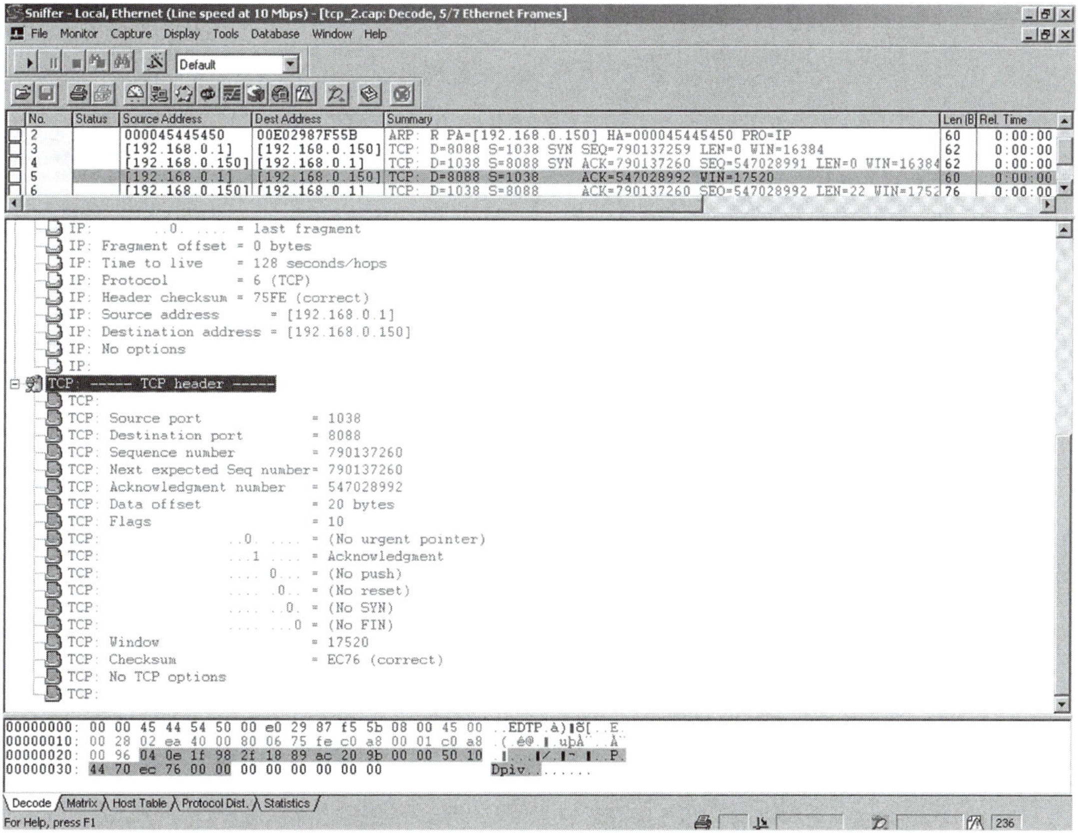

Sniffer Screen Capture 11.6: Use the Summary window and the detail area to gain a perspective on what's going on here.

The meat of Sniffer Screen Capture 11.6 is that the client acknowledges with 547028992 and satisfies the next sequence number expected by the server. The *Sequence number* and *Next expected Seq number* fields are identical, which means that there is no data in the client's acknowledgement. The lack of data kicks in the next piece of Easy Ethernet CS8900A TCP code laid out in Code Snippet 11.15.

```
//*********************************************************************
//*   TCP Function
//*   This function uses TCP protocol to act as a Telnet server on
//*   port 8088 decimal. The application function is called with
//*   every incoming character.
//*********************************************************************
   //If an ACK is received and the destination port address is valid
and no data is in the packet
   if(ACK_IN && portaddr == MY_PORT_ADDRESS && tcpdatalen_in == 0x00)
   {
      //assemble the acknowledgment number from the incoming packet
      incoming_ack =make32(packet[TCP_acknum],packet[TCP_acknum+1],
packet[TCP_acknum+2],
packet[TCP_acknum+3]);

   //if the incoming packet is a result of session establishment
   if(synflag)
   {
      //clear the SYN flag
      synflag = 0;

      //the incoming acknowledgment is my new sequence number
      my_seqnum = incoming_ack;

      //send the Telnet server banner
         j = sizeof(telnet_banner);
         for(i=0;i<j;++i)
            packet[TCP_data+i] = telnet_banner[i];
      //length of the banner message
         tcpdatalen_out = j;

      //expect to get an acknowledgment of the banner message
      expected_ack = my_seqnum + tcpdatalen_out;

      //send the TCP/IP packet
      send_tcp_packet();
   }
}
```

Code Snippet 11.15: If that Telnet banner shows, you'll be just as excited as you were when you saw your UDP data echoed earlier.

Let's break down the logic of the source code in Code Snippet 11.15. The first line of Code Snippet 11.15 is satisfied in that within the incoming TCP segment shown in Sniffer Screen Capture 11.6, the ACK flag bit is set, the port address matches the Easy Ethernet CS8900A's TCP port address (8088) and there is no data in the TCP segment.

```
//*******************************************************************
//*   TCP Function
//*   This function uses TCP protocol to act as a Telnet server on
//*   port 8088 decimal. The application function is called with
//*   every incoming character.
//*******************************************************************
   //assemble the acknowledgment number from the incoming packet
   incoming_ack =make32(packet[TCP_acknum],packet[TCP_acknum+1],
packet[TCP_acknum+2],
packet[TCP_acknum+3]);
```

Code Snippet 11.16: Here's yet another example of how the Custom Computer Services C Compiler built-in functions help us do TCP.

Depending on the upcoming circumstances, we may have to do some calculating with the acknowledgement number. So, in Code Snippet 11.16 the incoming acknowledgement number is reassembled into a 32-bit value from the values found in the packet array bytes that make up the incoming acknowledgement number.

```
//*******************************************************************
//*   TCP Function
//*   This function uses TCP protocol to act as a Telnet server on
//*   port 8088 decimal. The application function is called with
//*   every incoming character.
//*******************************************************************
   //if the incoming packet is a result of session establishment
   if(synflag)
   {
      //clear the SYN flag
      synflag = 0;

      //the incoming acknowledgment is my new sequence number
      my_seqnum = incoming_ack;
```

Code Snippet 11.17: We're ready to rock and roll now.

Recall that we set the *synflag* to signal to us that we are in the 3-way handshake/session establishment process if we passed through the *TCP* function again. Here we are. Part of Code Snippet 11.17 clears the *synflag*. The 3-way handshake process is now complete and the client and server are ready to exchange data. The incoming acknowledgement number we pulled from the client in Code Snippet 11.16 is the server's starting sequence number for data transmission.

```
//*********************************************************************
//*    TCP Function
//*    This function uses TCP protocol to act as a Telnet server on
//*    port 8088 decimal. The application function is called with
//*    every incoming character.
//*********************************************************************
      //send the Telnet server banner
         j = sizeof(telnet_banner);
         for(i=0;i<j;++i)
            packet[TCP_data+i] = telnet_banner[i];
      //length of the banner message
         tcpdatalen_out = j;
```

Code Snippet 11.18: The Telnet banner is stored in program memory instead of RAM.

Since the client has gotten to this point using Telnet, it would be nice to send a banner to the client, which would give the human using the client some positive feedback. The *telnet_banner* is hard-coded into the flash of the PIC16F877. We need to get the *telnet_banner* characters moved into the TCP data area in the PIC16F877's packet array RAM. We also need to tell the Easy Ethernet CS8900A TCP code how long the *telnet_banner* message is. Code Snippet 11.18 does all of that.

```
//*********************************************************************
//*    TCP Function
//*    This function uses TCP protocol to act as a Telnet server on
//*    port 8088 decimal. The application function is called with
//*    every incoming character.
//*********************************************************************
      //expect to get an acknowledgment of the banner message
      expected_ack = my_seqnum + tcpdatalen_out;
```

Code Snippet 11.19: This is what makes TCP/IP harder to code. You must keep up with every byte that is transferred. Our minimal TCP/IP firmware lacks many of the more complex error recovery procedures.

We already know how many bytes we'll be sending in our banner message and we know our starting sequence number. So, in Code Snippet 11.19, we calculate an expected acknowledgement number that the client will return after the banner message is delivered.

```
         //send the TCP/IP packet
         send_tcp_packet();
```

Code Snippet 11.20: Whoosh!

So far, we've only set up the data portion of the TCP segment, and Code Snippet 11.20 calls yet another TCP function, *send_TCP_packet.* You've been working very hard and your dump reading abilities are at the highest level. So, let's do the analysis of Code Snippet 11.21 backwards using Sniffer Screen Capture 11.7 as our guide.

```
//*******************************************************************
//*   Ethernet Header Layout
//*******************************************************************
int8  packet[96];       //50 bytes of UDP data available
#define  ip_pktlen            0x10  //packet length
#define  TCP_srcport          0x22  //TCP source port
#define  TCP_destport         0x24  //TCP destination port

//*******************************************************************
//*   PacketPage I/O Port Definitions
//*******************************************************************
#define  pageport_RxTxData0   0x00  //Receive/Transmit data Port 0
#define  pageport_TxCmd       0x04  //Transmit Command
#define  pageport_TxLen       0x06  //Transmit Length

int8 data_H,data_L;
int16 i,txlen,rxlen,chksum16,hdrlen,tcplen,tcpdatalen_in;
int16 tcpdatalen_out,ISN,portaddr,ip_packet_len;
int32 hdr_chksum,my_seqnum,client_seqnum,incoming_ack,expected_ack;
//*******************************************************************
//*   Send TCP Packet
//*   This routine assembles and sends a complete TCP/IP packet.
//*   40 bytes of IP and TCP header data is assumed.
//*******************************************************************
void send_tcp_packet()
{
   //count IP and TCP header bytes.. Total = 40 bytes
   ip_packet_len = 40 + tcpdatalen_out;
   packet[ip_pktlen] = make8(ip_packet_len,1);
   packet[ip_pktlen+1] = make8(ip_packet_len,0);
   setipaddrs();

   data_L = packet[TCP_srcport];
   packet[TCP_srcport] = packet[TCP_destport];
   packet[TCP_destport] = data_L;
   data_L = packet[TCP_srcport+1];
   packet[TCP_srcport+1] = packet[TCP_destport+1];
   packet[TCP_destport+1] = data_L;
```

```
assemble_ack();
set_packet32(TCP_seqnum,my_seqnum);

packet[TCP_hdrflags+1] = 0x00;
ACK_OUT;
if(finflag)
{
   FIN_OUT;
   finflag = 0;
}

packet[TCP_cksum] = 0x00;
packet[TCP_cksum+1] = 0x00;

hdr_chksum =0;
hdrlen = 0x08;
addr = &packet[ip_srcaddr];
cksum();
hdr_chksum = hdr_chksum + packet[ip_proto];
tcplen = ip_packet_len - ((packet[ip_vers_len] & 0x0F) * 4);
hdr_chksum = hdr_chksum + tcplen;
hdrlen = tcplen;
addr = &packet[TCP_srcport];
cksum();
chksum16= ~(hdr_chksum + ((hdr_chksum & 0xFFFF0000) >> 16));
packet[TCP_cksum] = make8(chksum16,1);
packet[TCP_cksum+1] = make8(chksum16,0);

txlen = ip_packet_len + 14;
if(txlen < 60)
   txlen = 60;
data_L = make8(txlen,0);
data_H = make8(txlen,1);

dataport_out;
WpppL(pageport_TxCmd,TXCMD_AFTER_ALL);
WpppH(pageport_TxCmd,0);
WpppL(pageport_TxLen,data_L);
WpppH(pageport_TxLen,data_H);
do{
      RPP(ppageBusStatus);
   }while(!(bit_test(data_H,BUSSTA_RDY4TXNOW_BIT)));
dataport_out;
for(i=0;i<txlen;i+=2)
```

```
    {
        WpppL(pageport_RxTxData0,packet[i]);
        WpppH(pageport_RxTxData0,packet[i+1]);
    }
```

Code Snippet 11.21: The txlen variable is checked to make sure it is at least 60 decimal. The CS8900A-CQ will automatically pad potential runt packets. The txlen code is a holdover from the code you'll see when we explore the RTL8019AS, which does not automatically pad a potentially runt packet.

Let's start our backwards analysis by first looking at the Summary window in Sniffer Screen Capture 11.7. We are working with frame 6 in the Summary window. Notice that just beyond the Summary data inside the Summary window is a field called *Len*. *Len* represents the total length of the frame in bytes. Let's look at the code and see if we can come up with 76 bytes.

```
Sniffer - Local, Ethernet (Line speed at 10 Mbps) - [tcp_2.cap: Decode, 6/7 Ethernet Frames]
File  Monitor  Capture  Display  Tools  Database  Window  Help
Default

No. Status Source Address   Dest Address      Summary                                                        Len (B) Rel. Time
3          [192.168.0.1]    [192.168.0.150]   TCP: D=8088 S=1038 SYN SEQ=790137259 LEN=0 WIN=16384               62  0:00:00
4          [192.168.0.150]  [192.168.0.1]     TCP: D=1038 S=8088 SYN ACK=790137260 SEQ=547028991 LEN=0 WIN=16384 62  0:00:00
5          [192.168.0.1]    [192.168.0.150]   TCP: D=8088 S=1038     ACK=547028992 WIN=17520                     60  0:00:00
6          [192.168.0.150]  [192.168.0.1]     TCP: D=1038 S=8088     ACK=790137260 SEQ=547028992 LEN=22 WIN=1752 76  0:00:00
7          [192.168.0.1]    [192.168.0.150]   TCP: D=8088 S=1038     ACK=547029014 WIN=17498                     60  0:00:00

IP: Fragment offset = 0 bytes
IP: Time to live    = 128 seconds/hops
IP: Protocol        = 6 (TCP)
IP: Header checksum = 75E8 (correct)
IP: Source address      = [192.168.0.150]
IP: Destination address = [192.168.0.1]
IP: No options
IP:
TCP: ----- TCP header -----
TCP:
TCP: Source port                = 8088
TCP: Destination port           = 1038
TCP: Sequence number            = 547028992
TCP: Next expected Seq number= 547029014
TCP: Acknowledgment number      = 790137260
TCP: Data offset                = 20 bytes
TCP: Flags                      = 10
TCP:               ..0. .... = (No urgent pointer)
TCP:               ...1 .... = Acknowledgment
TCP:               .... 0... = (No push)
TCP:               .... .0.. = (No reset)
TCP:               .... ..0. = (No SYN)
TCP:               .... ...0 = (No FIN)
TCP: Window                     = 17520
TCP: Checksum                   = 742D (correct)
TCP: No TCP options
TCP: [22 Bytes of data]
TCP:

00000000: 00 e0 29 87 f5 5b 00 00 45 44 54 50 08 00 45 00  .à)‡õ[..EDTP..E.
00000010: 00 3e 02 ea 40 00 80 06 75 e8 c0 a8 00 96 c0 a8  .>.ê@.€.uèÀ¨.–À¨
00000020: 00 01 1f 98 04 0e 20 9b 00 00 2f 18 89 ac 50 10  ......../.‰¬P.
00000030: 44 70 74 2d 00 00 0d 0a 45 44 54 50 20 54 65 6c  Dpt-....EDTP Tel
00000040: 6e 65 74 20 53 65 72 76 65 72 3e 00              net Server>.
Decode  Matrix  Host Table  Protocol Dist.  Statistics
For Help, press F1
```

Sniffer Screen Capture 11.7: You don't have to count them if you're using the Sniffer. Look in the Summary window, far right.

Since this implementation of TCP/IP is not by any means a complete implementation, I've cut to the chase and assumed no options or frills would be included in the TCP or IP headers. That would set the TCP and IP headers at a constant 20 bytes each. That's where the number 40 comes from in Code Snippet 11.22.

```
//******************************************************************
//*   Send TCP Packet
//*   This routine assembles and sends a complete TCP/IP packet.
//*   40 bytes of IP and TCP header data is assumed.
//******************************************************************
   //count IP and TCP header bytes.. Total = 40 bytes
   ip_packet_len = 40 + tcpdatalen_out;
   packet[ip_pktlen] = make8(ip_packet_len,1);
   packet[ip_pktlen+1] = make8(ip_packet_len,0);
```

Code Snippet 11.22: Some assumptions about the headers were made to keep from calculating the header length in detail every time we send a TCP segment.

The *tcpdatalen_out* value is equal to the length of our Telnet banner message. A look at the bottom of Sniffer Screen Capture 11.7 tells us that our banner message is 22 bytes long. Let's just count the bytes in Code Snippet 11.23.

```
//******************************************************************
//*   TELNET SERVER BANNER STATEMENT CONSTANT
//******************************************************************
int8 const telnet_banner[] = "\r\nEDTP Telnet Server>";
```

Code Snippet 11.23: Custom Computer Services C Compiler allows me to define an array without having to tell the compiler how long the array will be.

The carriage return and line feed characters (\r\n) count as 2 bytes. So, that's 2 (\r\n) + 4 (EDTP) + 7 (space Telnet) + 8 (space +Server>), which is equal to 21. What? A look back at the hex dump area of Sniffer Screen Capture 11.7 shows 0x00 at the end of our Telnet banner message. There are no padding messages in Sniffer Screen Capture 11.7, so what is that extra 0x00 doing in there?

```
Address                                              ASCII

0000   3000 008A 2C3A 0000 100A 108A 110A 0782 .0..:,.. ........
0008   340D 340A 3445 3444 3454 3450 3420 3454 .4.4E4D4 T4P4 4T4
0010   3465 346C 346E 3465 3474 3420 3453 3465 e4l4n4e4 t4 4S4e4
0018   3472 3476 3465 3472 343E 3400 100A 108A r4v4e4r4 >4.4....
```

Hex Dump 11.2: The "terminator"...

Let's see if we can find the 0x00. Since the Telnet banner is a constant string, that means it is somewhere in the PIC16F877's flash program memory. So, I loaded the MPLAB ICE 2000 with the Easy Ethernet CS8900A executable code and took a look at what was in the Easy Ethernet CS8900A's flash program memory. I found our 0x00 in Hex Dump 11.2.

Line	Address	Opcode	Disassembly	Fred's Translation
9	0008	340D	RETLW 0xd	carriage return
10	0009	340A	RETLW 0xa	line feed
11	000A	3445	RETLW 0x45	E
12	000B	3444	RETLW 0x44	D
13	000C	3454	RETLW 0x54	T
14	000D	3450	RETLW 0x50	P
15	000E	3420	RETLW 0x20	space
16	000F	3454	RETLW 0x54	T
17	0010	3465	RETLW 0x65	e
18	0011	346C	RETLW 0x6c	l
19	0012	346E	RETLW 0x6e	n
20	0013	3465	RETLW 0x65	e
21	0014	3474	RETLW 0x74	t
22	0015	3420	RETLW 0x20	space
23	0016	3453	RETLW 0x53	S
24	0017	3465	RETLW 0x65	e
25	0018	3472	RETLW 0x72	r
26	0019	3476	RETLW 0x76	v
27	001A	3465	RETLW 0x65	e
28	001B	3472	RETLW 0x72	r
29	001C	343E	RETLW 0x3e	>
30	001D	3400	RETLW 0	string terminator

Memory Dump 11.1: I speak a number of bit-oriented languages.

However, it's a bit easier to decipher in Memory Dump 11.1. The 0x00 is a null character and is used to indicate the end of a string. Thus, the string terminator (0x00) is our 22nd byte. So far, our ip_packet_len is 62, which includes the TCP header length, the IP header length and our Telnet banner message. Do you know where the other 14 bytes are? They're in the Ethernet DLC header. How about 6 bytes for the hardware destination address. Add 6 more bytes for the hardware source address. That's 12 bytes and that puts us at 74 bytes. The last 2 bytes are the Type bytes. That rounds out the frame and give us our total of 76 bytes just like Sniffer Screen Capture 11.7.

Again, just looking at Sniffer Screen Capture 11.7's event number 6 in the Sniffer Summary window, we can see that our *setipaddrs* and associated address swapping code in Code Snippet 11.24 worked to perfection.

```
//***************************************************************
//*    Send TCP Packet
//*    This routine assembles and sends a complete TCP/IP packet.
//*    40 bytes of IP and TCP header data is assumed.
//***************************************************************
   setipaddrs();

   data_L = packet[TCP_srcport];
   packet[TCP_srcport] = packet[TCP_destport];
   packet[TCP_destport] = data_L;
   data_L = packet[TCP_srcport+1];
   packet[TCP_srcport+1] = packet[TCP_destport+1];
   packet[TCP_destport+1] = data_L;
```

Code Snippet 11.24: Here's that swapperoonie code again.

There was no data from the client for the server to acknowledge, and the acknowledgement number in Sniffer Screen Capture 11.7 hasn't changed to reflect that.

```
//***************************************************************
//*    Send TCP Packet
//*    This routine assembles and sends a complete TCP/IP packet.
//*    40 bytes of IP and TCP header data is assumed.
//***************************************************************
   assemble_ack();
   set_packet32(TCP_seqnum,my_seqnum);

   packet[TCP_hdrflags+1] = 0x00;
   ACK_OUT;
   if(finflag)
   {
      FIN_OUT;
      finflag = 0;
   }
```

Code Snippet 11.25: More math and more flag waving. That's how it goes in TCP/IP.

The server's *assemble_ack* function in Code Snippet 11.25 adds the total of incoming bytes to the client's sequence number it received (790137260). The resulting acknowledgement from the server should reflect the total number of bytes sent by the client added to the received client's sequence number. In this case, no data was received and zero was added to the received client's sequence number. Therefore, the client's sequence number was not changed. In a similar fashion, the server sends its sequence number (547028992) to the client expecting to get an acknowledgement of the 22 bytes in the Telnet banner message, which through the acknowledgement from the client will increase the server's sequence number by 22 bytes. The TCP header Acknowledge flag bit will always be set after the client/server connection is established. Being set, the Acknowledge flag bit vouches for the acknowledgement number's validity.

A sender always sends a sequence number indicating the number of the first byte of data being sent. Note that the *Next expected Seq number* field in the Sniffer screen captures is not a TCP header field. It's a convenience offered by the Sniffer software. After the connection was established, the server's (Easy Ethernet CS8900A's) sequence number was set at 547028992 and the client's (personal computer's) sequence number was initialized at 790137260. The server sent 22 bytes of Telnet banner message. The Sniffer calculates the sender's TCP data length and offers up the *Next expected Seq number* field to help you see what the acknowledgement from the receiver should be. According to the Sniffer, the sender should get an acknowledgement number of 547029014.

```
//*******************************************************************
//*   Send TCP Packet
//*   This routine assembles and sends a complete TCP/IP packet.
//*   40 bytes of IP and TCP header data is assumed.
//*******************************************************************
   packet[TCP_cksum] = 0x00;
   packet[TCP_cksum+1] = 0x00;

   hdr_chksum =0;
   hdrlen = 0x08;
   addr = &packet[ip_srcaddr];
   cksum();
   hdr_chksum = hdr_chksum + packet[ip_proto];
   tcplen = ip_packet_len - ((packet[ip_vers_len] & 0x0F) * 4);
   hdr_chksum = hdr_chksum + tcplen;
   hdrlen = tcplen;
   addr = &packet[TCP_srcport];
   cksum();
   chksum16= ~(hdr_chksum + ((hdr_chksum & 0xFFFF0000) >> 16));
   packet[TCP_cksum] = make8(chksum16,1);
   packet[TCP_cksum+1] = make8(chksum16,0);

   txlen = ip_packet_len + 14;
   if(txlen < 60)
      txlen = 60;
   data_L = make8(txlen,0);
   data_H = make8(txlen,1);
```

Code Snippet 11.26: This ensures that a runt packet will not be transmitted.

The client hasn't sent a FIN (Finish Flag Bit) to end the connection yet. So, the server won't set its FIN flag this time around. The TCP checksum is calculated and the frame length is checked and assured to be at least 60 bytes in length in Code Snippet 11.26. Since Sniffer Screen Capture 11.7 exists, all went well with the server's transmission in Code Snippet 11.27.

```
//*******************************************************************
//*   Send TCP Packet
//*   This routine assembles and sends a complete TCP/IP packet.
//*   40 bytes of IP and TCP header data is assumed.
//*******************************************************************
   dataport_out;
   WpppL(pageport_TxCmd,TXCMD_AFTER_ALL);
   WpppH(pageport_TxCmd,0);
   WpppL(pageport_TxLen,data_L);
   WpppH(pageport_TxLen,data_H);
   do{
         RPP(ppageBusStatus);
     }while(!(bit_test(data_H,BUSSTA_RDY4TXNOW_BIT)));
   dataport_out;
   for(i=0;i<txlen;i+=2)
   {
   WpppL(pageport_RxTxData0,packet[i]);
   WpppH(pageport_RxTxData0,packet[i+1]);
```

Code Snippet 11.27: Send it...

The client (personal computer) has yet to send a byte of data, but it did acknowledge the 22 bytes of Telnet banner in Sniffer Screen Capture 11.8. I stopped the Sniffer trace at this point with the client's Telnet session window showing the Telnet banner transmitted by the Easy Ethernet CS8900A Telnet server in Sniffer Screen Capture 11.9.

We've written just enough code to set up a TCP/IP session between the Easy Ethernet CS8900A and the personal computer using Telnet. It should be pretty obvious now as to why TCP is known as a connection-oriented protocol, and UDP and IP are called connectionless protocols. Let's crank up the Sniffer and look at what happens when the client (personal computer) finally sends some data to the Easy Ethernet CS8900A Telnet server. Let's study the Easy Ethernet CS8900A's code first and then see if what we write as Easy Ethernet CS8900A source code makes sense as a Sniffer screen capture.

```
Sniffer - Local, Ethernet (Line speed at 10 Mbps) - [tcp_2.cap: Decode, 7/7 Ethernet Frames]
File  Monitor  Capture  Display  Tools  Database  Window  Help
Default

No. Status Source Address   Dest Address      Summary                                                                 Len (B) Rel. Time
4          [192.168.0.150] [192.168.0.1]     TCP: D=1038 S=8088 SYN ACK=790137260 SEQ=547028991 LEN=0 WIN=16384  62  0:00:00
5          [192.168.0.1]   [192.168.0.150]   TCP: D=8088 S=1038     ACK=547028992 WIN=17520                      60  0:00:00
6          [192.168.0.150] [192.168.0.1]     TCP: D=1038 S=8088     ACK=790137260 SEQ=547028992 LEN=22 WIN=1752  76  0:00:00
7          [192.168.0.1]   [192.168.0.150]   TCP: D=8088 S=1038     ACK=547029014 WIN=17498                      60  0:00:00

IP:        ..0. .... = last fragment
IP: Fragment offset = 0 bytes
IP: Time to live    = 128 seconds/hops
IP: Protocol        = 6 (TCP)
IP: Header checksum = 75FD (correct)
IP: Source address      = [192.168.0.1]
IP: Destination address = [192.168.0.150]
IP: No options
IP:
TCP: ----- TCP header -----
TCP:
TCP: Source port                = 1038
TCP: Destination port           = 8088
TCP: Sequence number            = 790137260
TCP: Next expected Seq number= 790137260
TCP: Acknowledgment number      = 547029014
TCP: Data offset                = 20 bytes
TCP: Flags                      = 10
TCP:              ..0. .... = (No urgent pointer)
TCP:              ...1 .... = Acknowledgment
TCP:              .... 0... = (No push)
TCP:              .... .0.. = (No reset)
TCP:              .... ..0. = (No SYN)
TCP:              .... ...0 = (No FIN)
TCP: Window                     = 17498
TCP: Checksum                   = EC76 (correct)
TCP: No TCP options
TCP:

00000000: 00 00 45 44 54 50 00 e0 29 87 f5 5b 08 00 45 00  ..EDTP.à)▮õ[..E.
00000010: 00 28 02 eb 40 00 80 06 75 fd c0 a8 00 01 c0 a8  .(.ë@.▮.uýÀ¨..À¨
00000020: 00 96 04 0e 1f 98 2f 18 89 ac 20 9b 00 16 50 10  .▮...▮/.▮¬ ▮..P.
00000030: 44 5a ec 76 00 00 00 00 00 00 00 00              DZìv........

Decode  Matrix  Host Table  Protocol Dist.  Statistics
For Help, press F1                                                   236
```

Sniffer Screen Capture 11.8: Notice the checksums we calculated are right on the money.

Sniffer Screen Capture 11.9: We've written a bunch of TCP/IP code just to get those 19 characters to appear.

```
//*********************************************************************
//*   TCP Function
//*   This function uses TCP protocol to act as a Telnet server on
//*   port 8088 decimal. The application function is called with
//*   every incoming character.
//*********************************************************************
   //if an ack is received and the port address is valid and there is
data in the incoming packet
   if(ACK_IN && portaddr == MY_PORT_ADDRESS && tcpdatalen_in)
      {
         for(i=0;i<tcpdatalen_in;++i)
      {
            //receive the data and put it into the incoming data buffer
            aux_data[i] = packet[TCP_data+i];

            //run the TCP application
            application_code();
         }
```

Code Snippet 11.28: The application code is run on a character-by-character basis.

We're still inside the *TCP* function, and Code Snippet 11.28 is an area within the code that responds when data arrives that is addressed to the Easy Ethernet CS8900A's Telnet server application. The Easy Ethernet CS8900A's Telnet application is character-based. So, every character that is received is acted upon. I've allocated a small 16-byte buffer (aux_data[]) just in case I ever wanted to receive and act on a block of characters. Code Snippet 11.28 is pretty easy to follow. So, let's explore the *application_code* function in Code Snippet 11.29.

```
//*********************************************************************
//*   TELNET SERVER BANNER STATEMENT CONSTANT
//*********************************************************************
int8 const telnet_banner[] = "\r\nEDTP Telnet Server>";

#define  latchdata   output_high(LE);  \
                     delay_us(1);      \
                     output_low(LE);

#define  set_hex  hexflag=1
#define  clr_hex  hexflag=0
int8 aux_data[16];   //tcp application received data area
int16 tcpdatalen_out,tcpdatalen_in;
int8  cntr,byteout;
```

```
//*****************************************************************
//*   Application Code
//*   Your application code goes here.
//*   Following a * this module writes the hex value that follows
//*   the * to the 74HCT573 latch..
//*   Use Telnet to connect.
//*   Example:  *55 writes 01010101 to the 74HCT573 latch
//*****************************************************************
void application_code()
{
   int8 i,j;

   ++cntr;

   if(aux_data[0] != 0x0A)
      tcpdatalen_out = tcpdatalen_in;
   if(aux_data[0] == 0x0A)
   {
      tcpdatalen_out = 0x00;
      clr_hex;
   }
      if(hexflag)
   {
      if(aux_data[0] >= '0' && aux_data[0] <= '9')
         aux_data[0] -= 0x30;
      else if(aux_data[0] >= 'A' && aux_data[0] <= 'F')
         aux_data[0] -= 0x37;
      else if(aux_data[0] >= 'a' && aux_data[0] <= 'f')
         aux_data[0] -= 0x67;
      else
      {
         cntr = 0x00;
         clr_hex;
      }

   if(cntr == 1)
      byteout = aux_data[0] << 4;
         if(cntr == 2)
   {
      byteout |= aux_data[0] & 0x0F;
      dataport_out;
      writedataport(byteout);
      latchdata;
      clr_hex;
      printf("Byte Latched = %x\r\n",byteout);
   }
   }
      if(aux_data[0] == '*')
```

```
{
   set_hex;
cntr=0;
}

if (aux_data[0] == 0x0D)
   {
      j = sizeof(telnet_banner);
            for(i=0;i<j;++i)
         packet[TCP_data+i] = telnet_banner[i];
      tcpdatalen_out = j;
   }
```

Code Snippet 11.29: Note that the hexflag in this application is actually a Boolean variable instead of a Boolean bit within a byte.

Believe it or not, we've already done the hard and dirty work. The TCP application is quite simple. Instead of popping your eyeballs out going back and forth, let's look at the code in Code Snippet 11.29 module by module, beginning with Code Snippet 11.30.

```
//********************************************************************
//*   Application Code
//*   Your application code goes here.
//*   Following a * this module writes the hex value that follows
//*   the * to the 74HCT573 latch..
//*   Use Telnet to connect.
//*   Example:  *55 writes 01010101 to the 74HCT573 latch
//********************************************************************
   ++cntr;

   if(aux_data[0] != 0x0A)
      tcpdatalen_out = tcpdatalen_in;
   if(aux_data[0] == 0x0A)
   {
      tcpdatalen_out = 0x00;
      clr_hex;
   }

   if (aux_data[0] == 0x0D)
      {
         j = sizeof(telnet_banner);
               for(i=0;i<j;++i)
            packet[TCP_data+i] = telnet_banner[i];
         tcpdatalen_out = j;
      }
```

Code Snippet 11.30: The cntr variable allows us to capture two consecutive ASCII bytes.

The variable *cntr* is a counter that keeps up with the number of bytes that have been processed. The *cntr* variable is only allowed to increment up to 2 before being reset.

As long as the incoming data is not a line feed character (0x0A), the incoming data is echoed. In other words, if you don't hit the ENTER key on the personal computer, the data you enter in the Telnet window will be echoed to your Telnet session. I've skipped to the bottom of the application in Code Snippet 11.30 to show you the code that responds to the carriage return character (0x0D) by throwing up the Telnet banner message.

The setting and clearing of the *hexflag* determines if the TCP application tickles the Easy Ethernet CS8900A's hardware or not. Notice that when a line feed character is sensed, the TCP data length is cleared to zero and the *hexflag* is cleared. When the *hexflag* is clear, only the data echo function will be active. Setting the *hexflag* activates the latched port output function of the TCP application. The *hexflag* gets set with a "*" (0x2A) character and is received from the Telnet client. As you can see in Code Snippet 11.31, the *cntr* variable is cleared right after the *hexflag* is set.

```
//******************************************************************
//*   Application Code
//*   Your application code goes here.
//*   Following a * this module writes the hex value that follows
//*   the * to the 74HCT573 latch..
//*   Use Telnet to connect.
//*   Example:  *55 writes 01010101 to the 74HCT573 latch
//******************************************************************
   if(aux_data[0] == '*')
   {
      set_hex;
      cntr=0;
   }
```

Code Snippet 11.31: Two ASCII bytes are collected following the ''.*

If you have read the application code banner, you already know that after the "*" is entered, the following two hexadecimal numbers will be combined into a byte and fed to the Easy Ethernet CS8900A's onboard 74HCT573 octal transparent latch, which presents the byte on a set of the Easy Ethernet CS8900A's I/O header pins.

OK, the "*" has been received. The *hexflag* gets set and the *cntr* variable is cleared. The TCP application only processes a character at a time and each character invokes the application code. The next character received from the client will flow through the TCP application code and increment the *cntr* variable to 1. The *hexflag* is set at this point and that kicks off the code in Code Snippet 11.32.

```
//*********************************************************************
//*    Application Code
//*    Your application code goes here.
//*    Following a * this module writes the hex value that follows
//*    the * to the 74HCT573 latch..
//*    Use Telnet to connect.
//*    Example:  *55 writes 01010101 to the 74HCT573 latch
//*********************************************************************
   if(hexflag)
      {
         if(aux_data[0] >= '0' && aux_data[0] <= '9')
            aux_data[0] -= 0x30;
         else if(aux_data[0] >= 'A' && aux_data[0] <= 'F')
            aux_data[0] -= 0x37;
         else if(aux_data[0] >= 'a' && aux_data[0] <= 'f')
            aux_data[0] -= 0x67;
         else
         {
            cntr = 0x00;
            clr_hex;
      }

   if(cntr == 1)
      byteout = aux_data[0] << 4;

   if(cntr == 2)
      {
         byteout |= aux_data[0] & 0x0F;
         dataport_out;
         writedataport(byteout);
         latchdata;
         clr_hex;
         printf("Byte Latched = %x\r\n",byteout);
      }
   }
```

Code Snippet 11.32: This is a simple ASCII-to-hex conversion routine. There are more elegant ways to do this, but we're not here for a programming glamour contest.

The incoming data from the client's Telnet session is ASCII data. Therefore, we have to convert the ASCII numbers to their numeric equivalents. ASCII character values 0 thru 9 run as shown in Table 11.1:

ASCII CHARACTER	ASCII HEX	NUMERIC HEX (DECIMAL)
0	0x30	0x00 (0)
1	0x31	0x01 (1)
2	0x32	0x02 (2)
3	0x33	0x03 (3)
4	0x34	0x04 (4)
5	0x35	0x05 (5)
6	0x36	0x06 (6)
7	0x37	0x07 (7)
8	0x38	0x08 (8)
9	0x39	0x09 (9)

Table 11.1: This is all very logical until you get to the number 10.

The code in Code Snippet 11.32 handles any ASCII character between *0* and *9* by simply subtracting 0x30 from the ASCII value. The tables turn in Table 11.2 when you have to figure out how to convert the rest of the hexadecimal numbers (A thru F) from ASCII to numeric.

ASCII CHARACTER	ASCII HEX	NUMERIC HEX (DECIMAL)
A	0x41	0x0A (10)
B	0x42	0x0B (11)
C	0x43	0x0C (12)
D	0x44	0x0D (13)
E	0x45	0x0E (14)
F	0x46	0x0F (15)
a	0x61	0x0A (10)
b	0x62	0x0B (11)
c	0x63	0x0C (12)
d	0x64	0x0D (13)
e	0x65	0x0E (14)
f	0x66	0x0F (15)

Table 11.2: Lower case, upper case...no difference when it comes to the numeric value.

The ASCII character representations of the hexadecimal characters are sequential. That means we can convert the ASCII alpha characters to numbers using a single conversion value. Subtracting 0x37 (decimal 55) from ASCII characters *A* thru *F* will give us the desired numeric value represented by the ASCII character. The same holds true for ASCII characters *a* thru *f* except the subtraction conversion value is 0x67 (103 decimal). If a "*" is followed by any ASCII character that cannot be converted directly to an equivalent hexadecimal value, the *hexflag* and the *cntr* variable are cleared.

After the ASCII is converted to its equivalent numeric value, the *cntr* variable is checked. If this is the high-order nibble of the byte to be latched out, the *cntr* variable will be equal to 0x01. If the value of *cntr* is 0x01, the high-order nibble of *aux_data[0]* is stored in the *byteout* variable.

The next character from the client should be the low-order nibble of the byte we want to latch out to the Easy Ethernet CS8900A's I/O pins. The *byteout* variable, which already contains the upper nibble of the byte to latch out, is combined via a bit-wise OR with the lower nibble garnered from the buffer array *aux_data[0]*. The completed byte within the *byteout* variable is then written to the 74HCT573 octal transparent latch. A serial port message is then fired off to confirm the value of the byte that was just latched out.

When the client wants to end the Telnet session, it sends a TCP segment with the FIN flag bit set in the TCP header. The Easy Ethernet CS8900A responds with a FIN-laden acknowledgement, and everything we worked so hard to build collapses into the bit bucket.

We've been studying TCP behavior frame-by-frame in relative slow motion, as I've been talking while the movie's on. So, let's look at a complete set of Sniffer screen captures from start to finish with minimal interruption by yours truly.

The following screen captures are shots of a Telnet session between the personal computer client and the Easy Ethernet CS8900A Telnet server implementation. The message is *ABC123*. Following the message the client ends the session by issuing a Telnet close, which throws a FIN TCP segment across to the Easy Ethernet CS8900A TCP application.

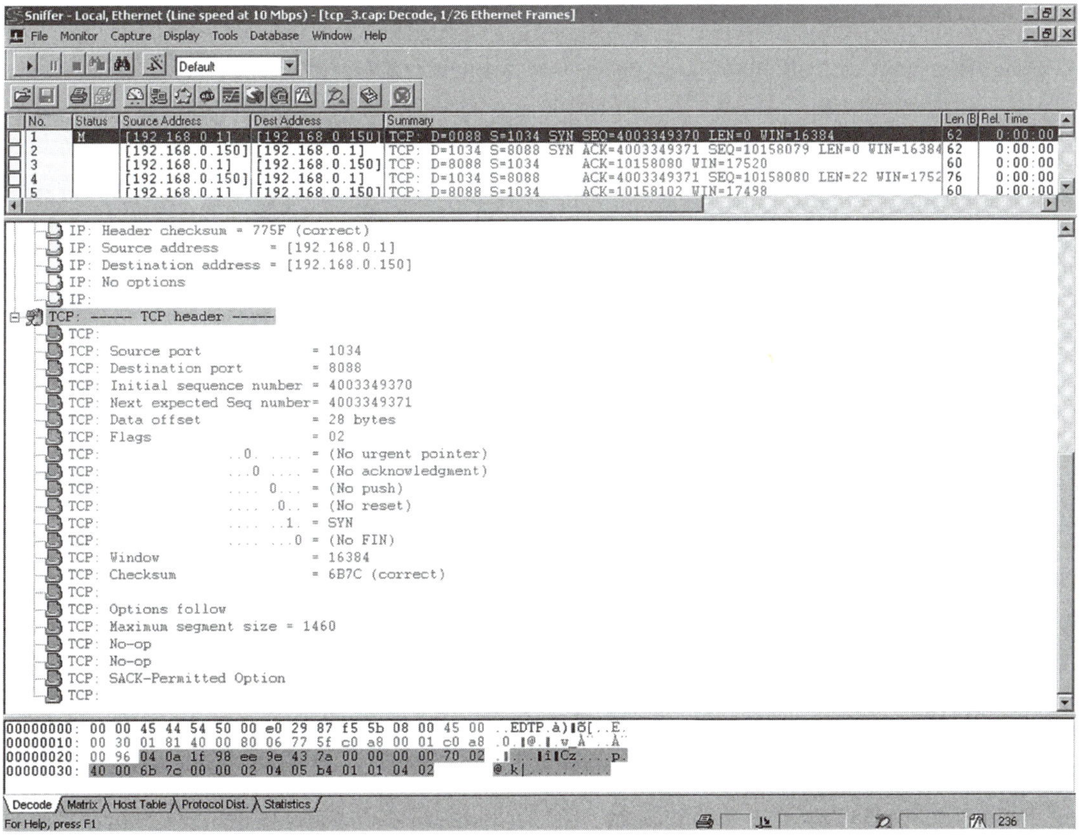

```
Sniffer - Local, Ethernet (Line speed at 10 Mbps) - [tcp_3.cap: Decode, 1/26 Ethernet Frames]
File  Monitor  Capture  Display  Tools  Database  Window  Help
Default

No. Status Source Address   Dest Address      Summary                                                                   Len (B) Rel. Time
1   M      [192.168.0.1]    [192.168.0.150]   TCP: D=0088 S=1034 SYN SEQ=4003349370 LEN=0 WIN=16384                     62  0:00:00
2          [192.168.0.150]  [192.168.0.1]     TCP: D=1034 S=8088 SYN ACK=4003349371 SEQ=10158079 LEN=0 WIN=16384        62  0:00:00
3          [192.168.0.1]    [192.168.0.150]   TCP: D=8088 S=1034     ACK=10158080 WIN=17520                             60  0:00:00
4          [192.168.0.150]  [192.168.0.1]     TCP: D=1034 S=8088     ACK=4003349371 SEQ=10158080 LEN=22 WIN=1752        76  0:00:00
5          [192.168.0.1]    [192.168.0.150]   TCP: D=8088 S=1034     ACK=10158102 WIN=17498                             60  0:00:00

IP: Header checksum = 775F (correct)
IP: Source address       = [192.168.0.1]
IP: Destination address = [192.168.0.150]
IP: No options
IP:
TCP: ----- TCP header -----
TCP:
TCP: Source port              = 1034
TCP: Destination port         = 8088
TCP: Initial sequence number = 4003349370
TCP: Next expected Seq number= 4003349371
TCP: Data offset              = 28 bytes
TCP: Flags                    = 02
TCP:                ..0. .... = (No urgent pointer)
TCP:                ...0 .... = (No acknowledgment)
TCP:                .... 0... = (No push)
TCP:                .... .0.. = (No reset)
TCP:                .... ..1. = SYN
TCP:                .... ...0 = (No FIN)
TCP: Window                   = 16384
TCP: Checksum                 = 6B7C (correct)
TCP:
TCP: Options follow
TCP: Maximum segment size = 1460
TCP: No-op
TCP: No-op
TCP: SACK-Permitted Option
TCP:

00000000: 00 00 45 44 54 50 00 e0 29 87 f5 5b 08 00 45 00  ..EDTP.à)▮õ[..E.
00000010: 00 30 01 81 40 00 80 06 77 5f c0 a8 00 01 c0 a8  .0.▮@.▮.w_À¨..À¨
00000020: 00 96 04 0a 1f 98 ee 9e 43 7a 00 00 00 00 70 02  .▮...▮î▮Cz....p.
00000030: 40 00 6b 7c 00 00 02 04 05 b4 01 01 04 02        @.k|......´....

Decode  Matrix  Host Table  Protocol Dist.  Statistics
For Help, press F1                                                        236
```

Sniffer Screen Capture 11.10: The client's human operator issues open 192.168.0.150 8088 *to kick off a Telnet session between the client (personal computer) and the server (Easy Ethernet CS8900A). The client sends this TCP segment with a client ISN of 4003349370. This is part 1 of the 3-way handshake.*

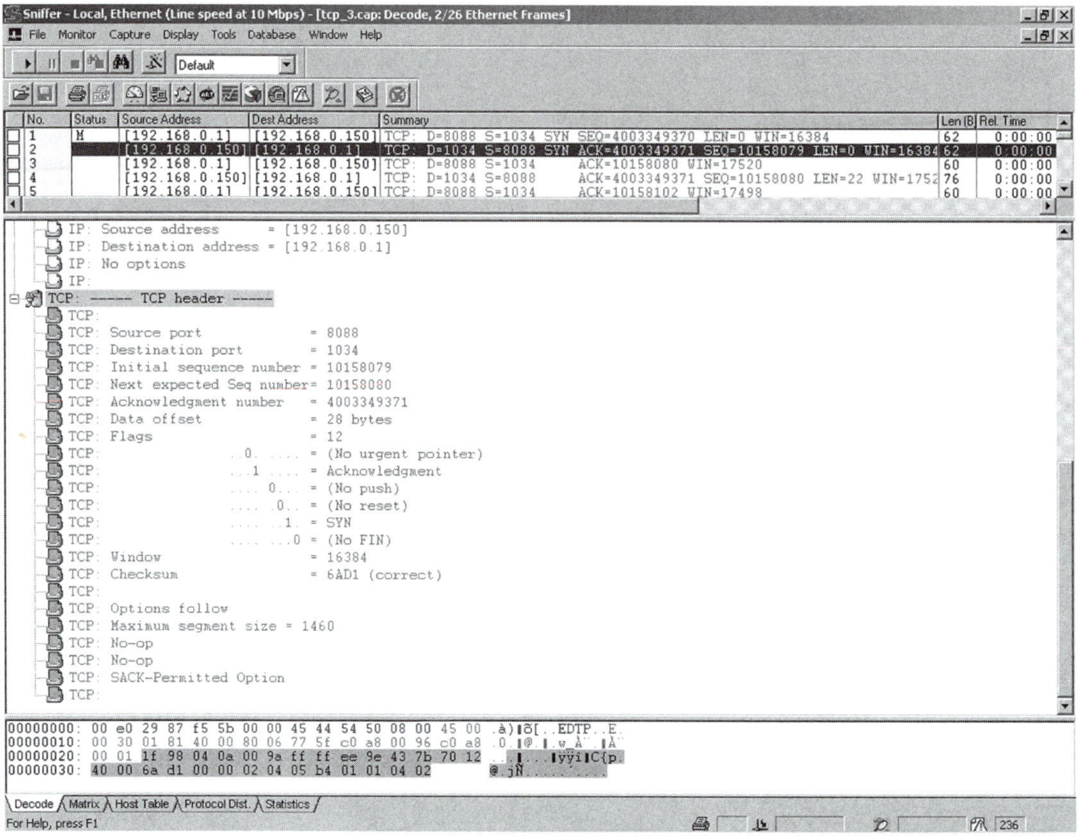

```
Sniffer - Local, Ethernet (Line speed at 10 Mbps) - [tcp_3.cap: Decode, 2/26 Ethernet Frames]
File  Monitor  Capture  Display  Tools  Database  Window  Help
Default

No. Status Source Address   Dest Address      Summary                                                         Len (B) Rel. Time
1   M      [192.168.0.1]    [192.168.0.150]   TCP: D=8088 S=1034 SYN SEQ=4003349370 LEN=0 WIN=16384                  62  0:00:00
2          [192.168.0.150]  [192.168.0.1]     TCP: D=1034 S=8088 SYN ACK=4003349371 SEQ=10158079 LEN=0 WIN=16384     62  0:00:00
3          [192.168.0.1]    [192.168.0.150]   TCP: D=8088 S=1034     ACK=10158080 WIN=17520                          60  0:00:00
4          [192.168.0.150]  [192.168.0.1]     TCP: D=1034 S=8088     ACK=4003349371 SEQ=10158080 LEN=22 WIN=1752     76  0:00:00
5          [192.168.0.1]    [192.168.0.150]   TCP: D=8088 S=1034     ACK=10158102 WIN=17498                          60  0:00:00

IP: Source address      = [192.168.0.150]
IP: Destination address = [192.168.0.1]
IP: No options
IP:
TCP: ----- TCP header -----
TCP:
TCP: Source port              = 8088
TCP: Destination port         = 1034
TCP: Initial sequence number  = 10158079
TCP: Next expected Seq number= 10158080
TCP: Acknowledgment number    = 4003349371
TCP: Data offset              = 28 bytes
TCP: Flags                    = 12
TCP:             ..0. .... = (No urgent pointer)
TCP:             ...1 .... = Acknowledgment
TCP:             .... 0... = (No push)
TCP:             .... .0.. = (No reset)
TCP:             .... ..1. = SYN
TCP:             .... ...0 = (No FIN)
TCP: Window                   = 16384
TCP: Checksum                 = 6AD1 (correct)
TCP:
TCP: Options follow
TCP: Maximum segment size = 1460
TCP: No-op
TCP: No-op
TCP: SACK-Permitted Option
TCP:

00000000: 00 e0 29 87 f5 5b 00 00 45 44 54 50 08 00 45 00  .à)▮õ[..EDTP..E.
00000010: 00 30 01 81 40 00 80 06 77 5f c0 a8 00 96 c0 a8  .0.▮@.▮.w_À¨.▮À¨
00000020: 00 01 1f 98 04 0a 00 9a ff ff ee 9e 43 7b 70 12  ...▮...▮ÿÿî▮C{p.
00000030: 40 00 6a d1 00 00 02 04 05 b4 01 01 04 02        @.jÑ..........

Decode  Matrix  Host Table  Protocol Dist.  Statistics
For Help, press F1                                                236
```

Sniffer Screen Capture 11.11: The Easy Ethernet CS8900A Telnet server acknowledges the client's SYN and issues a TCP segment with its ISN (10158079). Notice the ACK flag bit has been set indicating that the acknowledgement number field is now valid. This is part 2 of the 3-way handshake.

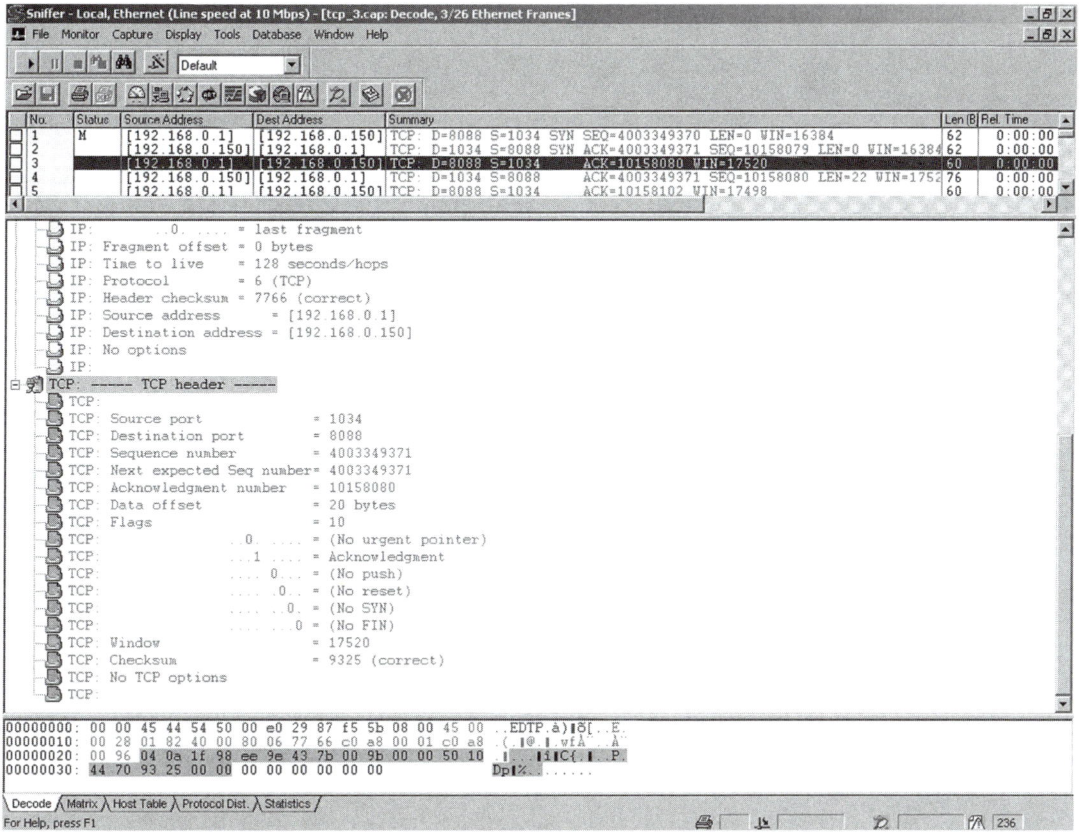

```
Sniffer - Local, Ethernet (Line speed at 10 Mbps) - [tcp_3.cap: Decode, 3/26 Ethernet Frames]
File  Monitor  Capture  Display  Tools  Database  Window  Help
Default

No. Status Source Address   Dest Address      Summary                                                              Len (B) Rel. Time
1   M      [192.168.0.1]    [192.168.0.150]   TCP: D=8088 S=1034 SYN SEQ=4003349370 LEN=0 WIN=16384                  62   0:00:00
2          [192.168.0.150]  [192.168.0.1]     TCP: D=1034 S=8088 SYN ACK=4003349371 SEQ=10158079 LEN=0 WIN=16384      62   0:00:00
3          [192.168.0.1]    [192.168.0.150]   TCP: D=8088 S=1034     ACK=10158080 WIN=17520                          60   0:00:00
4          [192.168.0.150]  [192.168.0.1]     TCP: D=1034 S=8088     ACK=4003349371 SEQ=10158080 LEN=22 WIN=1752     76   0:00:00
5          [192.168.0.1]    [192.168.0.150]   TCP: D=8088 S=1034     ACK=10158102 WIN=17498                          60   0:00:00

IP:         ..0. .... = last fragment
IP: Fragment offset = 0 bytes
IP: Time to live    = 128 seconds/hops
IP: Protocol        = 6 (TCP)
IP: Header checksum = 7766 (correct)
IP: Source address      = [192.168.0.1]
IP: Destination address = [192.168.0.150]
IP: No options
IP:
TCP: ----- TCP header -----
TCP:
TCP: Source port              = 1034
TCP: Destination port         = 8088
TCP: Sequence number          = 4003349371
TCP: Next expected Seq number= 4003349371
TCP: Acknowledgment number    = 10158080
TCP: Data offset              = 20 bytes
TCP: Flags                    = 10
TCP:              ..0. .... = (No urgent pointer)
TCP:              ...1 .... = Acknowledgment
TCP:              .... 0... = (No push)
TCP:              .... .0.. = (No reset)
TCP:              .... ..0. = (No SYN)
TCP:              .... ...0 = (No FIN)
TCP: Window                   = 17520
TCP: Checksum                 = 9325 (correct)
TCP: No TCP options
TCP:

00000000: 00 00 45 44 54 50 00 e0 29 87 f5 5b 08 00 45 00  ..EDTP.à)▌õ[..E.
00000010: 00 28 01 82 40 00 80 06 77 66 c0 a8 00 01 c0 a8  .(.▌@.▌.wfÀ¨..À¨
00000020: 00 96 04 0a 1f 98 ee 9e 43 7b 00 9b 00 00 50 10  .▌...▌î▌C{.▌..P.
00000030: 44 70 93 25 00 00 00 00 00 00 00 00              Dp▌%......

Decode / Matrix / Host Table / Protocol Dist. / Statistics
For Help, press F1                                                236
```

Sniffer Screen Capture 11.12: The client acknowledges the server's SYN segment, and both the client and server now have their beginning sequence numbers set. The ACK flag bit will remain set throughout the session indicating that the acknowledgement number fields are valid and should not be ignored. The 3-way handshake is complete and data can now be transferred.

```
Sniffer - Local, Ethernet (Line speed at 10 Mbps) - [tcp_3.cap: Decode, 4/26 Ethernet Frames]
File  Monitor  Capture  Display  Tools  Database  Window  Help
Default

No. Status Source Address   Dest Address      Summary                                                                  Len (B Rel. Time
1   M      [192.168.0.1]    [192.168.0.150]   TCP: D=8088 S=1034 SYN SEQ=4003349370 LEN=0 WIN=16384                    62     0:00:00
2          [192.168.0.150]  [192.168.0.1]     TCP: D=1034 S=8088 SYN ACK=4003349371 SEQ=10158079 LEN=0 WIN=16384       62     0:00:00
3          [192.168.0.1]    [192.168.0.150]   TCP: D=8088 S=1034     ACK=10158080 WIN=17520                            60     0:00:00
4          [192.168.0.150]  [192.168.0.1]     TCP: D=1034 S=8088     ACK=4003349371 SEQ=10158080 LEN=22 WIN=1752       76     0:00:00
5          [192.168.0.1]    [192.168.0.150]   TCP: D=8088 S=1034     ACK=10158102 WIN=17498                            60     0:00:00

IP: Fragment offset = 0 bytes
IP: Time to live    = 128 seconds/hops
IP: Protocol        = 6 (TCP)
IP: Header checksum = 7750 (correct)
IP: Source address      = [192.168.0.150]
IP: Destination address = [192.168.0.1]
IP: No options
IP:
TCP: ----- TCP header -----
TCP:
TCP: Source port                = 8088
TCP: Destination port           = 1034
TCP: Sequence number            = 10158080
TCP: Next expected Seq number= 10158102
TCP: Acknowledgment number      = 4003349371
TCP: Data offset                = 20 bytes
TCP: Flags                      = 10
TCP:               ..0. .... = (No urgent pointer)
TCP:               ...1 .... = Acknowledgment
TCP:               .... 0... = (No push)
TCP:               .... .0.. = (No reset)
TCP:               .... ..0. = (No SYN)
TCP:               .... ...0 = (No FIN)
TCP: Window                     = 17520
TCP: Checksum                   = 1ADC (correct)
TCP: No TCP options
TCP: [22 Bytes of data]
TCP:

00000000: 00 e0 29 87 f5 5b 00 00 45 44 54 50 08 00 45 00  .à)▮õ[..EDTP..E.
00000010: 00 3e 01 82 40 00 80 06 77 50 c0 a8 00 96 c0 a8  .>.▮@.▮.wPÀ¨.▮À¨
00000020: 00 01 1f 98 04 0a 00 9b 00 00 ee 9e 43 7b 50 10  ...▮...▮..î▮C{P.
00000030: 44 70 1a dc 00 00 0d 0a 45 44 54 50 20 54 65 6c  Dp.Ü....EDTP Tel
00000040: 6e 65 74 20 53 65 72 76 65 72 3e 00              net Server>.
Decode / Matrix / Host Table / Protocol Dist. / Statistics
For Help, press F1                                                  236
```

Sniffer Screen Capture 11.13: The Easy Ethernet CS8900A takes this opportunity to fire off the Telnet banner message, while the human operator is paused waiting for some sort of positive feedback. The Sniffer has predicted the client's acknowledgement in the Next expected Seq number field of the Sniffer screen capture.

```
Sniffer - Local, Ethernet (Line speed at 10 Mbps) - [tcp_3.cap: Decode, 5/26 Ethernet Frames]
File  Monitor  Capture  Display  Tools  Database  Window  Help
Default

No. Status Source Address   Dest Address      Summary                                            Len (B) Rel. Time     Delta Time  Abs. Time
3          [192.168.0.1]    [192.168.0.150]   TCP: D=8088 S=1034   ACK=10158080 WIN=17           60   0:00:00.001   0.000.104   08/
4          [192.168.0.150]  [192.168.0.1]     TCP: D=1034 S=8088   ACK=4003349371 SEQ=           76   0:00:00.003   0.001.812   08/
5          [192.168.0.1]    [192.168.0.150]   TCP: D=8088 S=1034   ACK=10158102 WIN=17           60   0:00:00.198   0.194.970   08/
6          [192.168.0.1]    [192.168.0.150]   TCP: D=8088 S=1034   ACK=10158102 SEQ=40           60   0:00:04.459   4.261.531   08/
7          [192.168.0.150]  [192.168.0.1]     TCP: D=1034 S=8088   ACK=4003349372 SEQ=           60   0:00:04.461   0.001.484   08/

IP:        ..0. .... = last fragment
IP: Fragment offset = 0 bytes
IP: Time to live    = 128 seconds/hops
IP: Protocol        = 6 (TCP)
IP: Header checksum = 7765 (correct)
IP: Source address      = [192.168.0.1]
IP: Destination address = [192.168.0.150]
IP: No options
IP:
TCP: ----- TCP header -----
TCP:
TCP: Source port              = 1034
TCP: Destination port         = 8088
TCP: Sequence number          = 4003349371
TCP: Next expected Seq number= 4003349371
TCP: Acknowledgment number    = 10158102
TCP: Data offset              = 20 bytes
TCP: Flags                    = 10
TCP:              ..0. .... = (No urgent pointer)
TCP:              ...1 .... = Acknowledgment
TCP:              .... 0... = (No push)
TCP:              .... .0.. = (No reset)
TCP:              .... ..0. = (No SYN)
TCP:              .... ...0 = (No FIN)
TCP: Window                   = 17498
TCP: Checksum                 = 9325 (correct)
TCP: No TCP options
TCP:

00000000: 00 00 45 44 54 50 00 e0 29 87 f5 5b 08 00 45 00  ..EDTP.à)‡õ[..E.
00000010: 00 28 01 83 40 00 80 06 77 65 c0 a8 00 01 c0 a8  .(.ƒ@.€.weÀ¨..À¨
00000020: 00 96 04 0a 1f 98 ee 9e 43 7b 00 9b 00 16 50 10  .–....îžC{.›..P.
00000030: 44 5a 93 25 00 00 00 00 00 00 00 00              DZ“%........

Decode / Matrix / Host Table / Protocol Dist. / Statistics
For Help, press F1                                                  236
```

Sniffer Screen Capture 11.14: The client has issued an acknowledgement number just as Sniffer Screen Capture 11.13 said it would. I've moved the Rel Time window into view in the Summary window area to show that around 4 seconds after this acknowledgement, the human operator at the client host wakes up and sends the first byte of data you see in Sniffer Screen Capture 11.15.

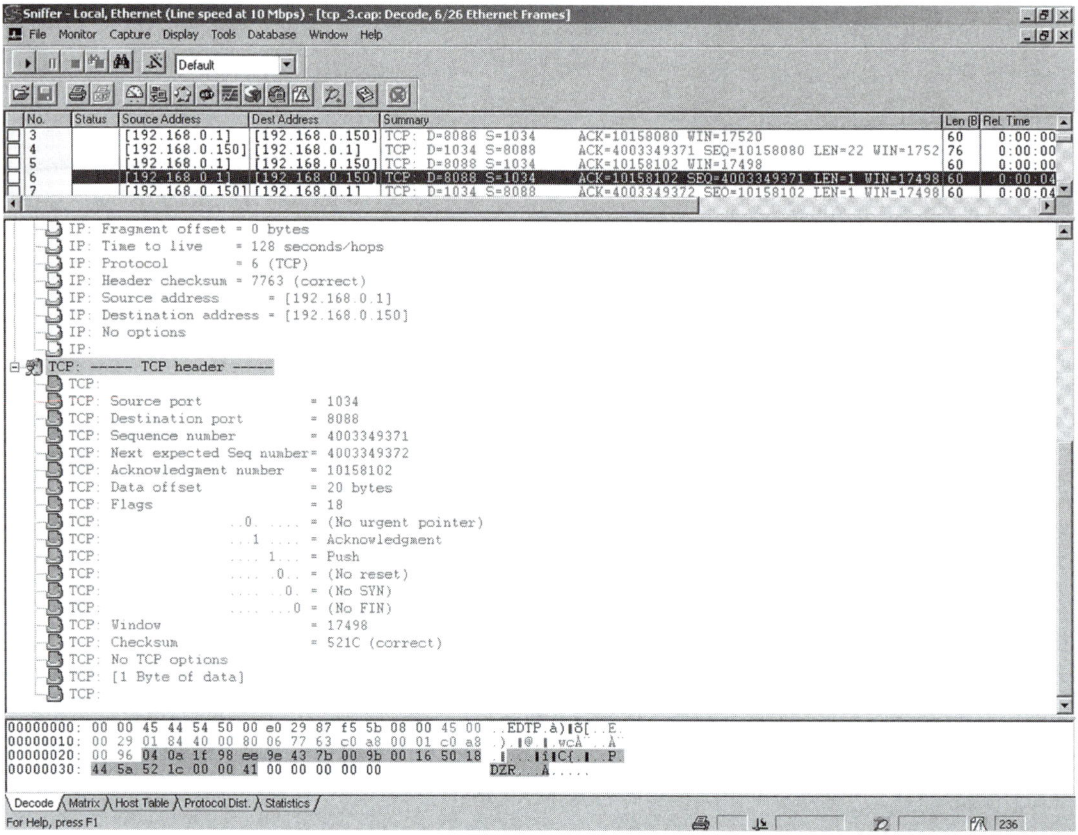

```
Sniffer - Local, Ethernet (Line speed at 10 Mbps) - [tcp_3.cap: Decode, 6/26 Ethernet Frames]
File  Monitor  Capture  Display  Tools  Database  Window  Help
Default

No. Status Source Address   Dest Address      Summary                                                          Len (B) Rel. Time
3          [192.168.0.1]    [192.168.0.150]   TCP: D=8088 S=1034   ACK=10158080 WIN=17520                       60   0:00:00
4          [192.168.0.150]  [192.168.0.1]     TCP: D=1034 S=8088   ACK=4003349371 SEQ=10158080 LEN=22 WIN=1752  76   0:00:00
5          [192.168.0.1]    [192.168.0.150]   TCP: D=8088 S=1034   ACK=10158102 WIN=17498                       60   0:00:00
6          [192.168.0.1]    [192.168.0.150]   TCP: D=8088 S=1034   ACK=10158102 SEQ=4003349371 LEN=1 WIN=17498  60   0:00:04
7          [192.168.0.150]  [192.168.0.1]     TCP: D=1034 S=8088   ACK=4003349372 SEQ=10158102 LEN=1 WIN=17498  60   0:00:04

IP: Fragment offset = 0 bytes
IP: Time to live    = 128 seconds/hops
IP: Protocol        = 6 (TCP)
IP: Header checksum = 7763 (correct)
IP: Source address      = [192.168.0.1]
IP: Destination address = [192.168.0.150]
IP: No options
IP:
TCP: ----- TCP header -----
TCP:
TCP: Source port               = 1034
TCP: Destination port          = 8088
TCP: Sequence number           = 4003349371
TCP: Next expected Seq number= 4003349372
TCP: Acknowledgment number     = 10158102
TCP: Data offset               = 20 bytes
TCP: Flags                     = 18
TCP:             ..0. .... = (No urgent pointer)
TCP:             ...1 .... = Acknowledgment
TCP:             .... 1... = Push
TCP:             .... .0.. = (No reset)
TCP:             .... ..0. = (No SYN)
TCP:             .... ...0 = (No FIN)
TCP: Window                    = 17498
TCP: Checksum                  = 521C (correct)
TCP: No TCP options
TCP: [1 Byte of data]
TCP:

00000000: 00 00 45 44 54 50 00 e0 29 87 f5 5b 08 00 45 00  ..EDTP.à)‡õ[..E.
00000010: 00 29 01 84 40 00 80 06 77 63 c0 a8 00 01 c0 a8  .)..@...wcÀ¨..À¨
00000020: 00 96 04 0a 1f 98 ee 9e 43 7b 00 9b 00 16 50 18  ........C{....P.
00000030: 44 5a 52 1c 00 00 41 00 00 00 00 00             DZR...A.....

Decode  Matrix  Host Table  Protocol Dist.  Statistics
For Help, press F1                                                   236
```

Sniffer Screen Capture 11.15: This TCP segment contains A. Also note that the Push flag bit is set. The Push flag bit tells the receiving host to "push" all of the buffered data to the application. Since we're running a minimal character-based TCP implementation on the Easy Ethernet CS8900A, the Push flag bit is ignored.

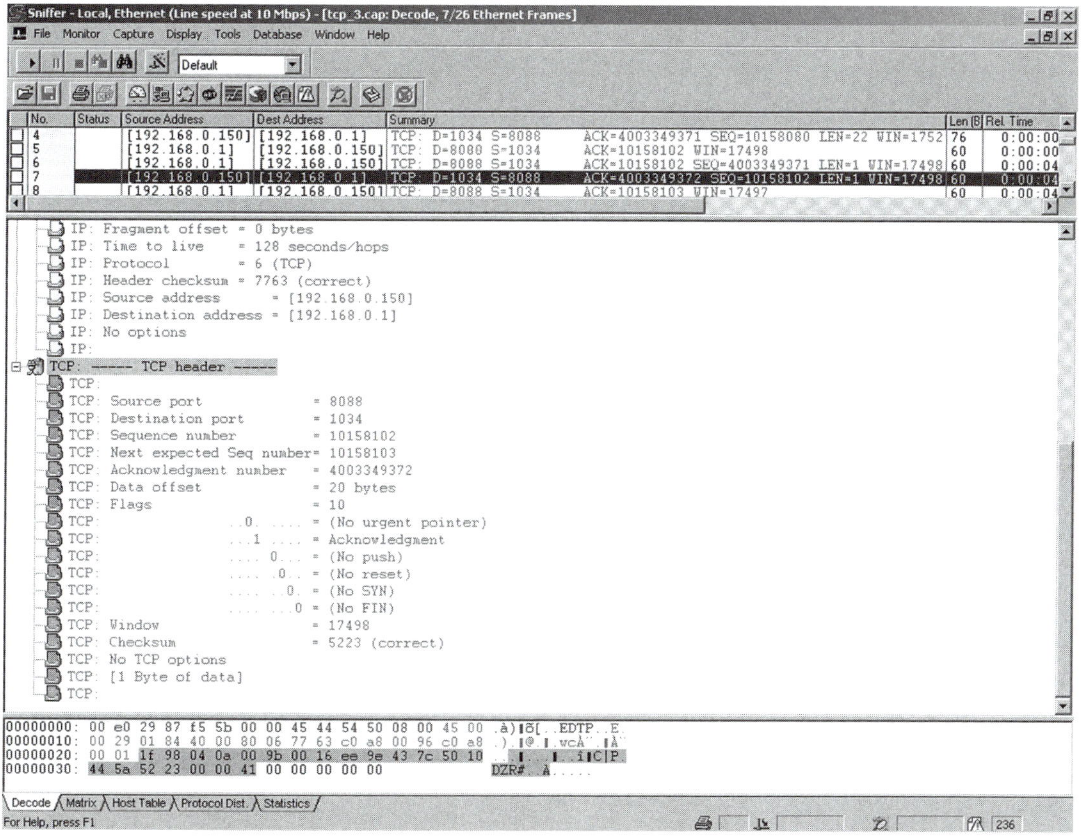

```
Sniffer - Local, Ethernet (Line speed at 10 Mbps) - [tcp_3.cap: Decode, 7/26 Ethernet Frames]
File  Monitor  Capture  Display  Tools  Database  Window  Help
Default

No. Status Source Address   Dest Address      Summary                                                           Len (B) Rel. Time
4          [192.168.0.150] [192.168.0.1]     TCP: D=1034 S=8088  ACK=4003349371 SEQ=10158080 LEN=22 WIN=1752    76      0:00:00
5          [192.168.0.1]   [192.168.0.150]   TCP: D=8088 S=1034  ACK=10158102 WIN=17498                         60      0:00:00
6          [192.168.0.1]   [192.168.0.150]   TCP: D=8088 S=1034  ACK=10158102 SEQ=4003349371 LEN=1 WIN=17498    60      0:00:04
7          [192.168.0.150] [192.168.0.1]     TCP: D=1034 S=8088  ACK=4003349372 SEQ=10158102 LEN=1 WIN=17498    60      0:00:04
8          [192.168.0.1]   [192.168.0.150]   TCP: D=8088 S=1034  ACK=10158103 WIN=17497                         60      0:00:04

IP: Fragment offset = 0 bytes
IP: Time to live    = 128 seconds/hops
IP: Protocol        = 6 (TCP)
IP: Header checksum = 7763 (correct)
IP: Source address       = [192.168.0.150]
IP: Destination address = [192.168.0.1]
IP: No options
IP:
TCP: ----- TCP header -----
TCP:
TCP: Source port              = 8088
TCP: Destination port         = 1034
TCP: Sequence number          = 10158102
TCP: Next expected Seq number= 10158103
TCP: Acknowledgment number    = 4003349372
TCP: Data offset              = 20 bytes
TCP: Flags                    = 10
TCP:            ..0. .... = (No urgent pointer)
TCP:            ...1 .... = Acknowledgment
TCP:            .... 0... = (No push)
TCP:            .... .0.. = (No reset)
TCP:            .... ..0. = (No SYN)
TCP:            .... ...0 = (No FIN)
TCP: Window                   = 17498
TCP: Checksum                 = 5223 (correct)
TCP: No TCP options
TCP: [1 Byte of data]
TCP:

00000000: 00 e0 29 87 f5 5b 00 00 45 44 54 50 08 00 45 00  .à)‚õ[..EDTP..E.
00000010: 00 29 01 84 40 00 80 06 77 63 c0 a8 00 96 c0 a8  .).„@.€.wcÀ¨.–À¨
00000020: 00 01 1f 98 04 0a 00 9b 00 16 ee 9e 43 7c 50 10  ...˜...›..îžC|P.
00000030: 44 5a 52 23 00 00 41 00 00 00 00 00              DZR#..A.....

Decode  Matrix  Host Table  Protocol Dist.  Statistics
For Help, press F1                                                236
```

Sniffer Screen Capture 11.16: The Easy Ethernet CS8900A echoes back every character to the client so the human Telnet operator can see what has been entered and sent. An Ethernet packet has to be at least 64 bytes in length. The Easy Ethernet CS8900A's code always sends a 60-byte message with the last 4 bytes being added as CRC bytes by the CS8900A-CQ. The 0x00 bytes that finish up the required 60-byte length requirement are simply what was in the PIC16F877's packet array at those locations. Remember the data length is determined by subtracting 40 decimal from the total length of the IP datagram that contains the TCP segment. In this screen shot, we have 20 bytes of IP header, 20 bytes of TCP header, 12 bytes of Ethernet DLC header, 2 bytes of Ethernet Type data and 6 bytes of data in the TCP data field. The extra 5 bytes in the data area prevent the packet from being too small or a runt. The TCP application knows to pick up only 1 byte of data, and the CS8900A-CQ was instructed to pad the outgoing Ethernet packet and not allow a runt packet to be transmitted. The CS8900A-CQ need not add padding as the Easy Ethernet CS8900A firmware always sends a 60-byte message.

```
Sniffer - Local, Ethernet (Line speed at 10 Mbps) - [tcp_3.cap: Decode, 8/26 Ethernet Frames]
File  Monitor  Capture  Display  Tools  Database  Window  Help
Default

No. Status Source Address    Dest Address       Summary                                                                   Len (B) Rel. Time
6          [192.168.0.1]     [192.168.0.150]    TCP: D=8088 S=1034   ACK=10158102 SEQ=4003349371 LEN=1 WIN=17498  60   0:00:04
7          [192.168.0.150]   [192.168.0.1]      TCP: D=1034 S=8088   ACK=4003349372 SEQ=10158102 LEN=1 WIN=17498  60   0:00:04
8          [192.168.0.1]     [192.168.0.150]    TCP: D=8088 S=1034   ACK=10158103 WIN=17497                       60   0:00:04
9          [192.168.0.1]     [192.168.0.150]    TCP: D=8088 S=1034   ACK=10158103 SEQ=4003349372 LEN=1 WIN=17497  60   0:00:04
10         [192.168.0.150]   [192.168.0.1]      TCP: D=1034 S=8088   ACK=4003349373 SEQ=10158103 LEN=1 WIN=17497  60   0:00:04

IP:           ..0. .... = last fragment
IP: Fragment offset = 0 bytes
IP: Time to live    = 128 seconds/hops
IP: Protocol        = 6 (TCP)
IP: Header checksum = 7763 (correct)
IP: Source address       = [192.168.0.1]
IP: Destination address = [192.168.0.150]
IP: No options
IP:
TCP: ----- TCP header -----
TCP:
TCP: Source port              = 1034
TCP: Destination port         = 8088
TCP: Sequence number          = 4003349372
TCP: Next expected Seq number= 4003349372
TCP: Acknowledgment number    = 10158103
TCP: Data offset              = 20 bytes
TCP: Flags                    = 10
TCP:               ..0. .... = (No urgent pointer)
TCP:               ...1 .... = Acknowledgment
TCP:               .... 0... = (No push)
TCP:               .... .0.. = (No reset)
TCP:               .... ..0. = (No SYN)
TCP:               .... ...0 = (No FIN)
TCP: Window                   = 17497
TCP: Checksum                 = 9324 (correct)
TCP: No TCP options
TCP:

00000000: 00 00 45 44 54 50 00 e0 29 87 f5 5b 08 00 45 00  ..EDTP.à)‚õ[..E.
00000010: 00 28 01 85 40 00 80 06 77 63 c0 a8 00 01 c0 a8  .(.‚@.‚.wcÀ¨..À¨
00000020: 00 96 04 0a 1f 98 ee 9e 43 7c 00 9b 00 17 50 10  .‚....î‚C|.‚..P.
00000030: 44 59 93 24 00 00 00 00 00 00 00 00              DY‚$......

Decode / Matrix / Host Table / Protocol Dist. / Statistics
For Help, press F1                                                         236
```

Sniffer Screen Capture 11.17: An acknowledgement to the echoed character. I think you've got the hang of this. The next Sniffer shot would look just like Sniffer Screen Capture 11.15 with the sequence numbers changed due to the acknowledgements being bantered back and forth between the client and the server and a B in the TCP data area. The B would get echoed by the Easy Ethernet CS8900A and you would end up here to do it all over again. Let's skip by the BC123 data entry and jump to the human Telnet operator issuing a Telnet close.

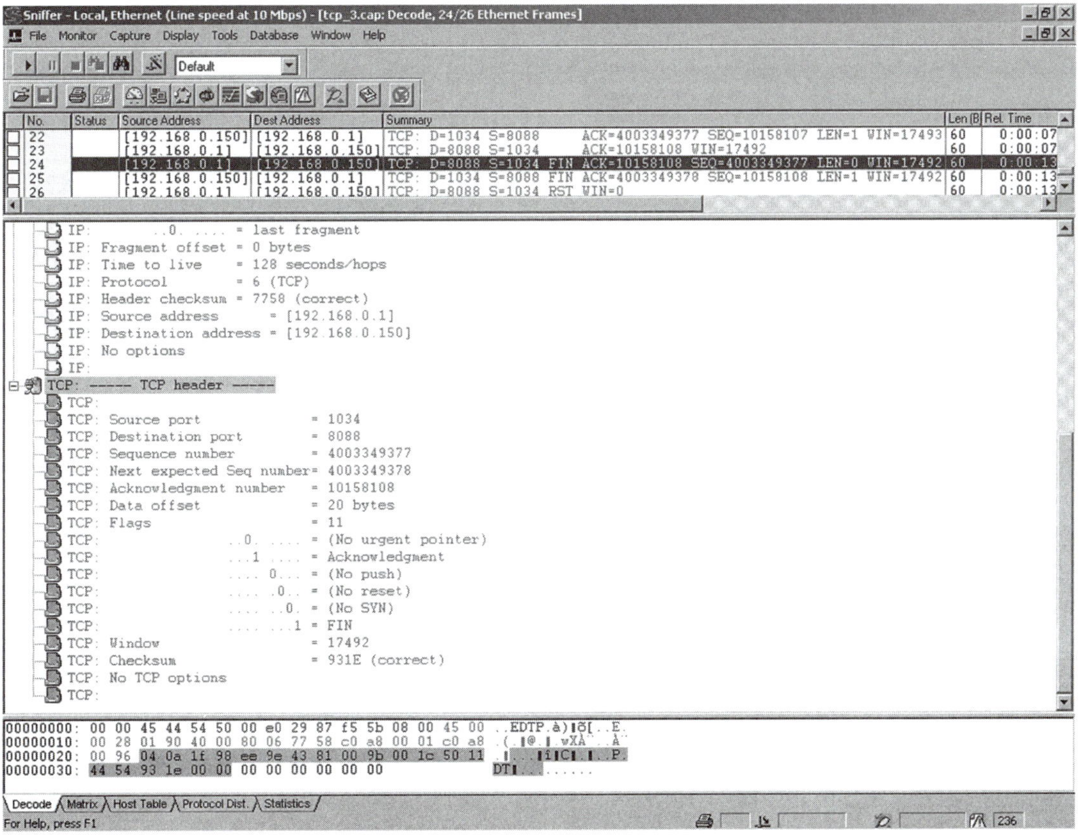

```
Sniffer - Local, Ethernet (Line speed at 10 Mbps) - [tcp_3.cap: Decode, 24/26 Ethernet Frames]
File  Monitor  Capture  Display  Tools  Database  Window  Help
Default

No. Status Source Address     Dest Address        Summary                                                                      Len (B) Rel. Time
22         [192.168.0.150]    [192.168.0.1]       TCP: D=1034 S=8088       ACK=4003349377 SEQ=10158107 LEN=1 WIN=17493        60      0:00:07
23         [192.168.0.1]      [192.168.0.150]     TCP: D=8088 S=1034       ACK=10158108 WIN=17492                             60      0:00:07
24         [192.168.0.1]      [192.168.0.150]     TCP: D=8088 S=1034 FIN ACK=10158108 SEQ=4003349377 LEN=0 WIN=17492          60      0:00:13
25         [192.168.0.150]    [192.168.0.1]       TCP: D=1034 S=8088 FIN ACK=4003349378 SEQ=10158108 LEN=1 WIN=17492          60      0:00:13
26         [192.168.0.1]      [192.168.0.150]     TCP: D=8088 S=1034 RST WIN=0                                                60      0:00:13

IP:        ..0. .... = last fragment
IP: Fragment offset = 0 bytes
IP: Time to live    = 128 seconds/hops
IP: Protocol        = 6 (TCP)
IP: Header checksum = 7758 (correct)
IP: Source address      = [192.168.0.1]
IP: Destination address = [192.168.0.150]
IP: No options
IP:
TCP: ----- TCP header -----
TCP:
TCP: Source port                = 1034
TCP: Destination port           = 8088
TCP: Sequence number            = 4003349377
TCP: Next expected Seq number= 4003349378
TCP: Acknowledgment number      = 10158108
TCP: Data offset                = 20 bytes
TCP: Flags                      = 11
TCP:               ..0. .... = (No urgent pointer)
TCP:               ...1 .... = Acknowledgment
TCP:               .... 0... = (No push)
TCP:               .... .0.. = (No reset)
TCP:               .... ..0. = (No SYN)
TCP:               .... ...1 = FIN
TCP: Window                     = 17492
TCP: Checksum                   = 931E (correct)
TCP: No TCP options
TCP:

00000000: 00 00 45 44 54 50 00 e0 29 87 f5 5b 08 00 45 00  ..EDTP.à)‡õ[..E.
00000010: 00 28 01 90 40 00 80 06 77 58 c0 a8 00 01 c0 a8  .(..@...wXÀ¨..À¨
00000020: 00 96 04 0a 1f 98 ee 9e 43 81 00 9b 00 1c 50 11  ........C.....P.
00000030: 44 54 93 1e 00 00 00 00 00 00 00 00              DT..........

Decode  Matrix  Host Table  Protocol Dist.  Statistics
For Help, press F1                                                        236
```

Sniffer Screen Capture 11.18: In this shot, the human Telnet operator has issued the Telnet close *command, which kicks out a TCP segment with the FIN flag bit set. This tells the server to expect no more data from the client. It is possible for the server to send more data to the client, but in our minimal TCP/IP implementation that is not possible.*

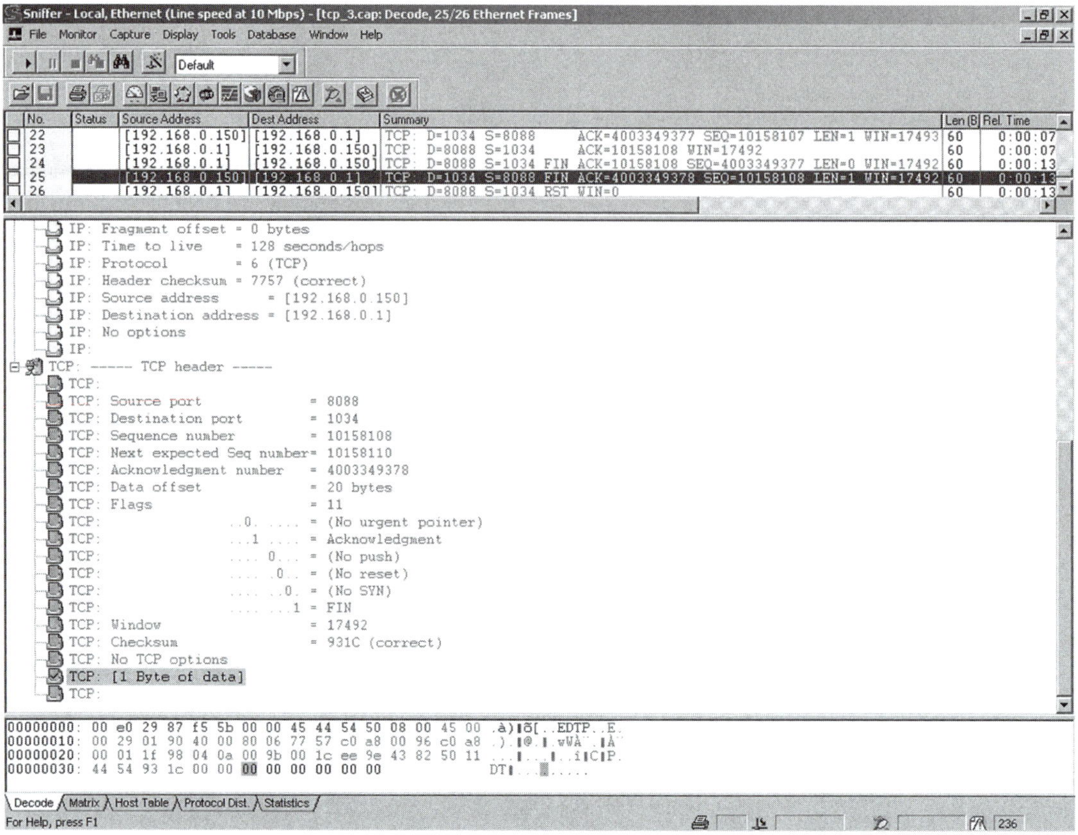

```
Sniffer - Local, Ethernet (Line speed at 10 Mbps) - [tcp_3.cap: Decode, 25/26 Ethernet Frames]
File  Monitor  Capture  Display  Tools  Database  Window  Help
Default

No. Status Source Address  Dest Address     Summary                                                              Len (B) Rel. Time
22         [192.168.0.150] [192.168.0.1]    TCP: D=1034 S=8088     ACK=4003349377 SEQ=10158107 LEN=1 WIN=17493  60  0:00:07
23         [192.168.0.1]   [192.168.0.150]  TCP: D=8088 S=1034     ACK=10158108 WIN=17492                       60  0:00:07
24         [192.168.0.1]   [192.168.0.150]  TCP: D=8088 S=1034 FIN ACK=10158108 SEQ=4003349377 LEN=0 WIN=17492  60  0:00:13
25         [192.168.0.150] [192.168.0.1]    TCP: D=1034 S=8088 FIN ACK=4003349378 SEQ=10158108 LEN=1 WIN=17492  60  0:00:13
26         [192.168.0.1]   [192.168.0.150]  TCP: D=8088 S=1034 RST WIN=0                                        60  0:00:13

IP: Fragment offset = 0 bytes
IP: Time to live    = 128 seconds/hops
IP: Protocol        = 6 (TCP)
IP: Header checksum = 7757 (correct)
IP: Source address       = [192.168.0.150]
IP: Destination address = [192.168.0.1]
IP: No options
IP:
TCP: ----- TCP header -----
TCP:
TCP: Source port                = 8088
TCP: Destination port           = 1034
TCP: Sequence number            = 10158108
TCP: Next expected Seq number= 10158110
TCP: Acknowledgment number      = 4003349378
TCP: Data offset                = 20 bytes
TCP: Flags                      = 11
TCP:              ..0. .... = (No urgent pointer)
TCP:              ...1 .... = Acknowledgment
TCP:              .... 0... = (No push)
TCP:              .... .0.. = (No reset)
TCP:              .... ..0. = (No SYN)
TCP:              .... ...1 = FIN
TCP: Window                     = 17492
TCP: Checksum                   = 931C (correct)
TCP: No TCP options
TCP: [1 Byte of data]
TCP:

00000000: 00 e0 29 87 f5 5b 00 00 45 44 54 50 08 00 45 00  .à)▮õ[..EDTP..E.
00000010: 00 29 01 90 40 00 80 06 77 57 c0 a8 00 96 c0 a8  .).▮@.▮.wWÀ¨.▮À¨
00000020: 00 01 1f 98 04 0a 00 9b 00 1c ee 9e 43 82 50 11  ...▮...▮..î▮C▮P.
00000030: 44 54 93 1c 00 00 00 00 00 00 00 00              DT▮.........

Decode  Matrix  Host Table  Protocol Dist.  Statistics
For Help, press F1                                                    236
```

Sniffer Screen Capture 11.19: The Easy Ethernet CS8900A sees the incoming FIN segment and immediately fires back an acknowledgement to the client's FIN and includes a FIN flag bit of its own. The session is totally closed at this point.

Code Snippet 11.33 is a fitting finish to the words we've devoted to the Easy Ethernet CS8900A hardware and firmware.

```
//*******************************************************************
//*   TCP Function
//*   This function uses TCP protocol to act as a Telnet server on
//*   port 8088 decimal. The application function is called with
//*   every incoming character.
//*******************************************************************
   //this code segment processes a FIN from the Telnet client
   //and acknowledges the FIN and any incoming data.
   if(FIN_IN && portaddr == MY_PORT_ADDRESS)
   {
      if(tcpdatalen_in)
         {
            for(i=0;i<tcpdatalen_in;++i)
            {
               aux_data[i] = packet[TCP_data+i];
               application_code();
            }
         }

      finflag = 1;

      ++tcpdatalen_in;

//*******************************************************************
//*   Send TCP Packet
//*   This routine assembles and sends a complete TCP/IP packet.
//*   40 bytes of IP and TCP header data is assumed.
//*******************************************************************
   packet[TCP_hdrflags+1] = 0x00;
   ACK_OUT;
   if(finflag)
   {
      FIN_OUT;
      finflag = 0;
   }
```

Code Snippet 11.33: That's all folks...

The Easy Ethernet CS8900A TCP code attempts to make sure that any data that may have arrived with the FIN segment is processed by the Easy Ethernet CS8900A's TCP application. The *finflag* is set and a byte is added to the outgoing acknowledgement number (++tcpdatalen_in). When the code falls through to the *send_tcp_packet* function, the ACK and FIN flag bits are set in the outgoing TCP segment.

You now possess the skills to allow you to drive a microcontroller-powered CS8900A-CQ on a LAN or the Internet. Even though the conversation has been limited to a PIC16F877 and the CS8900A-CQ. you can apply what you've learned here to other microcontrollers and Ethernet engine ICs.

CHAPTER 12

Let's Do It Again

You can add the CS8900A-CQ, Ethernet and Internet coding to your resume. It's time for some extra credit coding. We've managed to recycle everything we've coded up to this point. So, let's take all we know now and apply it once again to a completely different Ethernet engine IC.

If you're not "chapter hopping," you've just eaten one big sandwich made of Ethernet bread and Internet protocol meat. The Ethernet sandwich you just ate was flavored with the CS8900A-CQ Ethernet IC.

Believe it or not, you already know enough to put a big bite into another Ethernet sandwich flavored with the Realtek RTL8019AS. In the text to follow, you and I will take all of the hardware and firmware tricks we learned building the Easy Ethernet CS8900A and apply them to a new project called the Easy Ethernet W(hacked).

Easy Ethernet Whacked??? What the...?

The Easy Ethernet W project is an upgraded offspring of a project called the Packet Whacker (Photo 12.1). The Packet Whacker is no more than a Realtek RTL8019AS perched on a printed circuit board with all of its pins terminated at convenient .1-inch-center header points. Add a 20-MHz crystal, some power supply bypass capacitors, a few choice resistors and some custom magnetics and you have a fully functional 10 Mbps Ethernet engine.

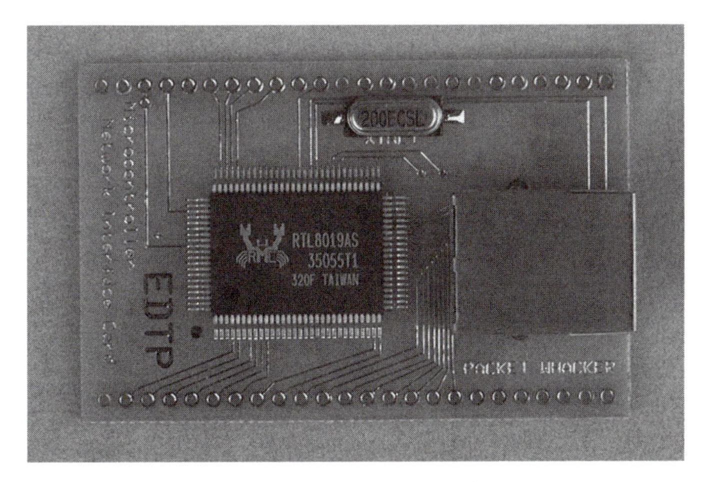

Photo 12.1: As far as Ethernet is concerned, this is as simple as it can get.

Originally, the Packet Whacker was designed to be integrated with a microcontroller of the designer's choice. If you continue reading this book, when you're finished you'll be able to interface the basic Packet Whacker design to a PIC16F877, a PIC18F452 and an Atmel ATmega16. The Atmel devices are so similar when it comes to coding them in C that you actually can use the ATmega16 code to Ethernet enable an ATmega32, ATmega64 and ATmega 128.

Do I have your attention now? Good. Let's start on our new Easy Ethernet W project by talking about the Realtek RTL8019AS.

The Realtek RTL8019AS

The Realtek RTL8019AS is an NE2000-compatible IC that is easily integrated into just about any microcontroller project that is being designed to use Ethernet connectivity. The Realtek RTL8019AS is based on the National DP8390 Network Interface Controller, which like the CS8900A-CQ, provides all the Media Access Control layer functions required for transmission and reception of packets in accordance with the IEEE 802.3 CSMA/CD (Carrier Sense Multiple Access/Collision Detection) standard.

The Realtek RTL8019AS has many of the features you are familiar with from the CS8900A-CQ. Like the CS8900A-CQ, the RTL8019AS provides interface auto-detect capability and can choose between an integrated 10Base-T transceiver, a BNC or an AUI interface. And, again like the CS8900A-CQ, the Realtek RTL8019AS's on-chip 10Base-T transceiver can automatically correct the polarity at its receiving cable pair.

In 8-bit mode, both the CS8900A-CQ and the RTL8019AS use a 4 Kb on-chip RAM area despite the RTL8019AS's 16 Kb specification, which only applies to 16-bit mode. In NE2000 fashion, the RTL8019AS's buffer memory is configured as a ring, while CS8900A-CQ PacketPage buffer memory is treated as a flat 4 Kb of buffer area.

In our exploration of the Easy Ethernet CS8900A's onboard CS8900A-CQ, we discovered that using the CS8900A-CQ's on-chip DMA resources to transfer data from the CS8900A-CQ's on-chip buffer memory to microcontroller memory in 8-bit mode is illegal. I use the word illegal because we know it can be done, but we were told by the CS8900A-CQ IC designers that you're not supposed to do that because in 8-bit mode it is not a reliable way to move data.

Unlike the CS8900A-CQ, the RTL8019AS uses internal DMA resources to manage and move data between the RTL8019AS's FIFO and the Realtek RTL8019AS's internal buffer memory. Within the Realtek RTL8019AS, the onboard FIFO (First In, First Out) and Local DMA (Direct Memory Access) channels work in conjunction to form a simple packet management scheme that provides up to 10 megabyte per second internal DMA transfers. The FIFO lies between the network interface and the Local DMA channel.

A second Realtek RTL8019AS Remote DMA channel is included on-chip to get data out of the RTL8019AS's internal Buffer Ring and into microcontroller memory for processing and vice versa. It's important to remember that the Local DMA channel moves data between the Realtek RTL8019AS's internal FIFO and the RTL8019AS's Buffer Ring, and the Remote RTL8019AS DMA channel moves data between the Realtek RTL8019AS's Buffer Ring and the microcontroller's working memory.

Despite the difference in the way the RTL8019AS and the CS8900A-CQ handle data internally, physically the Realtek RTL8019AS and the CS8900A-CQ interface with their external support circuitry and the network in a similar manner. Just like the CS8900A-CQ, the Realtek RTL8019AS was originally designed for major Ethernet applications in desktop personal computers and some of the RTL8019AS's functionality will be useless to the hardware used on our Easy Ethernet W. The Realtek RTL8019AS side of the Easy Ethernet W can be seen in Schematic 12.1.

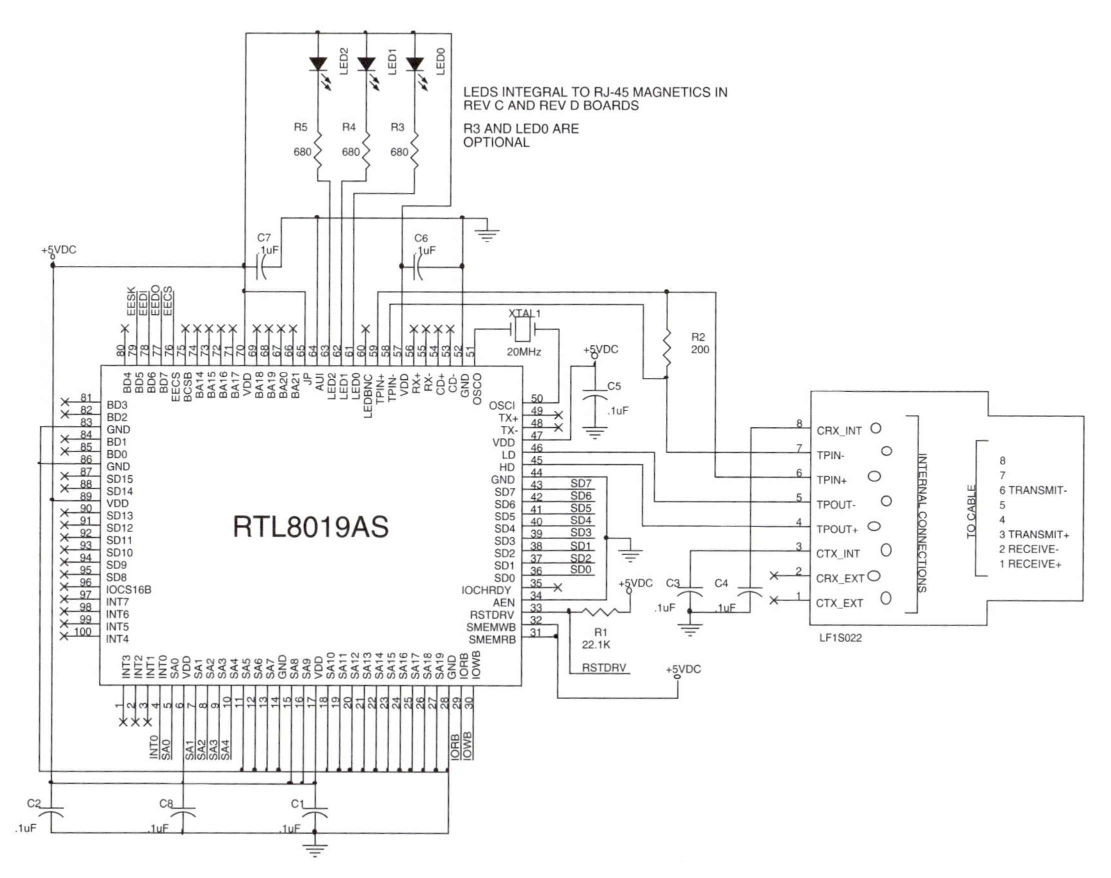

Schematic 12.1: Although designed to work as a personal computer Ethernet interface, the Realtek RTL8019AS is well-suited for work with 8-bit microcontrollers.

The Realtek RTL8019AS is controlled through an array of on-chip registers similar to those found in the CS8900A-CQ. The RTL8019AS doesn't use CS8900A-CQ PacketPage technology, but as you come to find out, the RTL8019AS's register operations are logically identical to the CS8900A-CQ's register operations.

Again, just like the CS8900A-CQ, the Realtek RTL8019AS registers are used during initialization, packet transmission and reception. There are also registers for Remote DMA operations on the Realtek RTL8019AS that don't exist on the CS8900A-CQ. Basically, using the Realtek RTL8019AS internal registers we can perform the same logical operations that are performed using the CS8900A-CQ's PacketPage registers. The basic operations include defining the hardware physical address, setting the receive parameters and setting the transmission parameters. For the Realtek RTL8019AS, add configuring DMA channels and allocating transmit and receive Buffer Ring areas to the aforementioned list of operations.

In that DMA is an integral part of the Realtek RTL8019AS's microcontroller interface, there must be a control mechanism or register to act as the traffic cop for the data flow between the RTL8019AS's buffers and the microcontroller memory and the RTL8019AS's MAC engine to Ethernet interface. That control register for the Realtek RTL8019AS is the Command Register (CR), which is used to initiate Remote DMA operations as well as data transmission. Remember, Remote DMA operations are used to move data between the RTL8019AS's buffer and the microcontroller's memory. As you examine the Easy Ethernet W's source code, you'll come to the conclusion that the RTL8019AS CR register is a well-used register. So, I've put its bits into graphical format in Figure 12.1.

CR

7	6	5	4	3	2	1	0
PS1	PS0	RD2	RD1	RD0	TXP	STA	STP

Figure 12.1: You're going to get to know this register very well.

An early version of the Easy Ethernet CS8900A used the CS8900A-CQ's internal interrupt structure to sense the presence of a valid frame in the CS8900A-CQ's receive buffer. The Easy Ethernet W detects a valid frame in a similar manner. The Easy Ethernet W's PIC16F877 microcontroller checks for a valid frame by polling an interrupt pin (INT0) on the RTL8019AS. Once a valid level is sensed on the INT0 pin, the PIC16F877 interrogates the Realtek RTL8019AS's Interrupt Status Register (ISR) to determine what type of interrupt has occurred.

We already know that both the CS8900A-CQ and the Realtek RTL8019AS transmit packets in accordance with the CSMA/CD protocol standards. Both the Realtek RTL8019AS and the CS8900A-CQ schedule retransmission of packets up to 15 times on collisions according to the truncated binary exponential backoff algorithm. The CS8900A-CQ datasheet calls this the Standard Backoff algorithm. Once you cut the transmit process loose, both Ethernet ICs run the show until the transmission cycle is aborted or completed.

Here's where things you already know about the Easy Ethernet CS8900A and CS8900A-CQ will be applied to the Easy Ethernet W and the RTL8019AS. Assuming buffer memory is allocated and free, transmitting packets with the Realtek RTL8019AS entails setting up an IEEE 802.3 frame in memory with:

- 6 bytes of the destination address (DA)
- 6 bytes of the source address (SA)
- The data length in bytes
- The data

Unless you're "chapter hopping," you may recognize the above bulleted sequence of bytes as a standard Ethernet frame. Once the required frame items are built in the microcontroller's packet array memory area, the Realtek RTL8019AS register TPSR (Transmit Page Start Register) is loaded with the frame starting address and the TBCR0 (Transmit Byte Count 0) and TBCR1 (Transmit Byte Count 1) registers are filled with the length of the frame. It's easier to visualize if you have some structure to view. You can see the structure of the TPSR and TBCR register set in Figure 12.2.

TPSR

7	6	5	4	3	2	1	0
A15	A14	A13	A12	A11	A10	A9	A8

TBCR0

7	6	5	4	3	2	1	0
TBC7	TBC6	TBC5	TBC4	TBC3	TBC2	TBC1	TBC0

TBCR1

7	6	5	4	3	2	1	0
TBC15	TBC14	TBC13	TBC12	TBC11	TBC10	TBC9	TBC8

Figure 12.2: The TPSR represents the upper byte of a 16-bit address. If that doesn't register right now, it will make more sense later when we talk about allocating buffer area within the RTL8019AS.

To initiate the transmission of a packet, the TXP (transmit packet) bit of the Realtek RTL8019AS Command Register is set. If the total length of the Ethernet packet is less than 46 bytes, the Realtek RTL8019AS cannot be instructed to automatically pad the packet to avoid sending a runt packet onto the network. Therefore, we must as programmers make sure we don't generate any runt packets. The TCP/IP section of the Easy Ethernet W's code checks for runts. The ARP, ICMP and UDP routines use the length of the incoming packets as their guide. Since we will setup the RTL8019AS to not accept runt packets, the UDP and ICMP packets received will always meet the minimum length requirement. The ARP code builds a 60-byte ARP reply packet.

Your first encounter with runt packet avoidance was in the TCP/IP section of the Easy Ethernet CS8900A source code. If you have a reason to break the rules for research or you're just playing around, by configuring the right bits in the right registers, you can tell either the CS8900A-CQ or the Realtek RTL8019AS to send and receive runt packets.

You're beginning to see how we can use knowledge acquired in our Easy Ethernet CS8900A project to quickly get our Easy Ethernet W project up and running. There are lots of similarities in the operation of the RTL8019AS and the CS8900A-CQ Ethernet ICs.

When it comes to allocating transmit buffer space, the CS8900A-CQ is a little different in that the on-chip buffer space is asked for, and permission is granted to load the buffer area before any data is transferred for transmission. This is called a "bid" for buffer space. The Realtek RTL8019AS transmit buffer area is allocated according to the contents of an RTL8019AS register. The Realtek RTL8019AS datasheet stresses that if the Buffer Ring area of the Realtek RTL8019AS is set up correctly at initialization, there should never be any contention for transmit buffer memory under normal operating conditions.

The act of transmitting data to the ether is a parallel process for both the CS8900A-CQ and RTL8019AS. Before jumping onto the ether, both the CS8900A-CQ and the RTL8019AS will check themselves internally to see if they are receiving data from the network. If the all clear is sounded, the CS8900A-CQ starts transmission of the 8-byte preamble once a specified number of bytes (5, 381,1021 or ALL) are loaded into the transmit buffer. You'll recall that this begin transmission threshold is determined by bit settings in the CS8900A-CQ Transmit Command Register.

The Realtek RTL8019AS uses its Local DMA channel and FIFO to follow the RTL8019AS-generated preamble with valid data. The Realtek RTL8019AS's Local DMA bursts data to the FIFO, which is then serialized out onto the network as clocked NRZ data.

If you stop and think about this, every Ethernet IC on a network has to conform to these standards in order to communicate with each other. You saw the NRZ data trick when we discussed the CS8900A-CQ MAC engine. The Realtek RTL8019AS's Local DMA refreshes the FIFO when the FIFO "send more" threshold is reached. The FIFO "send more" threshold is programmable. In both cases, the RTL8019AS and the CS8900A-CQ continue the transmission as long as the transmission byte count in the byte count registers is greater than zero. Once all bytes are sent, the CRC is calculated by both the CS8900A-CQ and RTL8019AS and is sent to complete the packet.

For either the CS8900A-CQ or the RTL8019AS, if a collision occurs during transmission, the transmission is stopped and 32 ones (a jam sequence) are transmitted to make sure everybody on the network segment knows a collision just took place. Both the CS8900A-CQ and the Realtek RTL8019AS execute the Standard Backoff algorithm and the transmission is retried. When the transmission completes, both the RTL8019AS and the CS8900A-CQ have transmit status registers that can be queried to see how the transmission went.

As you've probably already concluded, we're working with a bunch of standards that allow differing Ethernet IC manufacturers to design and build products that can communicate with each other over a common medium called Ethernet. Now with that in mind, transmitting data and receiving data from the ether is a similar process for the RTL8019AS and CS8900A-CQ as well. Both the CS8900A-CQ and the RTL8019AS listen to the wire sense a carrier and start syncing up with the alternating 1/0 preamble that starts a 10 Mbps Ethernet packet. Once the two consecutive ones of the SFD (Start of Frame Delimiter) are sensed, the preamble ends and the MAC engines within the RTL8019AS and CS8900A-CQ expect everything behind the set of SFD ones to be valid data.

Both the CS8900A-CQ and RTL8019AS check the destination address (DA) to see if the incoming packet is addressed to them. If it is not, it is not moved into buffer memory and the packet is discarded. On the other hand, if the packet destination address matches the Ethernet IC's address filter setting (hashed or individual), in the case of the RTL8019AS or CS8900A-CQ, the frame is moved into the Ethernet IC's on-chip buffer memory so it can be transferred to the microcontroller's RAM (Random Access Memory) for processing. If everything goes OK during the receive cycle, both the RTL8019AS and the CS8900A-CQ post receive status in their respective receive status registers. The RTL8019AS raises an interrupt I/O line, while the CS8900A-CQ must have its register bank polled to alert the microcontroller of a valid frame in the receive buffer.

Earlier, I mentioned that the Realtek RTL8019AS uses NE2000 conventions and thus buffers its data in a ring. The Realtek RTL8019AS differs from the CS8900A-CQ in that the data coming into the Realtek RTL8019AS from the network is put into a receive Buffer Ring; whereas the CS8900A-CQ stuffs the data into a flat predefined buffer area. To help you visualize a ring versus a flat memory area, think like sailors of ancient times. Those old salts thought the world was flat, and therefore, if they sailed far enough they would simply fall of the edge of the world. That's a flat memory area. Later, some enterprising sailors figured out that they could sail and sail and as long as they navigated correctly, they could end up where they started without falling off the edge. That's a ring.

The CS8900A-CQ flat memory model method is valid, and there's nothing special about the Realtek RTL8019AS's ring buffer. It's a classic circular, head and tail buffer scheme with four pointers controlling the activity in the Buffer Ring:

- PSTART
- PSTOP
- CURR
- BNRY

A graphical representation of the RTL8019AS Buffer Ring is shown in Figure 12.3.

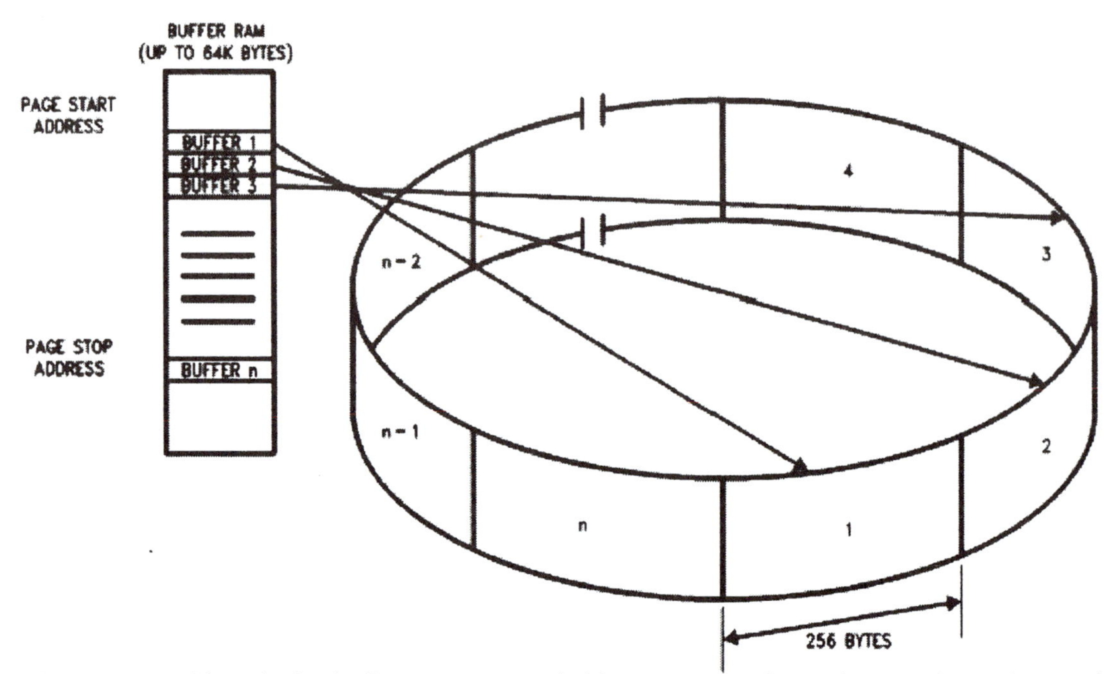

Figure 12.3: Although the buffer pointers can hold a 64K value (0xFFFF), note that only 8K of buffer RAM is available in the RTL8019AS.

PSTART (Page Start) is the beginning address of the Buffer Ring. PSTOP (Page Stop) is the address of the end of the Buffer Ring. The Buffer Ring size is determined by the number of bytes between PSTART and PSTOP. For the RTL8019AS, PSTART and PSTOP are loaded at initialization time. There are no CS8900A-CQ counterparts for PSTART and PSTOP. CURR, the Current Page Pointer, points to the next available buffer area for the next incoming frame. BNRY, or the Boundary Pointer, points to the next frame to be unloaded from the Buffer Ring. Think of the CURR as the write pointer and the BNRY as the read pointer for the Buffer Ring. As frames come in, the CURR pointer moves ahead of the BNRY pointer around the ring. If CURR reaches BNRY, the Buffer Ring is full. All receptions are aborted, and missed packet registers within the RTL8019AS are updated until this condition is cleared. The RTL8019AS's Remote DMA channel is the mechanism that removes frames from the Buffer Ring. Figure 12.4 is a representation of an initialized Buffer Ring.

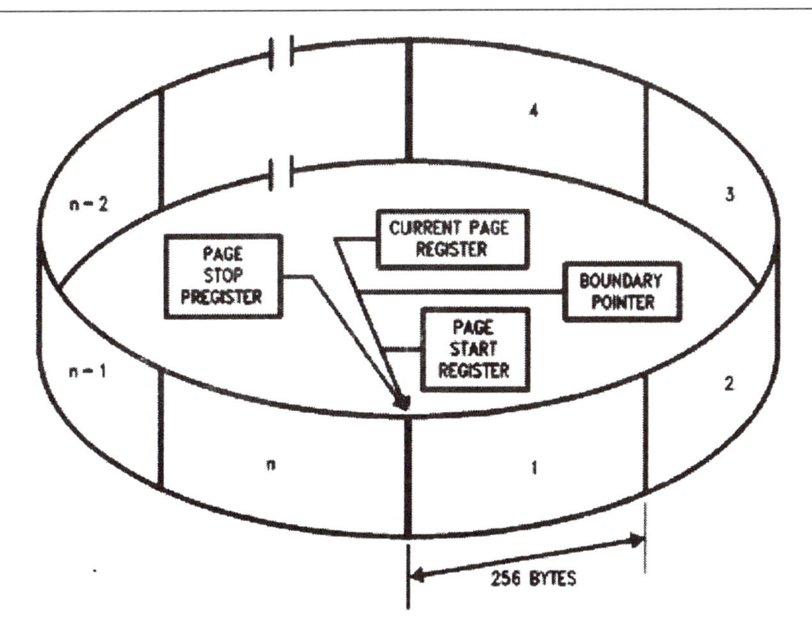

Figure 12.4: An empty ring is signaled by the CURR and BNRY pointers being equal.

Each Realtek RTL8019AS ring buffer segment in Figure 12.4 is 256 bytes in length. A valid received frame is placed at location CURR plus a 4-byte offset. Buffer segments are automatically linked together to receive frames larger than 256 bytes. When all the bytes are loaded, the RSR (Receive Status Register) status, a pointer to the next frame and the byte count of the current frame are written into the 4-byte offset. That's basically how the Realtek RTL8019AS and any other NE2000-compatible Ethernet IC works. A visual of a frame inside the RTL8019AS's Buffer Ring is shown in Figure 12.5.

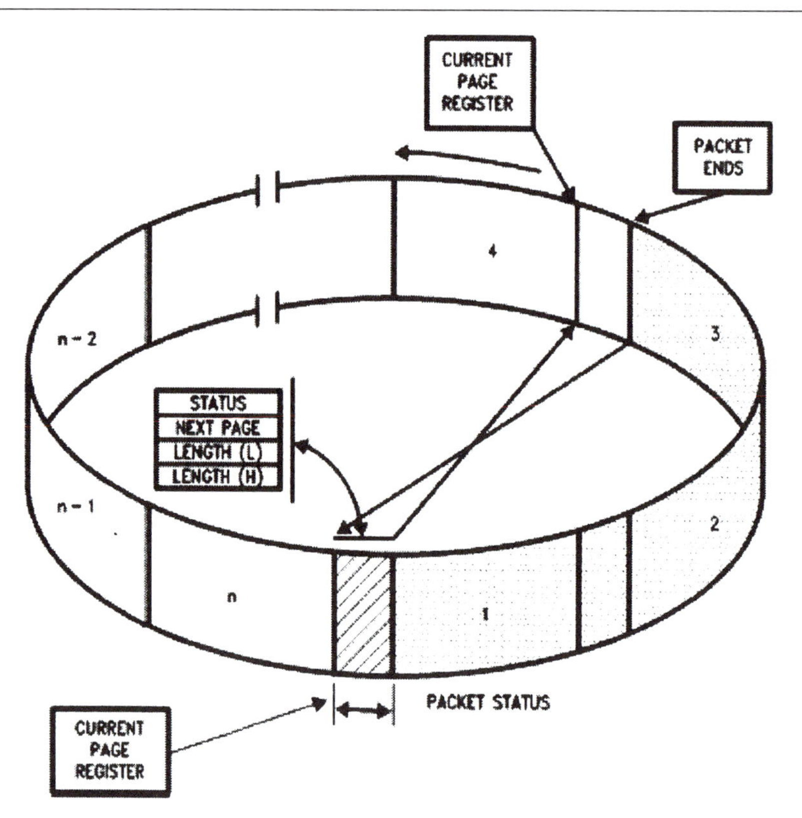

Figure 12.5: This is a graphic from the original National Semiconductor datasheet. The term "packet" is used loosely in the figure.

The Easy Ethernet W Hardware

There are only a couple of differences in the Easy Ethernet W hardware represented schematically in Schematic 12.2 and the Easy Ethernet CS8900A hardware. The bank of pull-up resistors is missing from the Easy Ethernet W schematic. I experimented with removing the pull-ups from the Easy Ethernet CS8900A and the Easy Ethernet CS8900A performed as expected with no problems. However, as a precaution, I left pads for the pull-up resistors on the Easy Ethernet CS8900A printed circuit board as the MPLAB ICE 2000 required them to operate properly. If you use the MPLAB ICE 2000, you'll need those pull-up resistors on the Easy Ethernet CS8900A printed circuit board. The only other physical difference in the Easy Ethernet CS8900A and the Easy Ethernet W are the pin assignments of the PIC16F877 microcontroller's I/O.

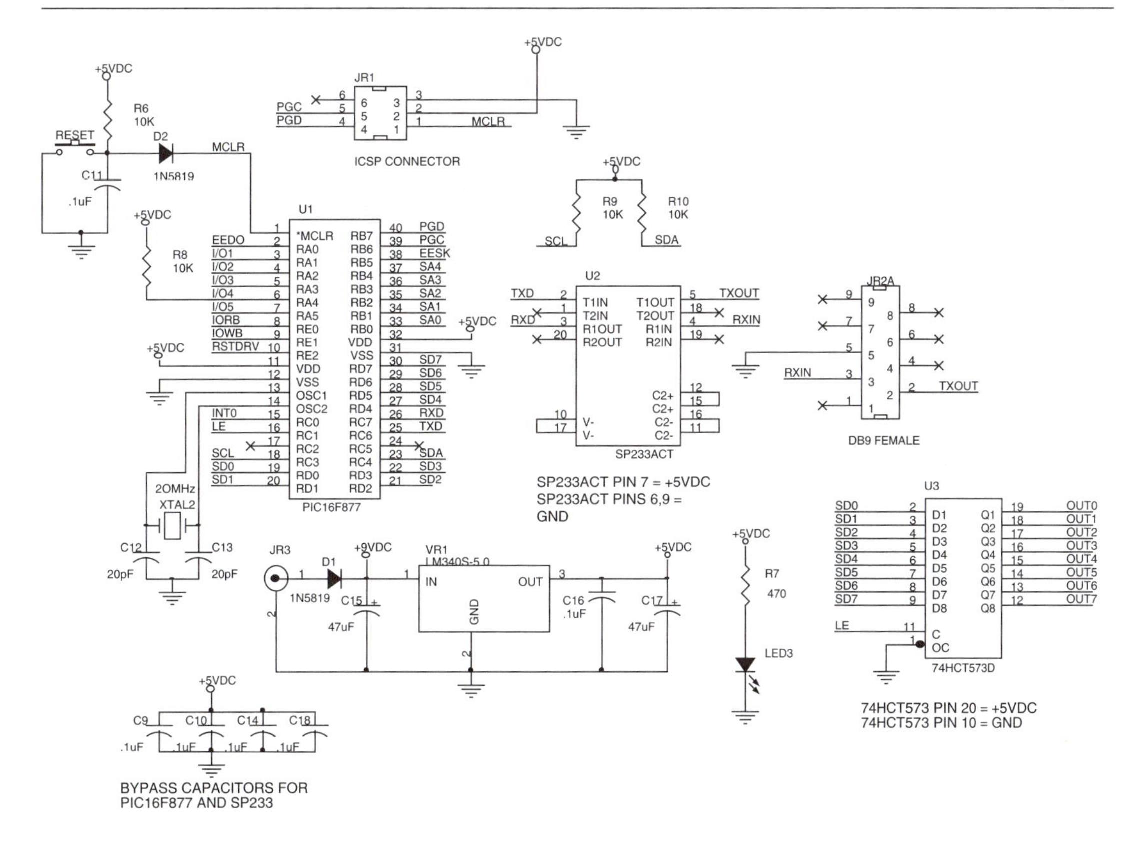

Schematic 12.2: We've even recycled the hardware.

There is another difference that isn't evident in Schematic 12.2. The PIC16F877 is loaded with CS8900A-CQ firmware that has been modified to control the RTL8019AS. Photo 12.2 gives us a view of a fully assembled Easy Ethernet W.

Photo12.2: This device is the result of crossbreeding a Packet Whacker, our RS-232 circuitry, our I^2C circuitry, a 74HCT573 latch (bottom of the board) and all of the supporting power supply and PIC programming components.

The Easy Ethernet W Firmware

You are already familiar with the CSMA/CD protocol, the Ethernet frame and packet and the Internet protocols TCP, UDP, IP, ARP and ICMP. Guess what? You don't have to "relearn" anything to build and code an Easy Ethernet W of your own. All of that knowledge you gained from building the Easy Ethernet CS8900A applies here for the Easy Ethernet W, and with the hardware changes from the Easy Ethernet CS8900A to the Easy Ethernet W being minimal, you're already checked out on the hardware and we can start coding the PIC16F877 firmware for the Easy Ethernet W right now.

Thanks to the portability of C, lots of the code we used in the Easy Ethernet CS8900A project will be reused in the Easy Ethernet W project without modification. The original CS8900A-CQ packet array memory area and the Ethernet and Internet protocol layouts remain unchanged in the RTL8019AS code. The IP address and MAC addresses for the Easy Ethernet W can easily be changed in your version of the project; however, as you see in Code Snippet 12.1, to avoid unneeded confusion I'll keep them just as they were in the Easy Ethernet CS8900A project. The only change I made in the Ethernet and Internet protocols area was the wording in the Telnet banner.

```
//*******************************************************************
//*   TELNET SERVER BANNER STATEMENT CONSTANT
//*******************************************************************
int8 const telnet_banner[] = "\r\nWhacked Easy Ethernet>";
//*******************************************************************
//*   IP ADDRESS DEFINITION
//*   This is the Ethernet Module IP address.
//*   You may change this to any valid address.
//*******************************************************************
int8 MYIP[4] = { 192,168,0,150 };
//*******************************************************************
//*   HARDWARE (MAC) ADDRESS DEFINITION
//*   This is the Ethernet Module hardware address.
//*   You may change this to any valid address.
//*******************************************************************
char MYMAC[6] = { 0,0,'E','D','T','P' };
```

Code Snippet 12.1: Using the same MAC and IP addresses for the Easy Ethernet W makes relating to the original CS8900A-CQ code and concepts a bit easier.

Earlier, I mentioned that a 4-byte header preceded each frame in the RTL8019AS's ring buffer (Figure 12.5). That prompted a slight change in the pageheader array in Code Snippet 12.3. The Easy Ethernet CS8900A's pageheader array layout is shown in Code Snippet 12.2.

```
//*******************************************************************
//*   Ethernet Header Layout
//*******************************************************************
int8  pageheader[4];
#define  enetpacketstatusH 0x00
#define  enetpacketstatusL 0x01
#define  enetpacketLenH 0x02
#define  enetpacketLenL 0x03
```

Code Snippet 12.2: This is the CS8900A-CQ's 4-byte header layout.

```
//********************************************************************
//*   Receive Ring Buffer Header Layout
//*   This is the 4-byte header that resides infront of the
//*   data packet in the receive buffer.
//********************************************************************
int8  pageheader[4];
#define  enetpacketstatus  0x00
#define  nextblock_ptr     0x01
#define  enetpacketLenL    0x02
#define  enetpacketLenH    0x03
```

Code Snippet 12.3: The CS8900A-CQ and RTL8019AS frame headers perform the same function but are totally different in logic and nature. I've included the Easy Ethernet CS8900A's pageheader array definition for comparison and your viewing pleasure in Code Snippet 12.2.

It stands to reason that the internal definitions for the RTL8019AS in Code Snippet 12.4 will differ from those we laid out for the CS8900A-CQ. We don't need to know the details of every RTL8019AS register. And, since the concepts of sending and receiving packets is common for both the RTL8019AS and the CS8900A-CQ, I'll point out the differences and highlight the similarities of the RTL8019AS firmware as they relate to the CS8900A-CQ firmware.

```
//********************************************************************
//*   REALTEK CONTROL REGISTER OFFSETS
//*   All offsets in Page 0 unless otherwise specified
//********************************************************************
#define CR        0x00
#define PSTART    0x01
#define PAR0      0x01    // Page 1
#define CR9346    0x01    // Page 3
#define PSTOP     0x02
#define BNRY      0x03
#define TSR       0x04
#define TPSR      0x04
#define TBCR0     0x05
#define NCR       0x05
#define TBCR1     0x06
#define ISR       0x07
#define CURR      0x07     // Page 1
#define RSAR0     0x08
#define CRDA0     0x08
#define RSAR1     0x09
#define CRDAL     0x09
#define RBCR0     0x0A
#define RBCR1     0x0B
```

```
#define RSR         0x0C
#define RCR         0x0C
#define TCR         0x0D
#define CNTR0       0x0D
#define DCR         0x0E
#define CNTR1       0x0E
#define IMR         0x0F
#define CNTR2       0x0F
#define RDMAPORT    0X10
#define RSTPORT     0x18
```

Code Snippet 12.4: If you're wondering what the "Page" comments are all about, we'll get to that soon. The creg *term you will see throughout the Easy Ethernet W source code refers to this Code Snippet. When you see* creg *think of that as short for Control Registers.*

You already have a bit of familiarity with the RTL8019AS, as I provided an overview of how the RTL8019AS worked earlier in the text. Rather than fill you with theory up front, let's do something different. Let's dive into the code and work our way through the changes needed to make the RTL8019AS our Ethernet IC of choice for the Easy Ethernet W.

Initializing the Realtek RTL8019AS

The Easy Ethernet W firmware was written for the PIC16F877 using Custom Computer Services C Compiler. The very first instruction in the Easy Ethernet W firmware in Code Snippet 12.5 begins the RTL8019AS's initialization process. Again, if you're not "chapter hopping," you should already know what the *synflag* and *finflag* are and why we're initializing them here.

The Easy Ethernet banner serves the same purpose here as it did in the Easy Ethernet CS8900A code. It verifies the operation of the Easy Ethernet W's serial port and provides a visual of the Easy Ethernet W firmware version.

```
int1 synflag,finflag;
//*******************************************************************
//*   MAIN MAIN MAIN MAIN MAIN MAIN MAIN MAIN MAIN MAIN MAIN MAIN
//*******************************************************************
void main()
{
   init_RTL8019AS();
   synflag = 0;
   finflag = 0;
   printf("Easy Ethernet W Version For PIC16F877 03.08.12\r\n");
```

Code Snippet 12.5: The banner is optional but nice to have as it tells you if the Easy Ethernet W is alive when you first apply power.

The RTL8019AS initialization code is a bit lengthy and is like eating breakfast. It's the most important meal of the day. The whole RTL8019AS initialization breakfast enchilada is offered up in Code Snippet 12.6. We'll look at it a bite at a time.

```
//*****************************************************************
//*   REALTEK CONTROL REGISTER OFFSETS
//*   All offsets in Page 0 unless otherwise specified
//*****************************************************************
#define CR          0x00
#define RSTPORT     0x18
#define RBCR0       0x0A
#define RBCR1       0x0B
#define RCR         0x0C
#define TPSR        0x04
#define TCR         0x0D
#define PSTART      0x01
#define BNRY        0x03
#define PSTOP       0x02
#define CURR        0x07    // Page 1
//*****************************************************************
//*   RTL8016AS PIN DEFINITIONS
//*****************************************************************
#define  INT0       PORTC,0
#define  le_pin     PORTC,1
#define  ior_pin    PORTE,0
#define  iow_pin    PORTE,1
#define  rst_pin    PORTE,2
//*****************************************************************
//*   RTL8019AS 9346 EEPROM PIN DEFINITIONS
//*****************************************************************
#define  EEDO       PORTA,0
//*****************************************************************
//*   RTL8019AS I/O PORT DEFINITIONS
//*****************************************************************
#define  cregaddr PORTB
#define  cregdata PORTD
#define  tocreg   set_tris_D(0x00);
#define  fromcreg set_tris_D(0xFF);
//*****************************************************************
//*   RTL8019AS INITIAL REGISTER VALUES
//*****************************************************************
#define rcrval    0x04
#define tcrval    0x00
#define dcrval    0x58
#define imrval    0x11     // PRX and OVW interrupt enabled
#define txstart   0x40
```

```
#define rxstart   0x46
#define rxstop 0x60

int8 byte_read;
//******************************************************************
//*   Initialize the RTL8019AS
//******************************************************************
void init_RTL8019AS()
{
   ADCON1 = 0x06;                   //00000110  all digital to start
   ADCON0 = 0;
   set_tris_C(0x9D);
   set_tris_A(0x00);
   bit_clear(EEDO);
   bit_clear(le_pin);
   set_tris_B(0xE0);                // setup address lines
   cregaddr = 0x00;                 // clear address lines
   fromcreg;                        // address lines = input
   set_tris_E(0x00);                // setup IOW, IOR, RESET
   bit_set(iow_pin);                // disable IOW
   bit_set(ior_pin);                // disable IOR
   bit_set(rst_pin);                // put NIC in reset
   delay_ms(2);                     // delay at least 1.6ms
   bit_clear(rst_pin);              // disable reset line
   read_creg(RSTPORT);              // read contents of reset port
   write_creg(RSTPORT,byte_read);   // do soft reset
   delay_ms(10);                    // give it time
   read_creg(ISR);                  // check for good soft reset
   if(!bit_test(byte_read,RST)){
      while(1){
      printf("INIT FAILED\n\r");
      }
   }

   write_creg(CR,0x21);             // stop the NIC, abort DMA, page 0
   delay_ms(2);                     // make sure nothing is coming in
                                    or going out
   write_creg(DCR,dcrval);          // 0x58
   write_creg(RBCR0,0x00);
   write_creg(RBCR1,0x00);
   write_creg(RCR,rcrval);
   write_creg(TPSR,txstart);
   write_creg(TCR,0x02);
   write_creg(PSTART,rxstart);
   write_creg(BNRY,rxstart);
   write_creg(PSTOP,rxstop);
```

```
    write_creg(CR,0x61);
    write_creg(CURR,rxstart);
    for(i=0;i<6;++i)
       write_creg(PAR0+i, MYMAC[i]);

    write_creg(CR,0x22);
    write_creg(ISR,0xFF);
    write_creg(IMR,imrval);
    write_creg(TCR,tcrval);
}
```

Code Snippet 12.6: You can break the code into three parts. The first part of the code initializes the RTL8019AS's on-chip hardware resources. The second part of the code prepares the RTL8019AS's register set for operation. The third part of the code activates the RTL8019AS and brings it online to the network.

Some of the PIC16F877's I/O pins (PORTA and PORTE) work double-duty and can be configured as analog inputs. The PIC16F877 analog pins are set in the analog input mode by default on power up. The Easy Ethernet W's UDP application code will use the PORTA I/O as an output port. Also, the Easy Ethernet W uses the PORTE pins to drive some of the RTL8019AS's bus control pins. So, we need to disable all of the analog functionality, and the first two lines of code in Code Snippet 12.7 disable all analog I/O functions. The analog inputs of the PIC16F877 can be enabled if you desire to use them. However, we won't be using them in this text.

The *set_tris_X(0xXX)* functions are built-in Custom Computer Services C Compiler functions that determine the direction the I/O lines will take. For instance, in Code Snippet 12.5, PORTA of the PIC16F877 is configured as an output port. This is standard PIC setup stuff and you've seen this before in the Easy Ethernet CS8900A source code. A '1' in a bit position indicates that the I/O pin associated with that bit is an input, while a '0' denotes output for an I/O pin position.

Like the CS8900A-CQ, the RTL8019AS is capable of reading some of its initial configuration parameters from an external EEPROM and tries to do this automatically unless it's headed off at the pass. As with the CS8900A-CQ and Easy Ethernet CS8900A, the Easy Ethernet W doesn't require any data to be stored on an external EEPROM. To prevent the RTL8019AS from expecting data from an external EEPROM at startup, we tell the RTL8019AS that no EEPROM device exists by taking the RTL8019AS's EEDO (EEPROM Data Output) line low (*bit_clear(EEDO)*) and leaving it low forever. The *bit_clear* function that puts the logic level on the PIC16F877 EEDO I/O pin is another one of the many Custom Computer Services C Compiler built-in functions.

The rest of the intentions in Code Snippet 12.7 are pretty obvious. Basically, we want to put everything inside and outside of the RTL8019AS in an operable state and then reset the RTL8019AS. The "NIC" (Network Interface Controller) mentioned in the Easy Ethernet W source code is actually referring to the RTL8019AS IC.

```
//*********************************************************************
//*   Initialize the RTL8019AS
//*********************************************************************
   ADCON1 = 0x06;              //00000110  all digital to start
   ADCON0 = 0;
   set_tris_C(0x9D);
   set_tris_A(0x00);
   bit_clear(EEDO);
   bit_clear(le_pin);
   set_tris_B(0xE0);           // setup address lines
   cregaddr = 0x00;            // clear address lines
   fromcreg;                   // address lines = input
   set_tris_E(0x00);           // setup IOW, IOR, RESET

   bit_set(iow_pin);           // disable IOW
   bit_set(ior_pin);           // disable IOR
   bit_set(rst_pin);           // put NIC in reset
```

Code Snippet 12.7: This part of the Easy Ethernet W code is actually working on setting up the Easy Ethernet W's PIC16F877 microcontroller, which directly affects the RTL8019AS's I/O control and reset pins.

Once the RTL8019AS reset pin is activated, we must give the RTL8019AS at least 1.6 ms to perform the internal reset process. Once we've waited long enough and think that the hard reset is finished, a write to the RTL8019AS RSTPORT initiates a soft reset. To be safe, we simply read the RTL8019AS's Reset Port (RSTPORT) and then write the contents we read from it back into it. The idea is to make sure the RTL8019AS has actually entered the reset state successfully. If the RST bit of the ISR (Interrupt Status Register) is found to be set, the RTL8019AS is in reset state and we can continue. If the RTL8019AS fails to enter reset state, an endless loop in the Easy Ethernet W firmware is entered that informs the user that the initialization process has failed via the Easy Ethernet W's serial port.

The RST bit isn't the only bit in the ISR we'll be using in the Easy Ethernet W firmware. So, I decided to show you the ISR layout in Figure 12.3.

ISR

7	6	5	4	3	2	1	0
RST	RDC	CNT	OVW	TXE	RXE	PTX	PRX

Figure 12.3: This is a register we wish we could have used in the Easy Ethernet CS8900A firmware.

```
//*******************************************************************
//*   Initialize the RTL8019AS
//*******************************************************************
   delay_ms(2);        // delay at least 1.6mS
   bit_clear(rst_pin);       // disable reset line
   read_creg(RSTPORT);  // read contents of reset port
   write_creg(RSTPORT,byte_read);    // do soft reset
   delay_ms(10);       // give it time
   read_creg(ISR);        // check for good soft reset
   if(!bit_test(byte_read,RST)){
      while(1){
      printf("INIT FAILED\n\r");
      }
```

Code Snippet 12.8: This is a very critical piece of code. If things don't go right here, we're dead in the water.

Code Snippet 12.8 uses the *read_creg* and *write_creg* functions to load and read the contents of the RTL8019AS's internal registers. The *read_creg* and *write_creg* functions are going to become your friends, as they will be used extensively in the Easy Ethernet W firmware. Remember that any terms that contain *creg* will be dealing with the RTL8019AS Control Registers you see listed in Code Snippet 12.3.

The *read_creg* function in Code Snippet 12.9 is easy to follow. First, the *fromcreg* macro puts the PIC16F877 I/O pins assigned as the databus in input mode. The register address is then loaded onto the PIC16F877's I/O pins doing the address bus duty. The RTL8019AS's IORB pin is activated and the RTL8019AS's register data is presented to the PIC16F877's databus port I/O pins, which are currently configured as inputs. The IORB pin is deactivated, and the byte that was just read from the RTL8019AS's register is returned in the variable *byte_read*.

```
//*******************************************************************
//*   Read From NIC Control Register
//*******************************************************************
int8 read_creg(int regaddr)
{
   fromcreg;
   cregaddr = regaddr;
   bit_clear(ior_pin);
   byte_read = input_d();
   bit_set(ior_pin);
   return(byte_read);
}
```

```
//*******************************************************************
//*   Write to NIC Control Register
//*******************************************************************
void write_creg(int regaddr, int regdata)
{
   cregaddr = regaddr;
   cregdata = regdata;
   tocreg;
   bit_clear(iow_pin);
   delay_cycles(1);
   bit_set(iow_pin);
   fromcreg;
}
```

Code Snippet 12.9: It's a bit easier to read and write the RTL8019AS registers versus the CS8900A-CQ registers.

The *write_creg* function in Code Snippet 12.10 is very similar to the *read_creg* function. Register data and address information is presented to the PIC16F877's data and address busses. The PIC16F877 databus pins are configured as outputs by the *tocreg* macro, and the RTL8019AS's IOWB pin is toggled. After the write operation is completed, the PIC16F877's databus I/O pins are reconfigured as inputs by the *fromcreg* macro. This is done to make sure the databus is free for other devices that may need to use it. The *delay_cycles(1)* function is native to the Custom Computer Services C Compiler and wastes time for 1 instruction cycle, which is 200 nS with the Easy Ethernet W's 20 MHz microcontroller clock.

```
//*******************************************************************
//*   Initialize the RTL8019AS
//*******************************************************************
   write_creg(CR,0x21);     // stop the NIC, abort DMA, page 0
   delay_ms(2);             // make sure nothing is coming in or going out
   write_creg(DCR,dcrval); // 0x58
   write_creg(RBCR0,0x00);
   write_creg(RBCR1,0x00);
   write_creg(RCR,rcrval);
   write_creg(TPSR,txstart);
   write_creg(TCR,0x02);
   write_creg(PSTART,rxstart);
   write_creg(BNRY,rxstart);
   write_creg(PSTOP,rxstop);
   write_creg(CR,0x61);
   write_creg(CURR,rxstart);
   for(i=0;i<6;++i)
      write_creg(PAR0+i, MYMAC[i]);
```

```
show_regs();
   while(1);
```

Code Snippet 12.10: The show_regs() *function is a way to see inside the RTL8019AS's Control Register Bank.*

We briefly mentioned the RTL8019AS's CR register (Command Register) and I warned you that you would become very familiar with it before we finish this project. Well, here it is again in Figure 12.4 and we're writing a value to it in the first line of Code Snippet 12.10.

CR

7	6	5	4	3	2	1	0
PS1	PS0	RD2	RD1	RD0	TXP	STA	STP

Figure 12.4: Curiously, the STA bit does nothing as far as starting and stopping the RTL8019AS.

The first line of code in Code Snippet 12.10 is writing a 0x21 to the RTL8019AS's CR register. Let's break 0x21 down into binary (00100001) and superimpose the binary mask on the CR register's bits.

CR

7	6	5	4	3	2	1	0
PS1	PS0	RD2	RD1	RD0	TXP	STA	STP
0	0	1	0	0	0	0	1

Taking it from left to right, the PS1 and PS0 bits are called Page Select bits. The CR register is common to all three NE2000-compatible Register Pages and the fourth test page that contains data that is unique to the RTL8019AS. Don't worry about the fourth Register Page right now as it is manipulated from an external EEPROM and used for performing tests on the RTL8019AS. Once you have a good grasp of the inner workings of the RTL8019AS you can fiddle with the bits in that fourth page.

Getting back to the task at hand, PS1 and PS0 are both 0 (zero) and point to Register Page 0. If you look once again at Code Snippet 12.4, you'll see that I've assigned a page number to some of the RTL8019AS register definitions. The Page Select bits in the CR register determine which RTL8019AS Register Page is addressed, and the definitions in Code Snippet 12.3 are used by the *read_creg* and *write_creg* functions to select the register inside the Register Page that is to be addressed.

RD2, RD1 and RD0 are the Remote DMA Command bits. The '1' in the RD2 position commands the RTL8019AS to abort any DMA activity that may be in progress. This effectively stops packet generation and reception. We want the RTL8019AS to be focused on the initialization process and not out there trying to receive or transmit packets on the network.

The TXP bit must be set to initiate a transmission. We're trying to reset the RTL8019AS and so far we've issued an abort DMA command, which resets this bit internally. We don't want this bit to be set right now as we're not in any position to begin any type of transmission process.

STA is short for START. As far as the RTL8019AS is concerned, this bit actually does nothing at all. The logic level of the STOP bit controls starting and stopping the RTL8019AS.

STP is not the super slick lubricant you buy at auto stores. STP is short for STOP. When active, this bit takes the RTL8019AS offline. No packets will be received or transmitted by the RTL8019AS. If any reception or transmission is in progress, it will continue to completion before the reset state is entered.

To exit the stopped state, the STP bit must be cleared and the STA bit must be set. To perform a software reset, the STP bit should be set high. Notice that we did not explicitly issue a command to set the STP bit when we performed our soft reset earlier. That's because the STP defaults to a high state upon RTL8019AS power up. After executing commands to put the RTL8019AS into reset mode, we checked the validity of our software reset by checking the RST bit in the ISR register. Since it is possible for an operation to be in progress while we're issuing a STOP command, we wait an ample amount of time (*delay_ms(2))* to make sure that nothing is coming into or going out of the RTL8019AS.

After killing some time to ensure tranquility between the RTL8019AS and any attached network, we begin a series of *write_creg* instructions that are called out in this sequence by the National Semiconductor® NE2000 software documentation. The first, *write_creg(DCR,dcrval)*, loads the DCR (Data Configuration Register) register with 0x58.

DCR

7	6	5	4	3	2	1	0
-	FT1	FT0	ARM	LS	LAS	BOS	WTS
0	1	0	1	1	0	0	0

Figure 12.5: I've superimposed the 0x58 onto the description of the DCR's bits.

Bit 7 is not used, and we begin our discussion of the DCR with the FIFO Threshold Select bits. FT1 and FT0 determine how many bytes are in the RTL8019AS's FIFO before a call to the Local DMA engine is made. We are set for 8 bytes. The FIFO receive threshold ranges from 2 bytes to 12 bytes depending on the bit pattern set in FT1 and FT0.

The ARM bit is set and that allows us to use the Send Packet command and auto-initialize the Remote DMA to extract a frame from the Buffer Ring. Don't get confused by using the term "send" in a receive operation. Remember the RTL8019AS's Remote DMA channel moves data into and out of the RTL8019AS buffer queue and literally sends the data to an external microcontroller's memory.

I'm going to get ahead of us a bit with this explanation of the significance of the ARM bit, but we'll cover the now "unknown" registers and concepts later and all of this will make sense then. To help the words, I've put up Figure 12.6 for you.

By setting the ARM (Auto-Initialize Remote) bit in the DCR, the RTL8019AS's Remote DMA channel can be automatically initialized to transfer a single frame from the Receive Buffer Ring. The PIC16F877 begins this automated transfer by issuing a Send Packet command (writing a 0x1A to the CR Register). The RTL8019AS's DMA engine will be initialized to the value contained in the Boundary Pointer Register. Remember that the boundaries fall on 256 byte intervals 0x0100). Therefore, the lower byte of the Remote Start Address will always be 0x00. The Remote Byte Count Register pair (RBCR0, RBCR1) is initialized to the value held in the Receive Byte Count fields found in the frame's 4-byte Buffer Header, which are the same bytes inside the Easy Ethernet W's pageheader array (*enetpacketLenL and enetpacketLenH)*. After the data is transferred, the RTL8019AS Boundary Pointer is advanced to allow the RTL8019AS buffers to be used for new incoming frames. The Remote Read will terminate when the Byte Count reaches zero. When the current Remote Read operation is complete, the RTL8019AS's Remote DMA engine is then prepared to read the next frame from the RTL8019AS's Receive Buffer Ring. If the RTL8019AS's DMA pointer crosses the Page Stop Register, it is reset to the Page Start Address. This allows the RTL8019AS's Remote DMA engine to remove frames that have wrapped around to the top of the RTL8019AS's Receive Buffer Ring.

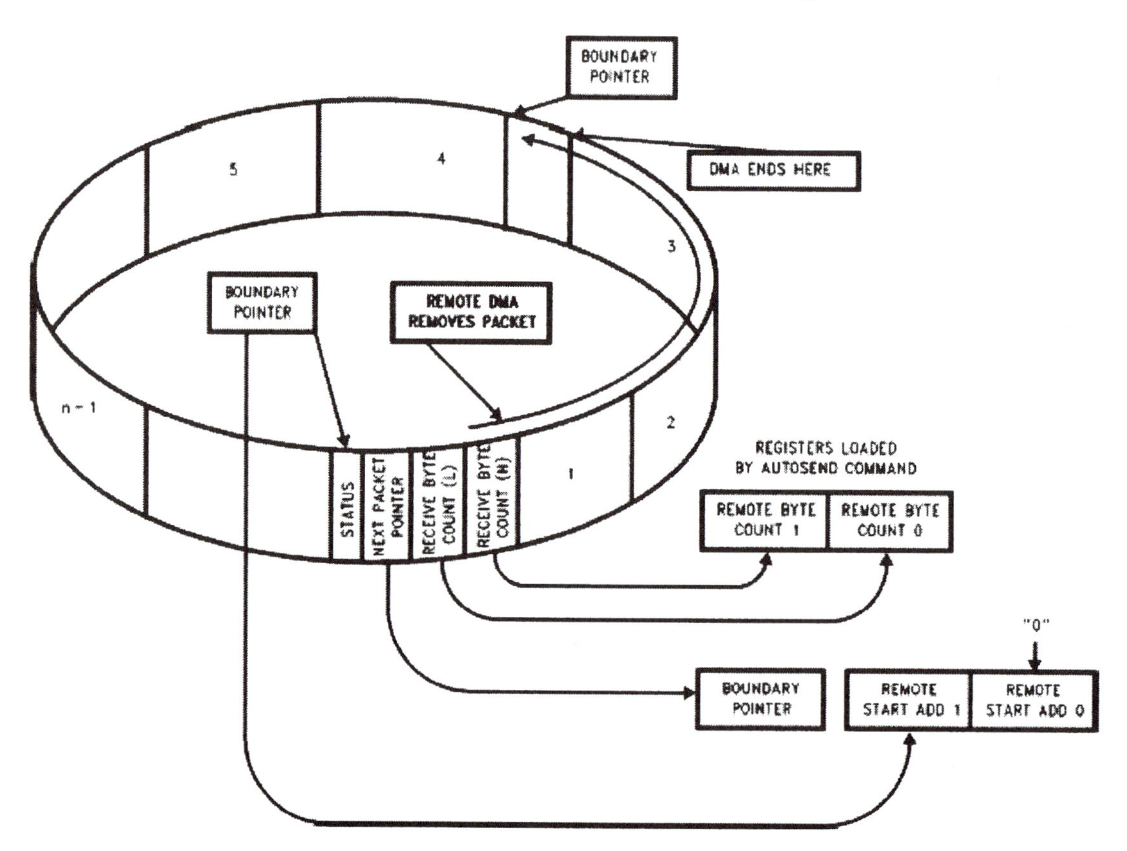

Figure 12.6: This drawing really helped me when I was first introduced to the RTL8019AS. Note again the loose use of the term "packet."

Getting back to our look at the bits we've loaded into the DCR, the LS (Loopback Select) bit is set at this point to allow the RTL8019AS to operate in normal mode. The LAS bit selects 32-bit or 16-bit dual DMA mode. The RTL8019AS requires that the LAS (Long Address Select) bit be cleared as the RTL8019AS only supports dual 16-bit DMA mode. The DCR is also used to program the RTL8019AS to operate in 8-bit or 16-bit DMA mode. The Easy Ethernet W needs to run in 8-bit DMA mode and this is accomplished by clearing the WTS (Word Transfer Select) bit in the DCR.

The DCR must be initialized prior to loading the Remote Byte Count Registers. Now that the DCR work is finished, we can safely clear both the RBCR0 and RBCR1 registers.

The RCR (Receiver Configuration Register) determines what packets to accept and whether or not to store them in the RTL8019AS's receive queue. We know from our CS8900A-CQ experience that we want to receive broadcast packets, and we need to make sure any packets other than broadcast packets handled by our RTL8019AS have the Easy Ethernet W's hardware address embedded within them. Our Easy Ethernet W code writes the value assigned to *rcrval* (0x04) to the RCR, and I've added the bit mask to the RCR graphic in Figure 12.7.

RCR

7	6	5	4	3	2	1	0
-	-	MON	PRO	AM	AB	AR	SEP
0	0	0	0	0	1	0	0

Figure 12.7: You can do some really cool things by bending the rules with bits in this register.

To effect our desired receiver configuration, we must set the AB (Accept Broadcast) bit and clear the PRO (Promiscuous Physical) bit. Putting the RTL8019AS into promiscuous mode allows any packet to be received regardless of its address. You already know that a Broadcast Packet is addressed in the Ethernet DLC area as *FF FF FF FF FF FF.*

To reject packets with errors, the SEP (Save Errored Packets) bit is cleared. We aren't interested in dealing with runt packets. So, the AR (Accept Runt Packets) bit is cleared to reject packets that are less than 64 bytes in length. The RTL8019AS on the Easy Ethernet W will not be using any multicast addressing schemes. So, the AM (Accept Multicast) bit is also cleared. And, we want to buffer any valid incoming frames. To enable frame buffering, the MON (Monitor Mode) bit is cleared by the Easy Ethernet W firmware.

The next line of code in Code Snippet 12.10 loads the predefined *txstart* value (0x40) into the TPSR (Transmit Page Start Register) shown in Figure 12.8.

TPSR

7	6	5	4	3	2	1	0
A15	A14	A13	A12	A11	A10	A9	A8
0	1	0	0	0	0	0	0

Figure 12.8: This 0x40 translates to a Transmit Page Start address of 0x4000.

The RTL8019AS datasheet tells us point blank to not exceed 0x60 as a PSTOP register value in 8-bit mode. That means our Buffer Ring memory area cannot go beyond physical address of 0x6000. The RTL8019AS has 8 Kb of on-chip memory area. In days before the RTL8019AS, that 8 Kb of memory was not on-chip, and the Buffer Ring addresses were dictated by the hardware. I took a look at the National Semiconductor NE2000 code and concluded that an effort was made to standardize the Buffer Ring addressing. I've included all of the original National Semiconductor documentation on the CD-ROM for you. With that little tidbit of history, let's work out our Buffer Ring strategy with the numbers we have been given.

We already have determined that PSTOP will be set for 0x60. Remember that this is the upper byte of a 16-bit address that always falls on a 256-byte boundary (0x0100). If you look ahead in Code Snippet 12.8, you'll see the PSTOP register is loaded with the *rxstop* value (0x60). Now that we have the top of our Buffer Ring, let's compute backwards towards the bottom. We know that we have 8 Kb of Buffer Ring area available on-chip with the RTL8019AS. If the maximum Buffer Ring memory address is 0x6000, and we subtract 8 Kb (0x2000) from the maximum address, that will give us our bottom Buffer Ring address, which is 0x4000 or PSTART. In the tradition of NE2000, we'll allocate the transmit buffer first. Since we only use the upper byte to designate the address, that puts 0x40 (the pre-defined value of *txstart*) in the TPSR.

While we're working on the transmit side of the RTL8019AS house, we need to make sure that the RTL8019AS's CRC checking and generation processes are enabled. The CRC functions are enabled in normal mode by clearing the CRC bit in the TCR (Transmit Configuration Register).

TCR

7	6	5	4	3	2	1	0
-	-	-	OFST	ATD	LB1	LB0	CRC
0	0	0	0	0	0	1	0

Figure 12.9: The loopback disables the RTL8019AS's Local DMA.

The RTL8019AS must be put into loopback mode before we set the remaining Buffer Ring pointers. Setting the LB0 bit in the TCR is Figure 12.9 puts the RTL8019AS in internal loopback mode. Now that we've stopped the RTL8019AS's Local DMA, we can set the PSTART, PSTOP and BNRY pointers.

Only one transmit frame will be present at any time in the RTL8019AS transmit buffer. Therefore, we only need enough space for one complete frame or 1518 bytes. 1518 bytes works out to 0x05EE, which doesn't fall on a 256-byte boundary. We can't round down. So, the next 256-byte boundary following 0x05EE is 0x0600, which will be the beginning address of the RTL8019AS's receive buffer relative to the beginning of the Buffer Ring. Doing a bit more memory map math gives us the starting address of the receive buffer:

TPSR (0x4000) + size of one Ethernet frame (0x0600) = Receive buffer starting address (0x4600)

Looking back at Figure 12.4, we know that the PSTART and BNRY pointers should be set to point at address 0x4600. So, we load the PSTART and BNRY registers with 0x46, which just happens to be the predefined value of *rxstart* in the Easy Ethernet W source code. We must also initialize the CURR pointer, which lies in Page 1. Writing 0x61 to the CR flips us into Page 1 where we write the *rxstart* value to the CURR register.

Remember the flip-flop we had to do in our Easy Ethernet CS8900A code to write the MAC address into the CS8900A-CQ? Well, the concept is the same for the RTL8019AS, but the actual code is nothing more than a *for* loop pushing the contents of the *MYMAC[]* array into the RTL8019AS's Physical Address Registers (PAR0:PAR5), which also happen to reside in Page 1.

We are just a C statement away from activating our Easy Ethernet W and putting it online. However, some of you see one C statement in particular that would prevent the RTL8019AS activation and, for that matter, any more code execution; *while(1)*. For those of you that may be C challenged, the *while(1)* statement is an endless loop statement.

I've halted the program execution here to show you a tool that is included in the Easy Ethernet W's firmware called *show_regs()*. We can't see into the RTL8019AS's Control Registers with the MPLAB ICE 2000 or the Sniffer. So, I wrote a simple function to dump the RTL8019AS Control Registers to a HyperTerminal window using the Easy Ethernet W's serial port. The *show_regs* code suite in Code Snippet 12.11 is quite simple.

```
//******************************************************************
//*   Converts Binary to Displayable Hex Characters
//*   ie.. 0x00 in gives 0x30 and 0x30 out
//******************************************************************
void bin2hex(binchar)
{
   high_nibble = (binchar & 0xF0) / 16;
   if(high_nibble > 0x09)
      high_char = high_nibble + 0x37;
   else
      high_char = high_nibble + 0x30;

   low_nibble = (binchar & 0x0F);
   if(low_nibble > 0x09)
      low_char = low_nibble + 0x37;
   else
      low_char = low_nibble + 0x30;
}
//******************************************************************
//*   Read/Write for show_regs
//*   This routine reads a NIC register and dumps it out to the
//*   serial port as ASCII.
//******************************************************************
```

```
//
void readwrite()
{
   read_creg(i);
   bin2hex(byte_read);
   printf("\t%c%c",high_char,low_char);
}
//**************************************************************
//*   Displays Control Registers in Pages 0, 1, 2 and 3
//*   This routine dumps all of the NIC internal registers
//*   to the serial port as ASCII characters.
//**************************************************************
void show_regs()
{
   write_creg(CR,0x21);
   cls();
   printf("\r\n");
   printf("    Realtek 8019AS Register Dump\n\n\r");
   printf("REG\tPage0\tPage1\tPage2\tPage3\n\r");

   for(i=0;i<16;++i)
   {
      bin2hex((int8) i);
      printf("%c%c",high_char,low_char);
      write_creg(CR,0x21);
      readwrite();
      write_creg(CR,0x61);
      readwrite();
      write_creg(CR,0xA1);
      readwrite();
      write_creg(CR,0xE1);
      readwrite();
      printf("\r\n");
   }
}
```

Code Snippet 12.11: Nothing to it…the code reads each register in each page and converts the binary register data to ASCII. The ASCII data is then displayed in an easy to read format.

Let's follow the order of register writes in Code Snippet 12.10 and see if we can find them in the *show_regs* screen capture in Figure 12.10.

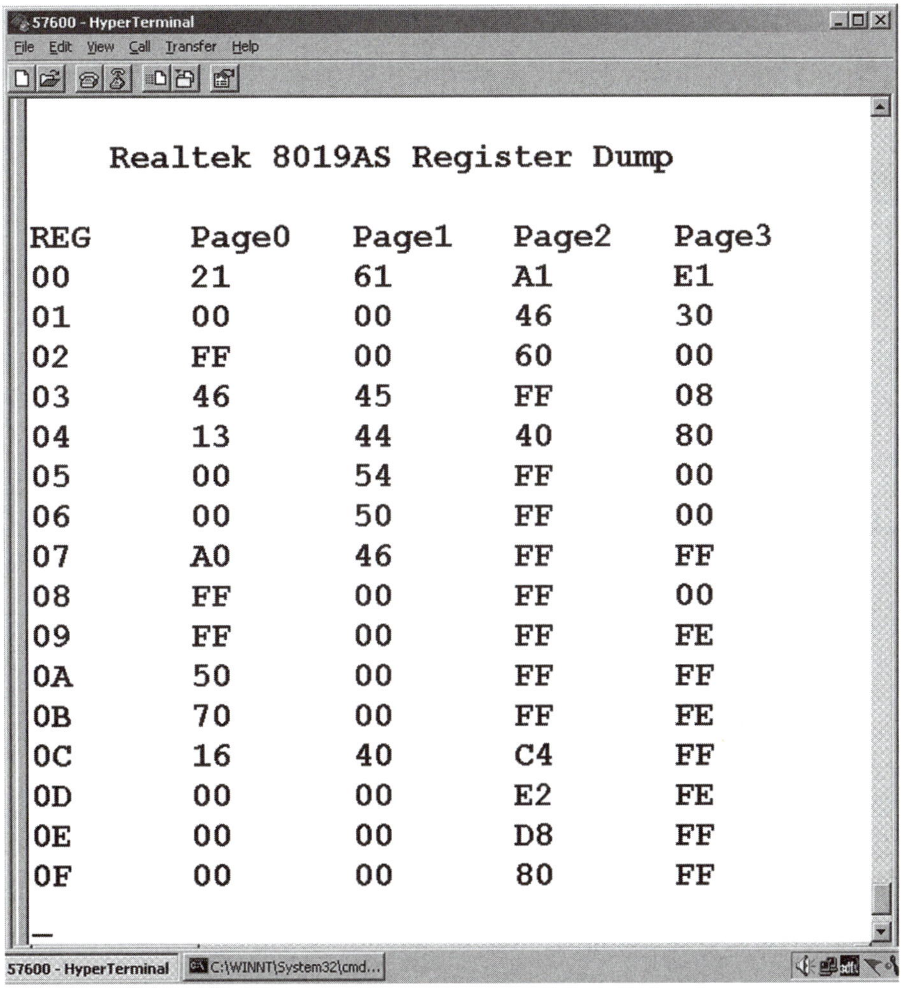

57600 - HyperTerminal

File Edit View Call Transfer Help

Realtek 8019AS Register Dump

REG	Page0	Page1	Page2	Page3
00	21	61	A1	E1
01	00	00	46	30
02	FF	00	60	00
03	46	45	FF	08
04	13	44	40	80
05	00	54	FF	00
06	00	50	FF	00
07	A0	46	FF	FF
08	FF	00	FF	00
09	FF	00	FF	FE
0A	50	00	FF	FF
0B	70	00	FF	FE
0C	16	40	C4	FF
0D	00	00	E2	FE
0E	00	00	D8	FF
0F	00	00	80	FF

57600 - HyperTerminal C:\WINNT\System32\cmd...

Figure 12.10: It may be helpful to reference the Control Register offsets in Code Snippet 12.3 while searching for the truth in this screen capture.

The CR (REG 00) spans across all of the RTL8019AS register pages. Each entry under a PageX header is the command given to go to that particular page. Let's see if we can find the DCR, which should contain a value of 0x58. The DCR is located at offset 0x0E in Page 0. There's 0x00 there. What gives? This is the quirk in the RTL8019AS Control Register set. The DCR is write-only at offset 0x0E in Page 0. To read the contents of the DCR we must travel to offset 0x0E in Page 2. The DCR value is read-only in Page 2. Found it? The value still isn't 0x58, is it?

DCR

7	6	5	4	3	2	1	0
-	FT1	FT0	ARM	LS	LAS	BOS	WTS
0	1	0	1	1	0	0	0

Figure 12.11: This value was written to the DCR.

Figure 12.11 is a reminder of what was written into the DCR before we ran the *show_regs* function. Bit 7 of the DCR is an unused bit. Let's perform a simple logical AND operation using the value we wrote to the DCR and the DCR value we read:

We wrote: 01011000 0x58

We read: 11011000 0xD8

Result: 01011000 0x58

Bit 7 of the DCR is a "don't care" bit and if we ignore it, we find our original value is stored in the DCR.

Now, let's find the RBCR0 and RBCR1 and check their values against what we loaded in the Easy Ethernet W source code. These registers should be located at offsets 0x0A and 0x0B, respectively, in Page 0. 0x50 and 0x70 don't look like 0x00 to me. What you see in the Control Register dump are the RTL8019AS ID bytes. The read-only data at offsets 0x0A and 0x0B are the RTL8019AS ID bytes. The RBCR0 and RBCR1 registers are write-only at these same offsets and there are no read-only locations for these bytes.

Let's try the RCR at offset 0x0C in Page 0. What you see in Figure 12.10 at offset 0x0C in Page 0 is not the contents of the RCR. What you see is the contents of the RSR (Receive Status Register). The contents of the RCR are found at the same offset in Page 2.

RCR

7	6	5	4	3	2	1	0
-	-	MON	PRO	AM	AB	AR	SEP
0	0	0	0	0	1	0	0

Figure 12.12: Can you see why 0xC4 is the value we read from the RCR?

However, again the RCR value in Page 2 doesn't match up with the 0x04 we originally put into the RCR. Take a look at Figure 12.12 and apply our "don't care" bits policy we used to find the true value of the DCR. I'm sure you can resolve the 0x04 in the RCR value we read.

You get the idea, I'm sure. The only read/write registers in Page 0 are the CR, BNRY and ISR. We haven't written to the ISR yet, but we can check the BNRY value. Take a look at offset 0x03 in Page 0. The Easy Ethernet W source code in Code Snippet 12.10 reveals that we wrote the value assigned to *rxstart* (0x46) to the BNRY, CURR and PSTART registers. Scanning the Control Register screen capture for the value of 0x46 gives us:

BNRY	offset 0x03	Page 0
CURR	offset 0x07	Page 1
PSTART	offset 0x04	Page 2

We know that the Easy Ethernet W MAC address (00EDTP) is loaded in the PAR0:PAR5 registers beginning at offset 0x01 in Page 1. The sequence is found in the Control Register screen capture under the Page 1 header as *00 00 45 44 54 50.*

The *show_regs* function is not intended to be a run-time tool as it will corrupt the data and give false readings. However, the *show_regs* function coupled with a *while(1)* statement is a great debugging tool. If you want to continue proving out values, consult the RTL8019AS datasheet for a complete listing of the Control Registers. A copy of the latest Realtek RTL8019AS datasheet is included on the CD-ROM.

As far as the RTL8019AS Control Registers are concerned, we've thus far established a place for everything, and everything is in its place. Code Snippet 12.12 issues a START command to the RTL8019AS, clears the ISR, enables the packet received OK and overflow interrupts and takes the RTL8019AS out of loopback mode enabling the RTL8019AS's Local DMA channel. Houston, the Easy Ethernet W is online.

CR

7	6	5	4	3	2	1	0
PS1	PS0	RD2	RD1	RD0	TXP	STA	STP
0	0	1	0	0	0	1	0

ISR

7	6	5	4	3	2	1	0
RST	RDC	CNT	OVW	TXE	RXE	PTX	PRX
1	1	1	1	1	1	1	1

IMR

7	6	5	4	3	2	1	0
-	RDC	CNT	OVW	TXE	RXE	PTX	PRX
-	0	0	1	0	0	0	1

TCR

7	6	5	4	3	2	1	0
-	-	-	OFST	ATD	LB1	LB0	CRC
0	0	0	0	0	0	0	0

Figure 12.13: The register values match up with the variables in Code Snippet 12.12.

```
//*******************************************************************
//*   Initialize the RTL8019AS
//*******************************************************************
   write_creg(CR,0x22);
   write_creg(ISR,0xFF);
   write_creg(IMR,imrval);
   write_creg(TCR,tcrval);
```

Code Snippet 12.12: This is akin to an automobile burnout. The accelerator is floored in the CR statement (the START command is given), and the brake is lifted in the TCR statement (the RTL8019AS Local DMA is activated).

Online with the Easy Ethernet W

It's all downhill from here. With Ethernet fundamentals under our belt, the only things we'll have to tackle to bring the Easy Ethernet W to Internet-protocol life are the differences in the way the RTL8019AS interfaces logically and physically. We still have to use the PIC16F877 microcontroller to sense the presence of a valid frame in the RTL8019AS Buffer Ring. The PIC must also initiate the fetch go get the data from the RTL8019AS's receive Buffer Ring to the PIC16F877's memory.

```
//*******************************************************************
//*   Look for a frame in the receive buffer ring
//*******************************************************************
   while(1)
   {
      //start the NIC
      write_creg(CR,0x22);

      //wait for a good frame
      while(!bit_test(INT0));

      //read the interrupt status register
      read_creg(ISR);

      //if the receive buffer has been overrun
      if(bit_test(byte_read,OVW))
         overrun();

      //if the receive buffer holds a good frame
      if(bit_test(byte_read,PRX))
         get_frame();
```

Code Snippet 12.13: Once the get_frame function is called by the Easy Ethernet W's PIC16F877 microcontroller, when compared to the Easy Ethernet CS8900A, there isn't very much difference in what happens next.

The presence of the *while(1)* tells us that this code runs forever, exiting only to retrieve or send a packet. The Easy Ethernet W frame retrieval code begins by starting the RTL8019AS. The RTL8019AS's INT0 line is polled. When INT0 goes high, the PIC16F877 reads the RTL8019AS's ISR. If the overrun bit (OVW) of the ISR is set, the *overrun* function is called.

The *overrun* function is actually a sequence called out by the NE2000 documentation that attempts to recover any latent frames that may be usable if the receive Buffer Ring blows up. We won't go into detail about the *overrun* function here. You can read about it in the NE2000 documentation I've included on the CD-ROM if you're interested in the details.

If the ISR's PRX (Packet Received OK) bit is set, the *get_frame* function is called. 0x1A is written to the CR to issue the Send Packet command. Since we "ARM"ed the RTL8019AS in the initialization procedure, the automatic Remote DMA functionality of the RTL8019AS is put into action to retrieve a frame from the receive Buffer Ring.

```
//******************************************************************
//*   Receive Ring Buffer Header Layout
//*   This is the 4-byte header that resides in front of the
//*   data packet in the receive buffer.
//******************************************************************
int8  pageheader[4];
#define  enetpacketstatus  0x00
#define  nextblock_ptr  0x01
#define  enetpacketLenL 0x02
#define  enetpacketLenH 0x03
//******************************************************************
//*   REALTEK CONTROL REGISTER OFFSETS
//*   All offsets in Page 0 unless otherwise specified
//******************************************************************
#define RDMAPORT  0X10
#define RSTPORT   0x18

int8 byte_read;
int16 i,rxlen;
//******************************************************************
//*   Get A Frame From the Ring
//*   This routine removes an Ethernet frame from the receive buffer
//*   ring.
//******************************************************************
void get_frame()
{
   //execute Send Packet command to retrieve the packet
   write_creg(CR,0x1A);
   for(i=0;i<4;++i)
      {
         read_creg(RDMAPORT);
```

```
            pageheader[i] = byte_read;
        }
            rxlen = make16(pageheader[enetpacketLenH],pageheader[enetpacketLenL]);
            for(i=0;i<rxlen;++i)
                {
                    read_creg(RDMAPORT);
                    //dump any bytes that will overrun the receive buffer
                    if(i < 96)
                        packet[i] = byte_read;
                }
    while(!bit_test(byte_read,RDC))
        read_creg(ISR);

    write_creg(ISR,0xFF);

    //process an ARP packet
    if(packet[enetpacketType0] == 0x08 && packet[enetpacketType1] == 0x06)
    {
        if(packet[arp_hwtype+1] == 0x01 &&
        packet[arp_prtype] == 0x08 && packet[arp_prtype+1] == 0x00 &&
        packet[arp_hwlen] == 0x06 && packet[arp_prlen] == 0x04 &&
        packet[arp_op+1] == 0x01 &&
        MYIP[0] == packet[arp_tipaddr] &&
        MYIP[1] == packet[arp_tipaddr+1] &&
        MYIP[2] == packet[arp_tipaddr+2] &&
        MYIP[3] == packet[arp_tipaddr+3] )
        arp();
    }
    //process an IP packet
    else if(packet[enetpacketType0] == 0x08 && packet[enetpacketType1] == 0x00
            && packet[ip_destaddr] == MYIP[0]
            && packet[ip_destaddr+1] == MYIP[1]
            && packet[ip_destaddr+2] == MYIP[2]
            && packet[ip_destaddr+3] == MYIP[3])
    {
        if(packet[ip_proto] == PROT_ICMP)
            icmp();
        else if(packet[ip_proto] == PROT_UDP)
            udp();
        else if(packet[ip_proto] == PROT_TCP)
            tcp();
    }

}
```

Code Snippet 12.14: As soon as the code starts to search for the frame type, the Easy Ethernet W's code looks exactly like the Easy Ethernet CS8900A's code.

Immediately following the Send Packet command, the Easy Ethernet W's code reads the 4-byte buffer header information into the *pageheader* array. The length of the buffered frame is computed and the bytes are read in from the RTL8019AS's Remote DMA port. The Remote DMA port is eight bytes long and begins at offset 0x10. Once all of the bytes have been transferred from the RTL8019AS receive Buffer Ring to the PIC16F877's *packet* array, the Easy Ethernet W waits for the Remote DMA done bit (RDC) in the ISR to signal the end of the Remote DMA operation. After verifying the end of the Remote DMA operation, writing 0xFF to the ISR clears the ISR register.

```
//*******************************************************************
//*   Look for a frame in the receive buffer ring
//*******************************************************************
   //make sure the receive buffer ring is empty
   //if BNRY = CURR, the buffer is empty
      read_creg(BNRY);
      data_L = byte_read;
      write_creg(CR,0x62);
      read_creg(CURR);
      data_H = byte_read;
      write_creg(CR,0x22);
   //buffer is not empty.. get next frame
      if(data_L != data_H)
         get_frame();

   //reset the interrupt bits
   write_creg(ISR,0xFF);
```

Code Snippet 12.15: It's a simple heads-and-tails game to determine if there are any more frames to be retrieved from the RTL8019AS's receive Buffer Ring.

The RTL8019AS receive Buffer Ring can hold more than one frame. So, when the code returns to the *main* function, the BNRY and CURR pointers are examined. If the BNRY and CURR pointers are not the same value, the receive Buffer Ring contains another frame that needs to be transferred and the *get_frame* function is called again. If the BNRY and CURR pointers match, the RTL8019AS receive Buffer Ring is empty. The ISR is cleared, and the code begins again as shown in Code Snippet 12.15.

Sending a Frame using the Easy Ethernet W

All of the logic that applied to filling arrays and building frames with the Easy Ethernet W is identical to what we wrote into the Easy Ethernet CS8900A project. The only significant differences in the two sets of code are the CS8900A-CQ and RTL8019AS transmit and receive processes. You've just seen how a frame is received and transferred using the RTL8019AS Remote DMA engine. Now, let's examine the code and procedures needed to transmit a packet using the RTL8019AS.

```
//*******************************************************************
//*   Echo Packet Function
//*   This routine does not modify the incoming packet size and
//*   thus echoes the original packet structure.
//*******************************************************************
void echo_packet()
{
   write_creg(CR,0x22);
   write_creg(TPSR,txstart);
   write_creg(RSAR0,0x00);
   write_creg(RSAR1,0x40);
   write_creg(ISR,0xFF);
   write_creg(RBCR0,pageheader[enetpacketLenL] - 4 );
   write_creg(RBCR1,pageheader[enetpacketLenH]);
   write_creg(CR,0x12);

   txlen = make16(pageheader[enetpacketLenH],pageheader[enetpacketLenL]) - 4;
   for(i=0;i<txlen;++i)
      write_creg(RDMAPORT,packet[enetpacketDest0+i]);

   byte_read = 0;
   while(!bit_test(byte_read,RDC))
      read_creg(ISR);

   write_creg(TBCR0,pageheader[enetpacketLenL] - 4);
   write_creg(TBCR1,pageheader[enetpacketLenH]);
   write_creg(CR,0x24);
}
```

Code Snippet 12.16: The basic rules of packet transmission for the RTL8019AS are found in this code.

After making sure the RTL8019AS is started (loading CR with 0x22), the TPSR must be initialized. This sets the beginning of the transmit buffer area. Since we will be transferring data from the Easy Ethernet W's PIC16F877 microcontroller to the RTL8019AS's transmit buffer area, we must set up the Remote DMA start address registers, RSAR0 and RSAR1, with the address of the transmit buffer and tell the Remote DMA engine how many bytes to transfer in the RBCR0 and RBCR1 register pair. In the meantime, we've also cleared the ISR to make sure we can detect the end of the Remote DMA operation.

Now that the RTL8019AS transmit buffer is defined and the Remote DMA knows how many bytes to move and where to put them, a Remote Read command (writing 0x12 to the CR) is issued. The RTL8019AS's Remote DMA port is bidirectional. So, using the same Remote DMA port we used to transfer bytes from the RTL8019AS to the PIC16F877, we write the number of bytes loaded into the RBCR0 and RBCR1 register set to the RTL8019AS's transmit buffer. Remember that the first 4 bytes in the RTL8019AS frame buffer are the pageheader bytes and must not be counted in our transfer length calculations.

When the PIC16F877 has transferred all of the required bytes, the ISR's RDC bit is interrogated to verify the completion of the Remote DMA operation. Even though we've specified the TPSR and loaded the RTL8019AS's transmit buffer, we haven't sent anything from the RTL8019AS's MAC engine yet. So, we must load the TBCR0 and TBCR1 (Transmit Byte Count Registers) and issue a Remote Write with the TXP bit set to the CR (0x24). The command (0x24) kicks off the Local DMA and bytes are transferred from the RTL8019AS's transmit buffer area to the FIFO through the MAC engine and out to the network. If you check the Easy Ethernet W's TCP/IP code in Code Snippet 12.17, you'll see that the transmit operation is logically identical to the *echo_packet* function used by UDP and ICMP.

```
//*******************************************************************
//*   Send TCP Packet
//*   This routine assembles and sends a complete TCP/IP packet.
//*   40 bytes of IP and TCP header data is assumed.
//*******************************************************************
void send_tcp_packet()
{
   //count IP and TCP header bytes.. Total = 40 bytes
   ip_packet_len = 40 + tcpdatalen_out;
   packet[ip_pktlen] = make8(ip_packet_len,1);
   packet[ip_pktlen+1] = make8(ip_packet_len,0);
   setipaddrs();

   data_L = packet[TCP_srcport];
   packet[TCP_srcport] = packet[TCP_destport];
   packet[TCP_destport] = data_L;
   data_L = packet[TCP_srcport+1];
   packet[TCP_srcport+1] = packet[TCP_destport+1];
   packet[TCP_destport+1] = data_L;

   assemble_ack();
   set_packet32(TCP_seqnum,my_seqnum);

   packet[TCP_hdrflags+1] = 0x00;
   ACK_OUT;
   if(finflag)
   {
      FIN_OUT;
      finflag = 0;
   }

   packet[TCP_cksum] = 0x00;
   packet[TCP_cksum+1] = 0x00;
```

```
    hdr_chksum =0;
    hdrlen = 0x08;
    addr = &packet[ip_srcaddr];
    cksum();
    hdr_chksum = hdr_chksum + packet[ip_proto];
    tcplen = ip_packet_len - ((packet[ip_vers_len] & 0x0F) * 4);
    hdr_chksum = hdr_chksum + tcplen;
    hdrlen = tcplen;
    addr = &packet[TCP_srcport];
    cksum();
    chksum16= ~(hdr_chksum + ((hdr_chksum & 0xFFFF0000) >> 16));
    packet[TCP_cksum] = make8(chksum16,1);
    packet[TCP_cksum+1] = make8(chksum16,0);

    txlen = ip_packet_len + 14;
    if(txlen < 60)
       txlen = 60;
    data_L = make8(txlen,0);
    data_H = make8(txlen,1);
    write_creg(CR,0x22);
    write_creg(TPSR,txstart);
    write_creg(RSAR0,0x00);
    write_creg(RSAR1,0x40);
    write_creg(ISR,0xFF);
    write_creg(RBCR0,data_L);
    write_creg(RBCR1,data_H);
    write_creg(CR,0x12);

    for(i=0;i<txlen;++i)
       write_creg(RDMAPORT,packet[enetpacketDest0+i]);

    byte_read = 0;
    while(!bit_test(byte_read,RDC))
       read_creg(ISR);

    write_creg(TBCR0,data_L);
    write_creg(TBCR1,data_H);
    write_creg(CR,0x24);
}
```

Code Snippet 12.17: Note the txlen code that ensures the outgoing packet is at least 60 bytes in length. This same piece of txlen code exists in the Easy Ethernet CS8900A firmware but is not really needed there, as the CS8900A-CQ can be instructed to automatically pad a potentially runt packet.

We can stick a fork in our discussion of the RTL8019AS. It's done. There's no need for any additional RTL8019AS Sniffer shots, as they would look exactly like their CS8900A-CQ counterparts. The inside-the-microcontroller views of the PIC16F877's *packet* array provided by the MPLAB ICE 2000 would also be identical for both the RTL8019AS and the CS8900A-CQ, as the data in the Ethernet frames is identical between the CS8900A-CQ and the RTL8019AS. The CS8900A-CQ and RTL8019AS TCP/IP and UDP applications perform identically, and the Test Panel application you saw in the CS8900A-CQ section of this book will work with the Easy Ethernet W in the exact same manner. As for the rest of the code, if it is bound to the PIC16F877 (checksum calculation, swapping IP and MAC addresses in the packet array, printing the Telnet banner, serial port operations, and so forth), it is identical code in both the Easy Ethernet CS8900A and Easy Ethernet W firmware.

You now have the skills necessary to implement two differing Ethernet IC technologies (RTL8019AS and CS8900A-CQ) over two types of Microchip microcontrollers (PIC16F877 and PIC18F452). Before we move on, there are a couple of tools I haven't mentioned that may help you with the development of your personal and unique Ethernet project.

Tools for Work and Play

I used the Sniffer and the MPLAB ICE 2000 extensively in my text to give you an in-depth look at how Ethernet is effected using small 8-bit microcontrollers. Before I had the luxury of owning a Sniffer or MPLAB ICE 2000, I used my programming powers to look into my work. You've already been introduced to the *show_regs* function. So, let me show you a function I wrote to mimic the MPLAB ICE 2000's ability to see into the PIC16F877's memory.

```
//*******************************************************************
//*   show_packet
//*   This routine is for diagnostic purposes and displays
//*   the Packet Buffer memory in the PIC.
//*******************************************************************
void show_packet()
{
   cls();
   printf("\r\n");
   data_L = 0x00;
   for(i=0;i<96;++i)
   {
      bin2hex(packet[i]);
      printf(" %c%c",high_char,low_char);
      if(++data_L == 0x10)
```

```
            {
                data_L = 0x00;
                printf("\r\n");
            }
    }
}
```

Code Snippet 12.18: This tool actually shows you the frame that is currently residing in the PIC16F877's packet *array.*

It is advantageous to be able to look at the contents of a frame in its raw form in the microcontroller memory. That's what the code in Code Snippet 12.18 does. Figure 12.14 is the fruit produced by Code Snippet 12.18. Let's see just how good you've gotten.

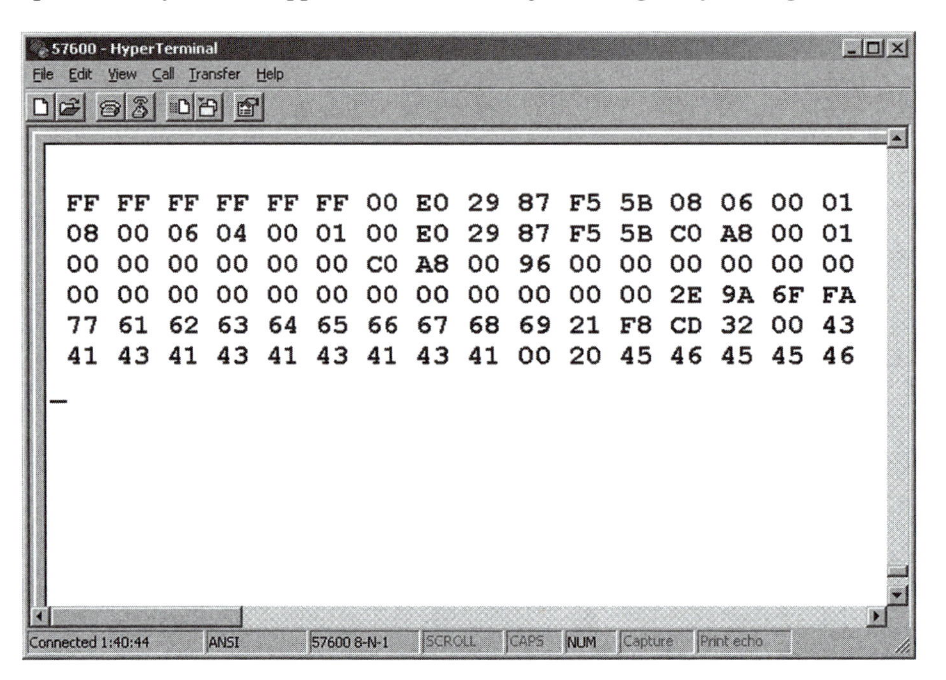

Figure 12.14: It doesn't have a fancy GUI interface, but it gives you what you want; a peek at the frame in the PIC16F877's memory.

Sure looks like a broadcast message to me. Those six bytes of *FF* in the DA should be very familiar to you. It also looks like I'm using that same personal computer to generate this frame. Yep, 00 E0 29 87 F5 5B matches up with the SA in Hex Dump 8.1. The next two bytes (08 06) tell us this is an ARP. Since the SA is a broadcast address you can bet this is an ARP request. Let's see if the Easy Ethernet W's MAC address is in there anywhere (00 00 45 44 54 50). Nope. Then that means it is an ARP request, and if we pegged the SA correctly the IP address should be 192.168.0.150 or C0 A8 00 01. It is. That's how I used to do it! After a while you get the feel for the locations of the fields, and by examining just a few fields you know what type of frame you're looking at.

There were times I needed to see the buffer area header that was captured into the PIC16F877's *pageheader* array. I wrote a small piece of code called aptly, *dump_header*, that you can see in Code Snippet 12.19.

```
//******************************************************************
//*   Dump Receive Ring Buffer Header
//*   This routine dumps the 4-byte receive buffer ring header
//*   to the serial port as ASCII characters.
//******************************************************************
void dump_header()
{
   cls();
   for(i=0;i<4;++i)
      {
         bin2hex(pageheader[i]);
         printf("\r\n%c%c",high_char,low_char);
      }
}
```

Code Snippet 12.19: The 4 bytes of the buffer header for the Easy Ethernet W are shown in Figure 12.15 and were generated with the hex dump shown in Figure 12.14. The first byte of the four is the Receive Status, and I've laid out the first byte's bit pattern in Figure 12.16. Let's figure out what these 4 bytes are telling us.

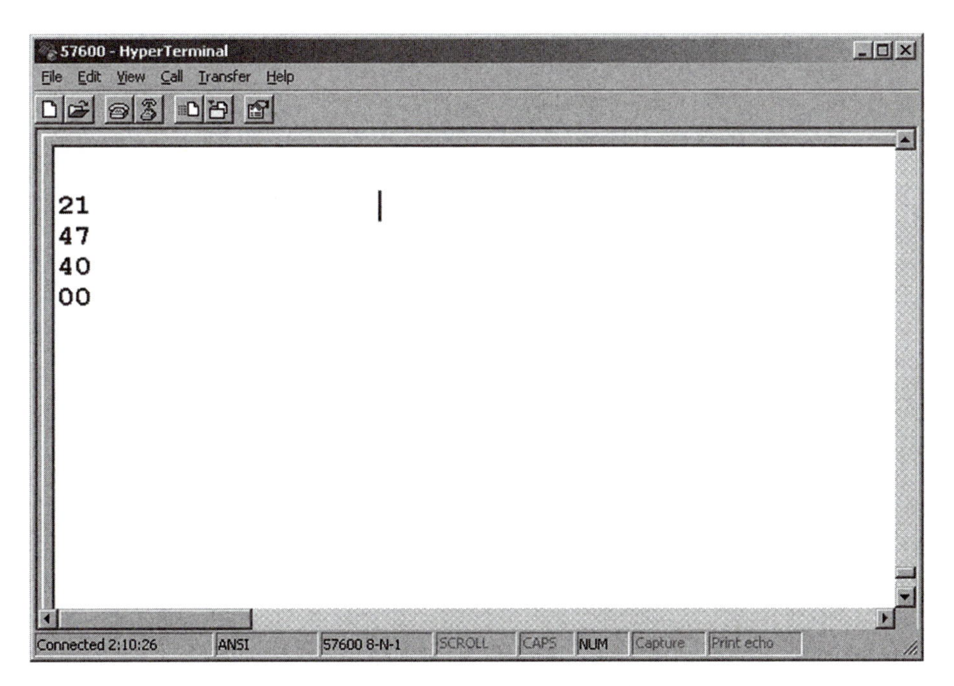

Figure 12.15: When you're in the debugging ditch, there's lots to be gained by knowing what's in these four little bytes.

The first byte is equivalent to the RSR (Receive Status Register). Looks like we have the PHY bit and the PRX bits set. The PHY bit being set tells us that the received packet contains a broadcast address. I know this to be true, as I generated the buffer header with a PING that generated an ARP request. The PRX bit says that the packet was received with no errors.

RSR

7	6	5	4	3	2	1	0
DFR	DIS	PHY	MPA	-	FAE	CRC	PRX
0	0	1	0	0	0	0	1

Fig 12.16: Interpreting this register can help dig you out of that debugging ditch.

The second byte of the buffer header points to the next buffer page. Remember that each buffer page is 256 bytes long and in hex that's 0x0100. Also, remember that each page falls on a 256-byte boundary and that we set the starting page at 0x46 when we initialized the RTL8019AS. You'll also remember that the 0x46 is the upper byte, and the lower byte must always fall on a 256-byte boundary. So, that give us a receive buffer starting address of 0x4600. Adding one page to 0x4600 gives us our next page of 0x4700, which is what the 0x47 represents.

The third byte of the buffer header is the low byte of the 16-bit length of the frame that is buffered. We must remember to subtract 4 from this value, as the 4 bytes of the buffer header are included in the count. So, 0x40 is equivalent to 64 decimal and taking into account our 4 bytes of header information, the frame is 60 bytes long. Anything coming in that is less than 60 bytes in length would be a runt, and the RTL8019AS has been instructed not to accept runt packets. The last of the buffer header bytes is the upper byte of the length value.

```
//******************************************************************
//*   Get A Frame From the Ring
//*   This routine removes an Ethernet frame from the receive buffer
//*   ring.
//******************************************************************
void get_frame()
{
   //execute Send Packet command to retrieve the packet
   write_creg(CR,0x1A);
   for(i=0;i<4;++i)
      {
         read_creg(RDMAPORT);
         pageheader[i] = byte_read;
      }
         rxlen = make16(pageheader[enetpacketLenH],pageheader
         [enetpacketLenL]);
         for(i=0;i<rxlen;++i)
```

```
            {
               read_creg(RDMAPORT);
               //dump any bytes that will overrun the receive buffer
               if(i < 96)
                  packet[i] = byte_read;
            }
   while(!bit_test(byte_read,RDC))
      read_creg(ISR);

   write_creg(ISR,0xFF);

show_packet();      //write the PIC16F877's packet array contents
    OR
dump_header();      // write the 4 receive buffer header bytes
while(1);           // wait here forever

   //process an ARP packet
   if(packet[enetpacketType0] == 0x08 && packet[enetpacketType1] == 0x06)
```

Code Snippet 12.20: Always use the while(1) statement to freeze the contents of the reads done by the show_packet or dump_header functions.

So, by using the software tools I've included in the Easy Ethernet W firmware, you can do some pretty heavy debugging by simply inserting *show_regs*, *show_packet* or *dump_header*, followed by a *while(1)* statement in the appropriate places in the Easy Ethernet W or Easy Ethernet CS8900A firmware.

CHAPTER 13

Putting the Easy Ethernet AVR Online

We've done wonderful things with Ethernet and Microchip's PICs. Now it's time for the Atmel AVR to shine. We have to use a totally different tool set and a completely different compiler package. However, we can still recycle 99% of everything we've done thus far, including the hardware.

This is going to be easy. All we have to do to get an Easy Ethernet AVR running is to replace the PIC16F877 or PIC18F452 on the Easy Ethernet W with an Atmel ATmega16 and change some coding behind the macros. Our new AVR-based project's schematic is shown in Schematic 13.1, and the fully assembled Easy Ethernet AVR can be seen in Photo 13.1.

Photo 13.1: Port C is left open to allow the attachment of the AVR JTAG ICE. You can also use Port C as a general purpose I/O.

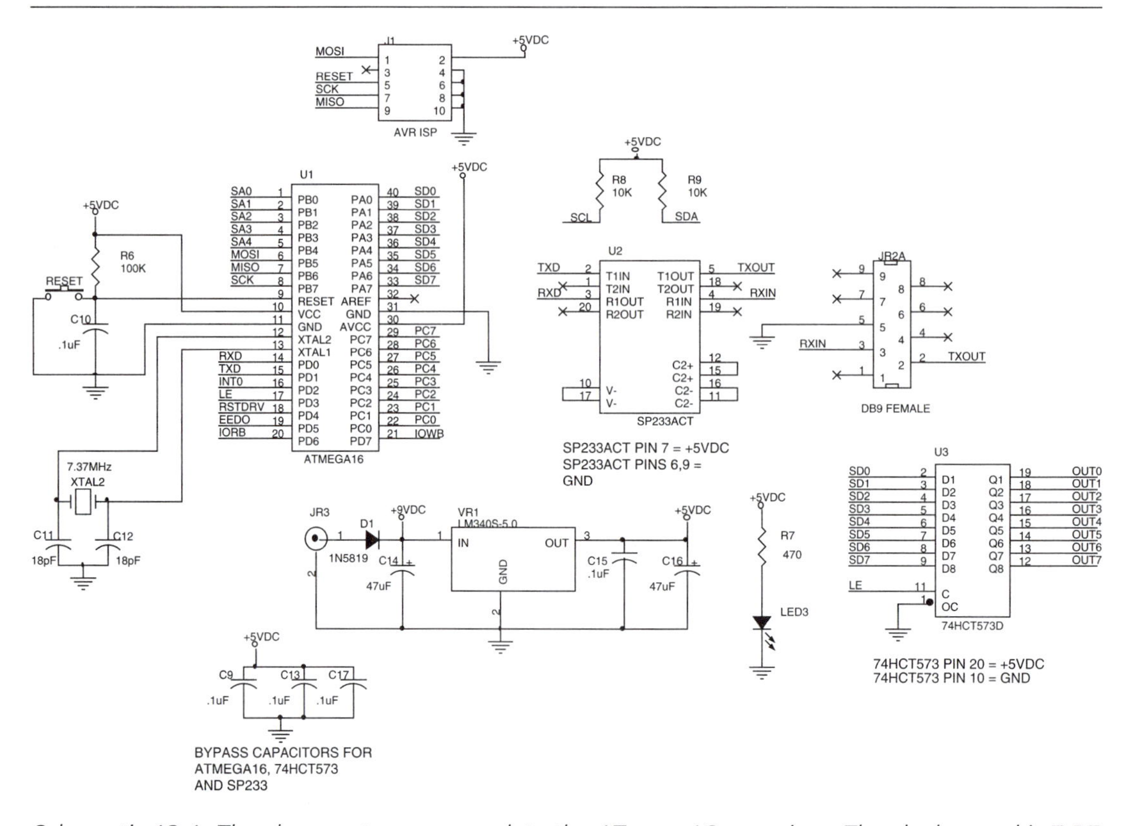

Schematic 13.1: The changes to accommodate the ATmega16 are minor. The clock speed is 7.37 MHz, and the AVR uses a 10-pin double-row header as the programming header instead of the RJ-11 phone jack programming interface on the Easy Ethernet CS8900A and Easy Ethernet W.

Most of the C statements will be identical to the Easy Ethernet W code. Some minor differences will have to be addressed, but most of the coding we will do will simply be adapting the logic to the new AVR hardware. For instance, the AVR uses a '1' to denote an I/O pin as an output. Conversely, the PIC uses a '1' to represent an I/O input pin. Let's take a look at Code Snippet 13.1 for an example.

```
//*********************************************************************
//*   Easy Ethernet W RTL8019AS I/O PORT DEFINITIONS
//*********************************************************************
#define  cregaddr    PORTB
#define  cregdata    PORTD
#define  tocreg      set_tris_D(0x00);
#define  fromcreg    set_tris_D(0xFF);
```

```
//******************************************************************
//*   Easy Ethernet AVR RTL8019AS DATA/ADDRESS PIN DEFINITIONS
//******************************************************************
#define  rtladdr      PORTB
#define  rtldata      PORTA
#define  tortl        DDRA = 0xFF
#define  fromrtl      DDRA = 0x00
//******************************************************************
//*   Initialize the RTL8019AS
//******************************************************************
void init_RTL8019AS()
{
   fromrtl;            // PORTA data lines = input
   PORTA = 0xFF;
   DDRB = 0xFF;
   rtladdr = 0x00;     // clear address lines
   DDRC = 0xFF;
   DDRD = 0xFA;        //setup IOW, IOR, EEPROM,RXD,TXD,CTS,LE

   PORTD = 0x05;       // enable pullups on input pins
```

Code Snippet 13.1: I've included some companion code from the Easy Ethernet W for comparison.

The Easy Ethernet AVR's *fromrtl* definition is equivalent to the Easy Ethernet W's *fromcreg* definition. Note that 0xFF is written to the port data direction register to indicate that the AVR I/O pins are output pins. That's opposite for the PIC controlling the Easy Ethernet W. Another difference found in the AVR is the ability to internally pull-up input pins by writing a '1' to the I/O pins that are designated as inputs.

Since most of the code changes occur behind the macros, let's look at some changes that must be made in Code Snippet 13.2.

```
//******************************************************************
//*   Easy Ethernet AVR PORT DEFINITIONS
//******************************************************************
#define iorwport   PORTD
//******************************************************************
//*   Easy Ethernet AVR RTL8019AS PIN DEFINITIONS
//******************************************************************
#define  iow_pin     0x80 //PORTD7 10000000
#define  rst_pin     0x10 //PORTD4 00010000
//******************************************************************
//*   Easy Ethernet W Initialize the RTL8019AS
//******************************************************************
bit_set(iow_pin)
bit_clear(rst_pin)
```

```
//**********************************************************************
//*   Easy Ethernet AVR Initialize the RTL8019AS
//**********************************************************************
#define set_iow_pin iorwport |= iow_pin
#define clr_rst_pin resetport &= ~rst_pin
```

Code Snippet 13.2: There are no AVR-specific built-in macros in the ImageCraft C Compiler package.

The *bit_set* and *bit_clear* functions used in the Easy Ethernet W are built into the Custom Computer Services C Compiler. To get the same functionality for the AVR, we have to build our own mini-macros. Let's translate the first AVR definition.

To set the RTL8019AS's IOWB pin, we code *set_iow_pin.* What really happens is:

iorwport = iorwport | iow_pin;

In effect, the AVR PORTD pin connected to the RTL8019AS's IOWB pin is set by ORing (|) PORTD with the *iow_pin* value (0x80). A look at Schematic 13.1 verifies the fact that the most significant bit of the ATmega16's PORTD is tied to the RTL8019AS IOWB pin.

To clear the reset pin (RSTDRV), we would code *clr_rst_pin,* which would equate to:

resetport = resetport & ~rst_pin;

The *rst_pin* value (0x10) is negated and becomes 0xEF. The negation flips the bit that represents the AVR *rst_pin* I/O pin. The AVR pin assigned to the RTL8019AS's RSTDRV will have no choice but to go low as ANDing (&) any value with a zero will result in zero. All of the set and clear functions used in the Easy Ethernet AVR firmware will follow the models we have just examined. Code Snippet 13.3 is a look at all of the set/clear mini-macros.

```
//**********************************************************************
//*   RTL8019AS 9346 EEPROM PIN DEFINITIONS
//**********************************************************************
#define  EEDO        0x20 //PORTD5 00100000
//**********************************************************************
//*   Flags
//**********************************************************************
#define synflag   0x01 //00000001
#define finflag   0x02 //00000010
#define hexflag   0x04 //00000100
#define synflag_bit  flags & synflag
#define finflag_bit  flags & finflag
#define hexflag_bit  flags & hexflag
//**********************************************************************
//*   PORT and LCD DEFINITIONS
//**********************************************************************
```

```
#define databus   PORTA
#define addrbus   PORTB
#define eeprom    PORTD
#define iorwport  PORTD
#define cport     PORTC
#define resetport PORTD
//******************************************************************
//*   RTL8019AS PIN DEFINITIONS
//******************************************************************
#define  ior_pin  0x40 //PORTD6 01000000
#define  iow_pin  0x80 //PORTD7 10000000
#define  rst_pin  0x10 //PORTD4 00010000
#define  INT0_pin 0x04 //PORTD2 00000100
#define  LE_pin      0x08 //PORTD3 00001000
//******************************************************************
//*   RTL8019AS PIN MACROS
//******************************************************************
#define set_ior_pin iorwport |= ior_pin
#define clr_ior_pin iorwport &= ~ior_pin
#define set_iow_pin iorwport |= iow_pin
#define clr_iow_pin iorwport &= ~iow_pin
#define set_rst_pin resetport |= rst_pin
#define clr_rst_pin resetport &= ~rst_pin
#define set_le_pin  iorwport |= LE_pin
#define clr_le_pin  iorwport &= ~LE_pin

#define set_cport_0 cport |= 0x01
#define set_cport_1 cport |= 0x02
#define set_cport_2 cport |= 0x04
#define set_cport_3 cport |= 0x08
#define set_cport_4 cport |= 0x10
#define set_cport_5 cport |= 0x20
#define set_cport_6 cport |= 0x40
#define set_cport_7 cport |= 0x80

#define clr_cport_0 cport &= ~0x01
#define clr_cport_1 cport &= ~0x02
#define clr_cport_2 cport &= ~0x04
#define clr_cport_3 cport &= ~0x05
#define clr_cport_4 cport &= ~0x10
#define clr_cport_5 cport &= ~0x20
#define clr_cport_6 cport &= ~0x40
#define clr_cport_7 cport &= ~0x80

#define clr_EEDO eeprom &= ~EEDO
#define set_EEDO eeprom |= EEDO
```

```
#define clr_synflag flags &= ~synflag
#define set_synflag flags |= synflag
#define clr_finflag flags &= ~finflag
#define set_finflag flags |= finflag

#define clr_hex flags &= ~hexflag
#define set_hex flags |= hexflag
```

Code Snippet 13.3: Using the AND NOT and OR logical operations, we can build an army of AVR bit set and clear mini-macros.

Millisecond and microsecond timing is another Custom Computer Services C Compiler built-in function that I took advantage of when using the PIC microcontroller for the Easy Ethernet W and Easy Ethernet CS8900A. Since the Ethernet code spans across three microcontrollers now, we must code some similar timing routines into our Easy Ethernet AVR firmware.

Before actually writing the code, one must figure out the values to put into the timing registers to get the desired timing intervals. Fortunately, a gentleman named Jack Tidwell wrote a very useful program called AVRCalc. As you can see in Figure 13.1, all I had to do was enter the AVR ATmega16 clock speed (7.37 MHz) and enter the timing interval I wanted (1 ms). The AVRCalc program does the rest. All I had to do was fill in the blanks in Code Snippet 13.4.

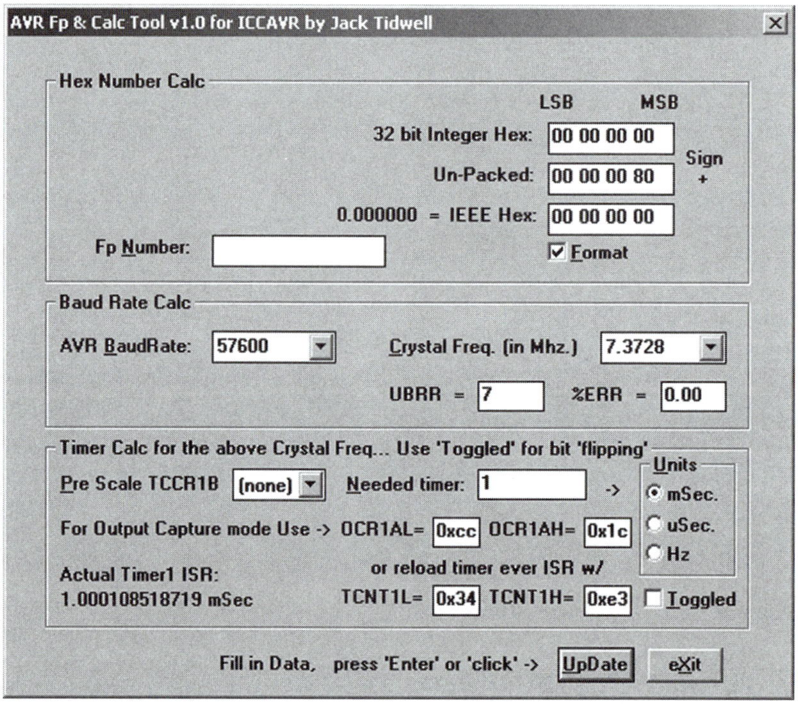

Figure 13.1: I use this tool quite often. It's great for "what if" work while you're developing code.

```
//*****************************************************************
//*   Delay millisecond Function
//*   This function uses Timer 1 and the A compare registers
//*   to produce millisecond delays.
//*
//*****************************************************************
void delay_ms(unsigned int delay)
{
   unsigned int i;
   OCR1AH = 0x1C;
   OCR1AL = 0xCC;
   TCCR1B = 0x00;           // Stop Timer1
   for(i=0;i<delay;++i)
   {
   TCCR1B = 0x00;           // Stop Timer1
   TCNT1H = 0x00;           // Clear Timer1
   TCNT1L = 0x00;
   TCCR1B = 0x09;           // Start Timer1 with clk/1
   while(!(TIFR & 0x10));
   TIFR |= 0x10;
   }
}
//*****************************************************************
//*   Delay microsecond Function
//*   This function uses Timer 1 and the A compare registers
//*   to produce microsecond delays.
//*
//*****************************************************************
void delay_us(unsigned int delay)
{
   unsigned int i;
   OCR1AH = 0x00;
   OCR1AL = 0x07;
   TCCR1B = 0x00;           // Stop Timer1
   for(i=0;i<delay;++i)
   {
   TCCR1B = 0x00;           // Stop Timer1
   TCNT1H = 0x00;           // Clear Timer1
   TCNT1L = 0x00;
   TCCR1B = 0x09;           // Start Timer1 with clk/1
   while(!(TIFR & 0x10));
   TIFR |= 0x10;
   }
}
```

Code Snippet 13.4: These hand-written AVR timing routines are very accurate. And, if I enter 1 μS as my timing interval, the AVRCalc program spits out an OCR1AL value of 7. Thanks, Jack.

The timing routines use a special compare mode that is driven by the AVR's internal timers and compare modules. When the count matches the value loaded in OCR1A, the OCF1A bit is set in the TIFR (Timer/Counter Interrupt Flag Register). Since we're not vectoring to an interrupt when the match occurs, we must write a '1' to the OCF1A bit to clear it (TIFR |= 0x10).

AVRCalc is also instrumental in calculating AVR baud rate values. Code Snippet 13.5 is the fill-in-the-blanks code I based on the baud rate calculation in Figure 13.1.

```
//*********************************************************************
//*   BAUD RATE NUMBERS FOR UBRR
//*********************************************************************
#define  b9600  47      // 7.3728MHz clock
#define  b19200 23
#define  b38400 11
#define  b57600 7
//*********************************************************************
//*   USART Function
//*
//*********************************************************************
void init_USART(unsigned int baud)
{
   UBRR = baud;
   UCSRB = 0x18;
}
```

Code Snippet 13.5: Once the baud rate value is derived, the rest is dead easy.

The Custom Computer Services C Compiler's "make" came in handy when I had to manipulate bytes within 32-bit and 16-bit numbers. If I wanted to be able to port the code across AVRs and PICs, I'd have to write some "make" macros like the ones in Code Snippet 13.6.

```
//*********************************************************************
//*   RTL8019AS PIN MACROS
//*********************************************************************
#define make8(var,offset)  (var >> (offset * 8)) & 0xFF

#define make16(varhigh,varlow)   ((varhigh & 0xFF)* 0x100) + (varlow & 0xFF)

#define make32(var1,var2,var3,var4) \
      ((unsigned long)var1<<24)+((unsigned long)var2<<16)+ \
      ((unsigned long)var3<<8)+((unsigned long)var4)
```

Code Snippet 13.6: Being able to mimic the Custom Computer Services C Compiler built-in functions allows easier porting of the original PIC code to ImageCraft C for the AVR devices.

There's nothing else I can show you with the Easy Ethernet AVR you haven't already seen in the Easy Ethernet CS8900A and Easy Ethernet W as far as how the code works. When you compare the Easy Ethernet AVR source code with the Easy Ethernet W source code, you'll notice that there are lots of places where the source code for each respective device reads exactly the same.

The reason you don't see any Easy Ethernet AVR Sniffer shots is because the logic within the Easy Ethernet AVR firmware, the Easy Ethernet CS8900A firmware and the Easy Ethernet W firmware is identical. Thus, the Sniffer screen captures you've already been introduced to apply to the Easy Ethernet AVR. In addition, the software tools I showed you in the Easy Ethernet W section of this book also apply to the Easy Ethernet AVR.

There is one last screen capture I'd like to show you before we leave the Easy Ethernet AVR.

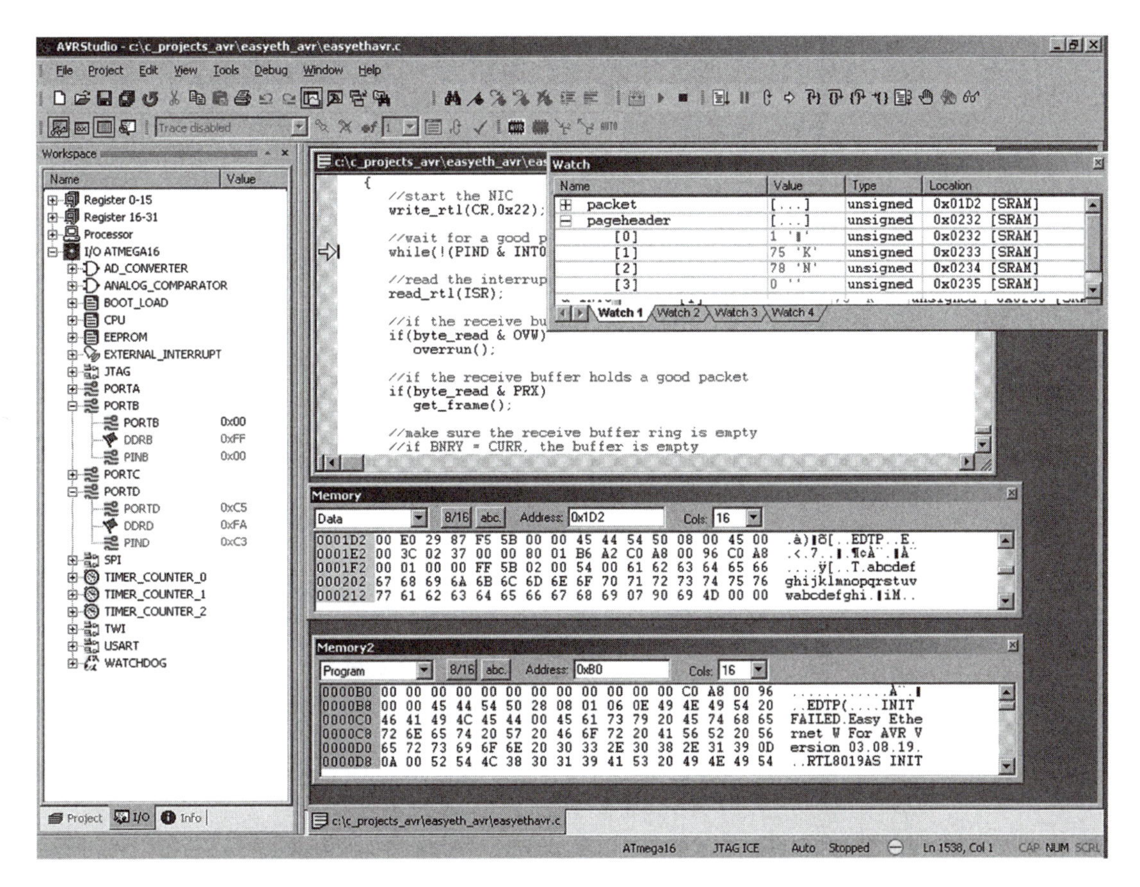

Figure 13.2: This is a screen shot of AVR Studio in debug mode with an AVR JTAG ICE attached. The source code you see is the firmware for the Easy Ethernet AVR. Can you see the ICMP frame in the Memory window? How about the Telnet banner in the Memory2 window?

This AVR Studio screenshot gives you an idea of just how powerful (and fun) coding for the Atmel AVR can be. Figure 13.2 was made possible by AVR Studio and an AVR JTAG ICE that is pinned into the JTAG interface of an Easy Ethernet AVR.

CHAPTER 14

Finale

You now have access to four Ethernet devices that can be used on a standard Ethernet LAN or on the Internet. The Packet Whacker is the most basic Ethernet package we have discussed and is based on the RTL8019AS. There's also a basic CS8900A-CQ configuration I would like to show you before you close this book. It's called NICki (Photo 14.1).

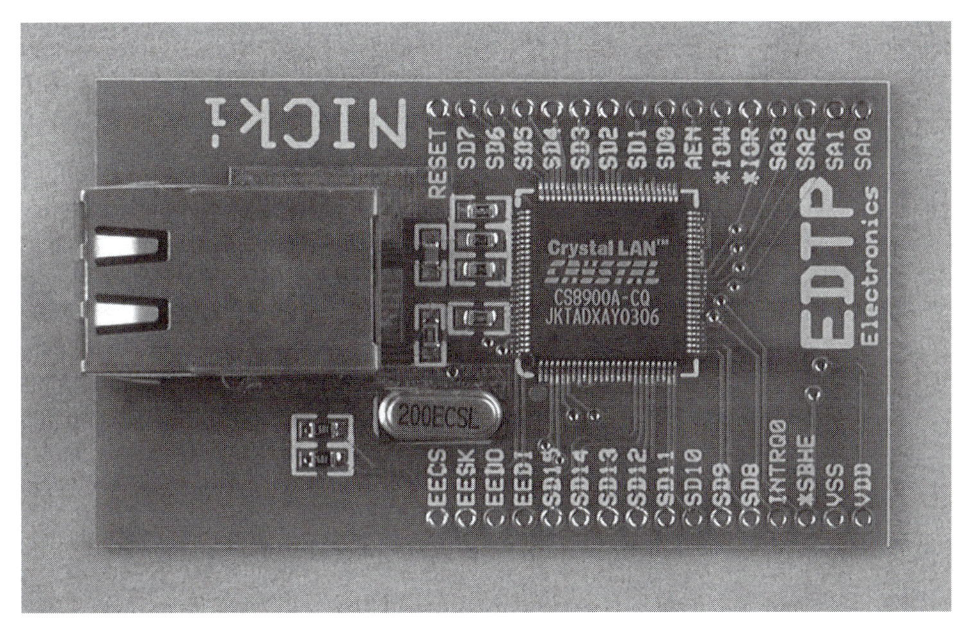

Photo 14.1: This represents the most basic CS8900A-CQ configuration.

NICki's circuitry is identical to the CS8900A-CQ circuitry found on the Easy Ethernet CS8900A. Like the Packet Whacker, NICki is designed to be interfaced to your favorite microcontroller.

Obtaining Easy Ethernet Devices

All of the Easy Ethernet devices can be purchased fully assembled or in kit form from EDTP Electronics. You can make Internet contact with EDTP Electronics at http://www.edtp.com. Once you are there, you can purchase the Easy Ethernet devices from the EDTP Electronics online store. You'll also find a wealth of information and source code relating to various Ethernet and Internet devices on the EDTP Electronics web site.

About the Author

For the past 18 years, Fred has written technical columns and articles for engineering journals and electronics magazines.

Fred is well versed in embedded system programming and his hardware expertise spans the embedded hardware spectrum including 8748, 8051, PIC and Atmel microcontrollers. Many of Fred's columns have also dealt with applications using the services of X86-based single board computers.

As an engineering consultant, Fred has implemented communications networks for the space program and designed firmware and hardware for the medical, retail and public utility industries. He currently designs and markets microcontroller-based hardware through his Internet-based online store.

Index

Numbers

A

B

C

D

E

F

This Agreement will remain in effect until terminated pursuant to the terms of this Agreement. You may terminate this Agreement at any time by removing from Your system and destroying the CD-ROM Product. Unauthorized copying of the CD-ROM Product, including without limitation, the Proprietary Material and documentation, or otherwise failing to comply with the terms and conditions of this Agreement shall result in automatic termination of this license and will make available to Elsevier Science legal remedies. Upon termination of this Agreement, the license granted herein will terminate and You must immediately destroy the CD-ROM Product and accompanying documentation. All provisions relating to proprietary rights shall survive termination of this Agreement.

LIMITED WARRANTY AND LIMITATION OF LIABILITY

NEITHER ELSEVIER SCIENCE NOR ITS LICENSORS REPRESENT OR WARRANT THAT THE INFORMATION CONTAINED IN THE PROPRIETARY MATERIALS IS COMPLETE OR FREE FROM ERROR, AND NEITHER ASSUMES, AND BOTH EXPRESSLY DISCLAIM, ANY LIABILITY TO ANY PERSON FOR ANY LOSS OR DAMAGE CAUSED BY ERRORS OR OMISSIONS IN THE PROPRIETARY MATERIAL, WHETHER SUCH ERRORS OR OMISSIONS RESULT FROM NEGLIGENCE, ACCIDENT, OR ANY OTHER CAUSE. IN ADDITION, NEITHER ELSEVIER SCIENCE NOR ITS LICENSORS MAKE ANY REPRESENTATIONS OR WARRANTIES, EITHER EXPRESS OR IMPLIED, REGARDING THE PERFORMANCE OF YOUR NETWORK OR COMPUTER SYSTEM WHEN USED IN CONJUNCTION WITH THE CD-ROM PRODUCT.

If this CD-ROM Product is defective, Elsevier Science will replace it at no charge if the defective CD-ROM Product is returned to Elsevier Science within sixty (60) days (or the greatest period allowable by applicable law) from the date of shipment.

Elsevier Science warrants that the software embodied in this CD-ROM Product will perform in substantial compliance with the documentation supplied in this CD-ROM Product. If You report significant defect in performance in writing to Elsevier Science, and Elsevier Science is not able to correct same within sixty (60) days after its receipt of Your notification, You may return this CD-ROM Product, including all copies and documentation, to Elsevier Science and Elsevier Science will refund Your money.

YOU UNDERSTAND THAT, EXCEPT FOR THE 60-DAY LIMITED WARRANTY RECITED ABOVE, ELSEVIER SCIENCE, ITS AFFILIATES, LICENSORS, SUPPLIERS AND AGENTS, MAKE NO WARRANTIES, EXPRESSED OR IMPLIED, WITH RESPECT TO THE CD-ROM PRODUCT, INCLUDING, WITHOUT LIMITATION THE PROPRIETARY MATERIAL, AN SPECIFICALLY DISCLAIM ANY WARRANTY OF MERCHANTABILITY OR FITNESS FOR A PARTICULAR PURPOSE.

If the information provided on this CD-ROM contains medical or health sciences information, it is intended for professional use within the medical field. Information about medical treatment or drug dosages is intended strictly for professional use, and because of rapid advances in the medical sciences, independent verification f diagnosis and drug dosages should be made.

IN NO EVENT WILL ELSEVIER SCIENCE, ITS AFFILIATES, LICENSORS, SUPPLIERS OR AGENTS, BE LIABLE TO YOU FOR ANY DAMAGES, INCLUDING, WITHOUT LIMITATION, ANY LOST PROFITS, LOST SAVINGS OR OTHER INCIDENTAL OR CONSEQUENTIAL DAMAGES, ARISING OUT OF YOUR USE OR INABILITY TO USE THE CD-ROM PRODUCT REGARDLESS OF WHETHER SUCH DAMAGES ARE FORESEEABLE OR WHETHER SUCH DAMAGES ARE DEEMED TO RESULT FROM THE FAILURE OR INADEQUACY OF ANY EXCLUSIVE OR OTHER REMEDY.

U.S. GOVERNMENT RESTRICTED RIGHTS

The CD-ROM Product and documentation are provided with restricted rights. Use, duplication or disclosure by the U.S. Government is subject to restrictions as set forth in subparagraphs (a) through (d) of the Commercial Computer Restricted Rights clause at FAR 52.22719 or in subparagraph (c)(1)(ii) of the Rights in Technical Data and Computer Software clause at DFARS 252.2277013, or at 252.2117015, as applicable. Contractor/Manufacturer is Elsevier Science Inc., 655 Avenue of the Americas, New York, NY 10010-5107 USA.

GOVERNING LAW

This Agreement shall be governed by the laws of the State of New York, USA. In any dispute arising out of this Agreement, you and Elsevier Science each consent to the exclusive personal jurisdiction and venue in the state and federal courts within New York County, New York, USA.